Problem Plants of Ohio

Problem Plants of Ohio

**Megan E. Griffiths,
Melissa A. Davis,
and David Ward**

The Kent State University Press

Kent, Ohio

© 2020 by The Kent State University Press

All rights reserved

Library of Congress Catalog Number 2020027331

ISBN 978-1-60635-402-5

Manufactured in Korea

Library of Congress Cataloging-in-Publication Data

Names: Griffiths, Megan E., author. | Davis, Melissa A., author. | Ward, David (Herrick Chair of Plant Biology), author.

Title: Problem plants of Ohio / Megan E. Griffiths, Melissa A. Davis, David Ward.

Description: Kent, Ohio : The Kent State University Press, [2020] | Includes bibliographical references and index.

Identifiers: LCCN 2020027331 | ISBN 9781606354025 (hardback) | ISBN 9781631014154 (epub)

Subjects: LCSH: Invasive plants--Ohio--Identification.

Classification: LCC SB613.5 .G58 2020 | DDC 333.95/3309771--dc23

LC record available at https://lccn.loc.gov/2020027331

24 23 22 21 20   5 4 3 2 1

Contents

Introduction 1

1 Grasses 21

2 Forbs 63

3 Creepers and Climbers 200

4 Shrubs 247

5 Trees 286

6 Aquatic and Wetland Plants 320

Acknowledgments 354

Appendix A Plant Species Designated as Invasive in Ohio 356

Appendix B Ohio Invasive Plants Council Plant Assessment Results 358

Appendix C Plant Species Designated as Prohibited Noxious Weeds in Ohio 361

Appendix D Herbicides 363

Appendix E Additional Synonyms of Scientific Names for Problem Plant Species 366

Glossary 371

Online Resources 373

Bibliography 375

Index 378

Introduction

Nonnative Plant Species and Problem Plants

There are many terms used to describe plant species occurring outside of their native ranges—*introduced, alien, exotic, nonindigenous, nonnative*—all of which indicate that these plant species have been moved out of the ecosystems in which they evolved. The introduction of nonnative plants to new ranges is almost always the result of human activity, whether intentional or unintentional. European settlers brought many plants to North America for use as food or medicine or for fiber production, forage, erosion control, or ornamental purposes. Other introductions have been accidental, through contamination of agricultural and nursery seed or the transfer of seeds or other reproductive plant parts (such as rhizomes) in ship ballast or packing material.

We examined data from the seven catalogs of vascular plants that have been compiled for the state of Ohio, as well as from comprehensive county and regional assessments (Fig. 1). Examining these historic records makes it clear that there has been a steady increase in the number of nonnative plant species in Ohio since the first records were made in 1840. At the time of the most recent statewide survey of vascular plants, Cooperrider, Cusick, and Kartesz (2001) found that there were 2,716 plant species growing in Ohio. Of these, 931 (34%) of the species were identified as nonnative. That represents a profound change in the composition of the flora in less than 200 years.

It is important to emphasize that not all nonnative species are equally problematic. Most have limited environmental impact aside from occupying space that might otherwise support native vegetation. In fact, many nonnative plants are economically important or even vital to our food supply. The majority of plant species used in agriculture or horticulture do not spread beyond the area where they have been planted. However, some nonnative plant species do escape cultivation and grow out of control. When nonnative plant species establish in natural areas and cause significant economic or environmental harm, they are called *invasive*.

Invasive plant species have profound negative effects on ecosystems by displacing native plant species, affecting wildlife that rely on native plant communities, and forming monocultures that reduce biological diversity overall. Because of this, invasive plant species are regulated at both the state and federal level in the United States. In Ohio, the Department of Agriculture (ODA) is the legal authority

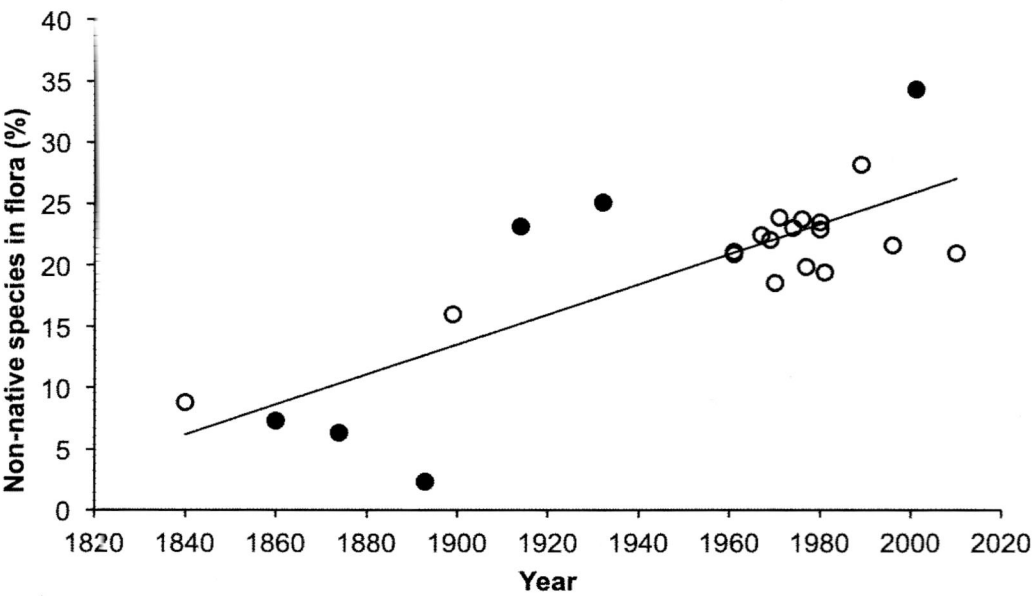

Fig. 1. Percentage of plant species classified as nonnative in Ohio, based on surveys for the entire state (in solid points) and counties or regions (in open points). Data were compiled from the seven state floral catalogs (Newberry 1860; Beardslee 1874; Kellerman and Werner 1893; Kellerman 1899; Schaffner 1914; Schaffner 1932; Cooperrider, Cusick, and Kartesz 2001) and county and regional assessments (Sullivant 1840; Selby 1899; Amann 1961; Hawver 1961; Cusick 1967; Anderson 1969; Silberhorn 1970; Weishaupt 1971; Wilson 1974; Pusey 1976; Cline 1977; Andreas 1980; Burns 1980; Emmitt 1981; Andreas 1989; Curtis 1996; Davis 2012).

that controls invasive plant species. The ODA has developed a list of 38 species designated as invasive plants in the state, and regulation of these species went into effect in January 2018 (ODA 901:5–30–01, Invasive Plant Species, available in appendix A of this volume [pp. 356–357]). All of the invasive species on this list cannot legally be sold, propagated, or distributed in Ohio. The ODA invasive species list is reviewed and updated every five years. In addition to this governmental regulation, the Ohio Invasive Plants Council (OIPC)—an independent coalition of agencies, organizations, and individuals—also helps to identify invasive plants and assesses their invasiveness using rigorous scientific methods. The OIPC list, available at the organization's website, is updated continually and should be consulted regularly to stay current as assessments are completed (https://www.oipc.info/assessment-re-

sults.html). We include the results of their plant assessments in appendix B [pp. 358–360].

There are also many plant species that cause significant economic harm in agricultural systems. These species are typically designated as prohibited noxious weeds at the state or federal level; see the US Department of Agriculture's Animal Health and Plant Inspection Service website for information (https://www.aphis.usda.gov). Under state or federal law, these species cannot be planted, sold, or transported, and they cannot be present as contaminants in seed sold for agricultural purposes. In almost all cases, prohibited noxious weeds are non-native species, although there are a few exceptions, such as *Conyza canadensis* (marestail) [pp. 104–105], which is native to much of North America, including Ohio. As with invasive species, the ODA designates prohibited noxious weeds in the state and this list is reviewed and updated every five years. The ODA updated the prohibited noxious weed list for Ohio in September 2018 (ODA 901:5–37–01; appendix C [pp. 361–362]).

In this book, we provide species accounts for all plant species designated as invasive by the ODA and the OIPC, all designated prohibited noxious weeds in Ohio (with the exception of seven species that were added to the list after the book was completed; these are indicated in appendix C [pp. 361–362]), and all federal noxious weeds that occur in Ohio. We have also included additional nonnative species that are commonly found growing as nuisance weeds in gardens, landscaping, and lawns. Most plants that fall under this last category tend to occur in highly disturbed human habitats and do not spread readily into natural areas. Collectively, all of these species are considered problem plants. We stress that any plants included in this book that do not have the designation of OIPC invasive, ODA invasive, or prohibited noxious weed should not be considered— or referred to as—invasive. They are merely aggressive species or those with weedy tendencies that we have observed to grow beyond cultivated areas.

Nonnative problem plant species have both direct and indirect effects on ecosystems. The most obvious is the replacement of diverse native plant communities by stands of a single invasive species. Many problem plants are successful invaders because they outcompete native species for resources such as water, nutrients, and sunlight. Some are hosts for pathogens, making them sources of disease for economically important plants like food crops. Others are allelopathic: they produce chemicals that are released into the soil and inhibit the growth of other plant species. Still others can alter fire regimes, hydrologic conditions, or soil chemistry. Some species, such as *Phalaris arundinacea* (reed canary grass) [pp. 46–47], produce large amounts of leaf litter that suppress the growth of native species. Species such as *Conium maculatum* (poison hemlock) [pp.

73–74] or *Euphorbia esula* (leafy spurge) [pp. 139–141] are toxic to livestock and degrade rangelands and pastures. Collectively, the economic damage caused by nonnative plant species in the United States is estimated to be $28 billion annually (Pimentel 2011). Because of these ecological and economic costs, it is important to be able to identify and control nonnative plant species.

This book is intended as a resource for anyone wanting to learn more about the plant species that are problems in Ohio. We have included distribution maps showing the counties for which there are records of each species being present. The problem plant species have been organized in chapters according to growth form or habitat ("Grasses," "Forbs," "Creepers and Climbers," "Shrubs," "Trees," and "Aquatic and Wetland Plants") and then alphabetically by family so related species are grouped together. To help with identification, each species has a full description and images. We have tried to limit the botanical terminology, and we have included a glossary to help define those terms that cannot be avoided. Each species description also includes a brief suggestion of control methods.

Where Do Nonnative Invasive Plants Come From?

The most serious problem plants are invasive plants, so we discuss invasive species in particular in the following sections. Invasions by nonnative species occur in three stages: first, an invasion begins with **dispersal**, which is followed by **establishment** and then **spread** of the species into a larger range. In many cases, establishment and spread can be viewed as parts of a continuum. Sources for nonnative plants are largely Europe and western Asia, where the climate is similar to Ohio's (Fig. 2). Some, particularly woody species that have been introduced for horticulture, have come from temperate areas of eastern Asia as well.

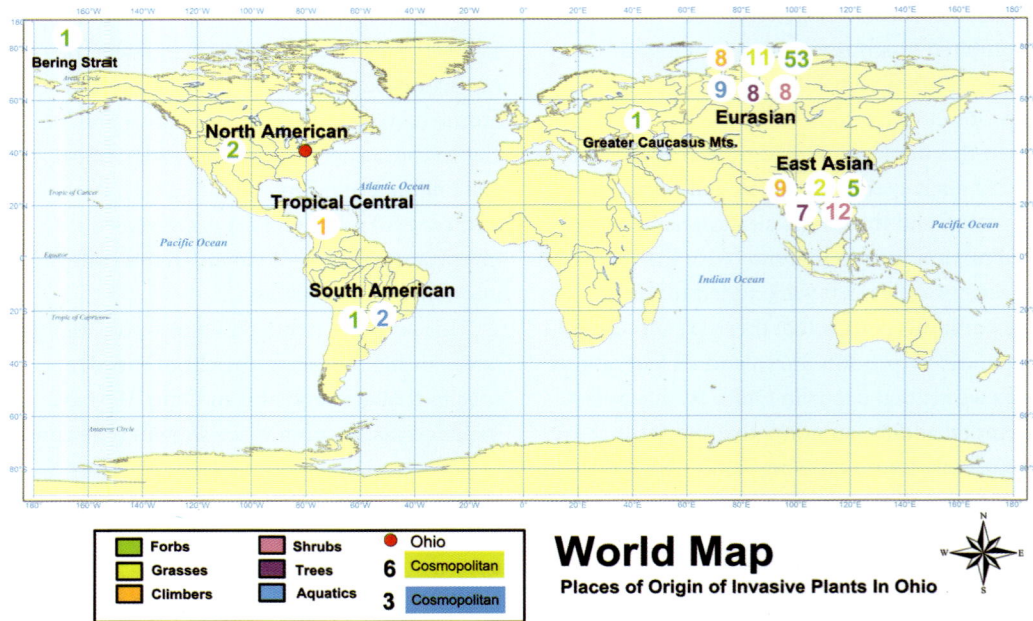

Fig. 2. World map with places of origin for the problem plants included in this book. The colored numbers represent the numbers of species in each plant form from a particular origin. Some of the Eurasian species are from predominantly temperate regions, but others have more Mediterranean origins and occur in northern Africa as well as in Europe and western Asia (Middle East). Similarly, eastern Asian species can be either temperate or tropical.

Only three species have come from tropical areas of South America, presumably because the climate is so different from Ohio's. No species with a native range of Australia or Africa (south of the Sahara) are currently problematic in Ohio, probably also a result of an incompatible climate and the less frequent and direct transportation of species from those regions to North America. Six cosmopolitan grass species and three aquatic species are found in Ohio; these are distributed in most areas of the world, with the exception of Antarctica.

Dispersal

There are three primary mechanisms of **dispersal** into a new region. The first is **release**, which can be either accidental or intentional. Seeds have been accidentally released, for example, through the transport and dumping of ballast soil, which was used in sailing ships prior to bal-

last water. This is thought to have been the route of entry for the beautiful but extremely invasive *Lythrum salicaria* (purple loosestrife) [pp. 156–158]. Another source of dispersal is **escape**, which often means that a species spreads from areas of cultivation. Finally, there can be **natural dispersal** from a neighboring region where the species is already established. Following an initial human-aided introduction, many species subsequently disperse into adjacent regions of their own accord. In general, terrestrial plants are more likely to be introduced by intentional release than aquatic plants, which often arrive as escapees or stowaways.

Establishment

Most of the plant species that disperse to other countries never become established. The primary reason nonnative species do not survive is that they are often not physiologically suited to their introduced environments. Establishment may not be successful because the climate is too cold or too hot (e.g., hard freezes and ice cover prevent *Eichhornia crassipes* [water hyacinth] from establishing in Ohio, whereas it is extremely successful in more southern states). In other instances, there might be inappropriate soil conditions or too little sunlight. Failed establishment of nonnative species is usually not detected or documented. We only notice new populations when they are abundant and long lasting and produce a noticeable impact.

Establishment is most likely to occur if the introduced species have existing traits that enable them to take advantage of available resources. Subsequent persistence is due more to adaptation to the new environment, which can be genetic or nongenetic. Nongenetic factors include phenotypic plasticity, the ability of an individual to take on different forms depending on environmental conditions (Pigliucci 2005; Dietz and Edwards 2006).

Occasionally, invasions may exhibit delayed establishment. For example, *Schinus molle* (Peruvian pepper tree) remained restricted to plantings in Florida for decades before it successfully established outside of cultivation. It then expanded rapidly into neighboring states and subsequently spread across to Texas, Mexico, and California. This is a good example of the importance of propagule pressure or introduction effort, a composite measure derived from the number of introductions and the number of individuals introduced (Lockwood, Cassey, and Blackburn 2005). Many species only become invasive when there is high propagule pressure. In other words, the more times a nonnative species is introduced and the more individuals that are introduced, the greater the likelihood that species will establish.

Spread

In many cases, within-region dispersal of introduced species is achieved independently of human activity. Plants are often dispersed by vectors such as water, wind, and animals.

Humans can also be an important dispersal vector; for example, hikers might inadvertently pick up seeds on their socks and boots and transfer them elsewhere, or vehicle tires may move them from site to site. In some cases, there is dispersal through mutualistic relationships: birds and large mammals ingest fruits and disperse the seeds elsewhere. In a common Ohio example, white-tailed deer regularly disperse invasive *Lonicera* species (shrub honeysuckles) [pp. 255–257] from suburbs to rural areas and probably back again (Vellend 2002). Similar deer-mediated effects have been recorded for *Berberis thunbergii* (Japanese barberry) [pp. 251–252], *Rubus phoenicolasius* (wineberry) [pp. 284–285], and *Microstegium vimineum* (Japanese stiltgrass) [pp. 42–43] (Shen et al. 2016).

Spread can also be influenced by propagule pressure, because species with more introductions and more individuals introduced will likely have higher genetic diversity in the population. When plant populations are small on arrival from another country, they may be subject to random changes that can cause them to become locally extinct (this is called *extirpation*). Repeated introductions allow these populations to sustain themselves. Moreover, higher genetic diversity often results in a higher likelihood of successful colonization under a broader range of ecological and physiological conditions. This higher genetic diversity can allow a nonnative species access to a greater range of habitats as it spreads. It is also possible, however, that there is no connection with genetic diversity. For example, *Egeria densa* (Brazilian waterweed) [pp. 335–336] has spread across the United States despite having only male individuals. This species was introduced for the freshwater aquarium industry because of its showy flowers, and it has become established as people repeatedly discard the contents of their aquariums into natural bodies of water.

Characteristics of Successful Invaders

Many studies have looked for common characteristics of invasive plants. Such studies are usually done by comparing successful invasive and noninvasive plants. For example, Rejmánek, Richardson, and Pyšek (2005) found that invasiveness in plants was strongly associated with three attributes: small seed mass, decreased time to reproductive maturity, and increased frequency of large seed crops. This last characteristic means that the number of seeds produced by a single invasive plant was much larger than for a noninvasive one. In another study, Daws et al. (2007) compared invasive and noninvasive species in the Asteraceae (aster family) and Poaceae (grass family). Among the Asteraceae, they found that invasive species produced seeds that were at least twice as big as those of noninvasive species (i.e., the opposite of what Rejmánek and colleagues found). Similarly, Daws and colleagues found that the invasive grass species produced seeds that were 68% bigger than noninvasive species. Thus, the results of these studies are inconsistent.

Pyšek and Richardson (2007) performed a study in which they compared invasive and noninvasive congeners (related species belonging to the same genus). They concluded that invasion success was consistently related to higher fecundity, vigorous vegetative growth, greater water- and nutrient-use efficiency, increased dispersal ability, and early and extended flowering. They also found that the importance of many traits was ambiguous. For example, the pollen vectors (wind or animal) were important in some comparisons of invasive and nonnative plants, but this was not always the case.

Some invasive species persist and spread through vegetative reproduction. There are distinct advantages to vegetative growth, which have to do primarily with the rapid increase in the size of a plant and its consequent occupation of a larger area. Larger plants may capture more resources and use the storage to minimize the negative effects of herbivory by shunting nutrients below ground. Another advantage is that there are none of the costs of sexual reproduction or hazards of recruitment from seed. However, vegetative reproduction also has distinct disadvantages. A plant that reproduces vegetatively may have fewer resources left over to produce seeds that can be dispersed away from the parent plant, where their probability of survival may be greater. Another issue is a lower genetic diversity, which could make the species more susceptible to disease (e.g., 95% of elms in the Netherlands died because they were descended from a single clone). Furthermore, clones may suffer from local crowding, and the resource costs of clonal growth may delay flowering.

Many successful nonnative plant species have characteristics that may give them a competitive advantage against native species. Problem plants may have specialized features, such as abundant lenticels as in *Frangula alnus* (glossy buckthorn) [pp. 298–299], that allow for oxygen exchange in anoxic growing conditions, copious seed production for dispersal as in *Lythrum salicaria* (purple loosestrife) [pp. 156–158], unique chemical defenses that prevent herbivory as in *Pastinaca sativa* (wild parsnip) [pp. 80–81], or allelopathy that inhibits the growth of neighboring plants as in *Centaurea stoebe* ssp. *micranthos* (spotted knapweed) [pp. 94–96]. These isolated characteristics help give individual species a competitive advantage under particular conditions, but they are not helpful in defining the general traits that predispose plant species to be invasive.

In some cases, native species can be more competitive in a community than nonnative species. For example, Seabloom et al. (2003) found that in a California grassland native perennials were better competitors for light, soil water, and soil nitrogen (usually the most important soil resource) than nonnative annuals. They also found that when planted together in a stand, native perennials reduced the abundance of nonnative annuals. One might reasonably ask how the nonnative annual species succeed in these communities? In this case, they are ultimately more successful because under natural conditions there are severe dispersal limitations in the native perennial species.

General Hypotheses of Invasion

A number of major hypotheses have been developed to explain and predict biological invasions, and they are not necessarily mutually exclusive. Most of these hypotheses provide logical explanations for general patterns observed in invasive species but many are still not supported by strong empirical evidence (Moles et al. 2012). However, it is worth outlining some of the more commonly cited ideas that relate directly to how and why plant species become invasive.

Enemy Release Hypothesis

One of the concepts with the most support in the literature is the enemy release hypothesis, which relates to specialist herbivores (Keane and Crawley 2002). In its native range, a plant species is often controlled by a native herbivore that consumes it. Typically, the herbivore is a specialist that only eats a single plant species. This keeps the abundance of the plant species low, so it is not able to outgrow or outbreed its potential competitors. When the plant species invades a new area, it is no longer suppressed by this specialist herbivore and the plant grows unchecked. Orians and Ward (2010) found that about 75% of plant invasions could be attributed to enemy release.

Biotic Resistance Hypothesis

The biotic resistance hypothesis relates to the ability of native species to resist or outcompete invasive species (Levine, Adler, and Yelenik 2004). Usually, the effect of such resistance is a reduction in the density of the invasive species that seldom, if ever, results in extinctions. Biotic resistance also tends to reduce the initial establishment of invaders but seldom causes complete repulsion of invasions.

Trait Diversity

Within a plant community, different types of plants perform different functions. For instance, a community with high functional diversity may have trees that draw water from deep down in the aquifer, legumes changing nitrogen so that it can be used by other plants, and grasses and forbs with shallow roots taking up rainwater and nutrients from the top layer of soil. Contrastingly, a shrubland with low functional diversity may have only one or two types of short plants occupying similar niches. Evidence is increasing that high functional diversity reduces the possibility of invasion because essentially all the functional niches are taken. Communities with high functional diversity are predicted to be more difficult for nonnative species to invade.

Empty Niches

An alternative idea concerns the role of empty niches, which presumes that the environment has spaces where new species can establish. In the 1950s, the biologist Charles Elton proposed that communities with more species (i.e., high-productivity environments with high rainfall and soil nutrients) are less likely to be invaded by other species because they have less space for new species to establish. In other words, it is predicted that there will be a negative correlation between the number of invasive species and the number of native species: an area with more native species will have fewer invasive species. However, Shea and Chesson (2002) argued that this hypothesis was overly simplistic. They found that for any given place, there often is a negative correlation between the number of native species and the number of invasive species. However, they observed that as one proceeds along a productivity gradient—for example, from the arid deserts of the US Southwest to the temperate deciduous forests of the Northeast—there is a positive correlation between the number of native species and the number of invasive species. That is, where there are more native species, there are correspondingly more invasive species because productive habitats are able to support more species overall. There is ample evidence that communities are not truly saturated with species and that as the productivity of the environment increases there are more niches for invasive plants to occupy.

Darwin's Naturalization Hypothesis

Charles Darwin (1859) observed that there was intense competition between invasive and native congeners (related members of the same genus). He postulated that newly arriving species would have a more difficult time establishing and persisting if they were more closely related to the native species than if they were unrelated to them. Darwin based this hypothesis on the assumption that related species were likely to require similar resources and would have increased interspecific competition. Some studies have found that species were more successful in establishing in new communities if they were less related to the species already there (Rejmánek 1998; Strauss, Webb, and Salamin 2006), supporting Darwin's naturalization hypothesis. However, other studies have been unable to find such support (Daehler 2001; Duncan and Williams 2002). In fact, being closely related could have both negative and positive effects on the establishment of invasive species (Mitchell et al. 2006). Because of their similarities, introduced species could experience increased competition, and they would likely share herbivores as well as pathogens. In contrast, invasive species could benefit from the closely related native plants in any given new community by sharing their pollinators and beneficial mycorrhizal fungi (which increase their uptake of key nutrients from the soil such as phosphorus and nitrogen). Another positive effect would be that their physiological similarities would make them more likely to be tolerant of the physical environment.

Novel Weapons Hypothesis

Another prominent theory is the novel weapons hypothesis (Callaway and Aschehoug 2000), which is related to plants' chemical defenses. Some plants are allelopathic, meaning that they exude chemicals from their roots that, in turn, inhibit the growth of other plant species. If an invasive plant uses allelochemicals in a plant community that is not accustomed to these exudates, the native plant species will often die out, allowing the invasive species to take over. A classic example of this is found in knapweeds—such as *Centaurea stoebe* ssp. *micranthos* (spotted knapweed) [pp. 94–96]—that have taken over large areas of the arid western United States.

Disturbance

An important issue with regard to the success of invasive species is the condition of the local ecosystem. In many cases, invasive species become successful because they are able to exploit disturbances. Such disturbances may occur before or upon immigration; invaders either require disturbance to facilitate their successful establishment or the native plants are not able to tolerate such disturbances. There are many examples in which disturbance facilitates the establishment of invasive plants. One need think only of road verges and residential developments and the scores of weedy species that quickly occupy those spaces before grass is able to grow. The same is true of old fields. Usually, disturbances involve physical movement of the soil, which creates open spaces for seedling establishment by invasive species. Similarly, disturbances can be created by selective grazing or fire, both of which create space for the seeds of invasive plants to establish.

Why Are Invasive Plants a Problem?

The redistribution of species around the world has often resulted in regional and local increases in species richness, as found with the increase in plant species in Ohio due to the addition of nonnative species to the flora. However, it is clear that there is also a homogenization of the world's biota. Biotic homogenization results from three interacting factors: introductions of nonnative species, the local elimination (extirpation) of native species, and habitat changes that facilitate both the introductions of nonnative species and the extirpation of native species.

One of the most negative impacts of invasive species is their effect on ecosystem services. These services are the processes through which natural ecosystems sustain human life—such as purification of air and water, drought and flood mitigation, and pollination of crops and natural vegetation. Disruptions to services include barriers to pollination and dispersal of seeds, alterations in nutrient cycling, changes in the hydrology of ecosystems, and changes in the presence of agricultural pests. Vilà and colleagues (2011) and Van Wilgen et al. (2009) have shown that one of the most important changes wrought by invasive species is the reduction in water availability to native plants. On the positive side, however, invasive species might reduce the effects of floods by acting as barriers, retarding the rate of water flow, or slowing down the rates of soil erosion.

Invasive species also have the potential to affect soil quality and productivity. Negative plant–soil feedback occurs when plants are less able to grow in soil that was previously occupied by a member of the same species. Often this occurs as a result of depletion of soil nutrients or because soil pathogens become more abundant where plant species are more abundant. An additional factor is allelopathy, where chemicals from the roots reduce soil quality for competing organisms (even of the same species). Negative plant–soil feedback is thought to be an important factor in helping plants to coexist because this will cause certain species to outcompete others. If a plant is very abundant, as is often the case with invasive species, then allelopathy, soil pathogens, and other negative factors will become common, diminishing the plants' growth. Diez and colleagues discovered that invasive plant species that became

established longer ago exhibited stronger negative feedbacks with the soil than recent invasive arrivals (2010). They also found negative soil feedbacks for invasive species that were dependent on time of arrival—those species that arrived earlier (and/or were more widespread) showed more negative soil feedbacks.

Invasive plants may ultimately cause the modification of some habitats, just as habitat modification drives species invasions. For example, dunes on the west coast of the United States have become stabilized by the invasive grass *Ammophila arenaria* (beach grass or marram grass). The invasive shrub *Berberis thunbergii* (Japanese barberry) [pp. 251–252] has been shown to alter the rate of nutrient cycling in the understory of deciduous forests in eastern North America (Kourtev, Ehrenfeld, and Häggblom 2003). The suppression of fire in habitats that evolved with fire—such as is the case in much of Ohio—can lead to enormous changes in habitats and concomitant changes in species composition, often accelerated by the presence of invasive species. Acid rain, a common problem in Ohio caused by pollution from coal-burning power plants, can also result in changed composition of species, many of which may be invasive.

Another issue of considerable importance is related to the coevolution between animal species and native plants. For instance, certain native plants are important host plants for insects. Where these are replaced by invasive species, this key interaction may be lost. A classic example of this can be seen with *Danaus plexippus* (monarch butterfly), which uses native species of the genus *Asclepias* as host plants for the development of its caterpillars (Fig. 3). In many parts of the United States, the *Asclepias* host plants are being replaced by invasive species, which is one of many factors contributing to the decline of this charismatic butterfly.

Another serious issue with regard to invasive species is the potential for interaction effects, many of which are not always immediately obvious. Plants are at the bottom of the terrestrial

Fig. 3. *Danaus plexippus* (monarch butterfly) caterpillar on its host plant, *Asclepias* sp. Image by Andrew Cannizzaro, used under a CC BY 2.0 license.

food web, which means that there are many more of them (in terms of both numbers of species and biomass of plants) than the herbivores that eat them, and fewer still of the carnivores that eat the herbivores. This is called a trophic pyramid because there are fewer species and less biomass in each level up the hierarchy. Invasive plant species often form dense monocultures that reduce the diversity of plants in an area, which means that the base of the pyramid gets narrower. This may result in effects higher up in the pyramid. Lower plant diversity can reduce the total number of herbivores, which limits the number of carnivorous organisms. These organisms include arthropod predators and parasitoids (usually wasps and related hymenopterans, which parasitize herbivorous insects), spiders, and insectivorous birds and mammals.

Invasive plants have negative financial impacts, as well as ecological ones. Nonna-

tive plants are spreading and invading approximately 700,000 hectares of habitat in the United States each year, costing an estimated $28 billion annually (Pimentel 2011). Crop weeds alone are responsible for a $20.5 billion loss, between a reduction in crop yield and expenses for herbicides and other control methods. Pimentel also calculated that pasture weeds cost about $6 billion each year. Furthermore, a single invasive species such as *Lythrum salicaria* (purple loosestrife) [pp. 156–158] costs as much as $45 million annually. Nonnative weeds in gardens, lawns, and golf courses have an additional expense of about $1.5 billion per year. These estimates are all based on 2011 values; the current costs are likely far higher.

Transgenic Crops and Biofuels

During the Green Revolution of the 1960s and 1970s, there was strong selection for high productivity in plants, usually based on total biomass for either the entire plant or just the fruit. However, this artificial selection was often associated with a loss of natural defense characteristics. Transgenic crops are currently being developed to reengineer the antiherbivore characteristics back into the plant so that crop plants can grow well and defend themselves against pests. While these traits are positive for farmers, there is also a real danger that these genetically altered plants could invade and spread into natural areas. There is also considerable concern over hybridization between transgenic species and native species. This issue might be particularly acute in cases where the hybrids have higher survival and/or fecundity than the native species. In the case of transgenic crops such as *Brassica napus* (canola) or biofuel species such as *Miscanthus giganteus* (silvergrass) that have been genetically modified to increase their productivity and herbivore resistance, there is concern that hybridization with invasive species may result in even more invasive strains that are difficult to control.

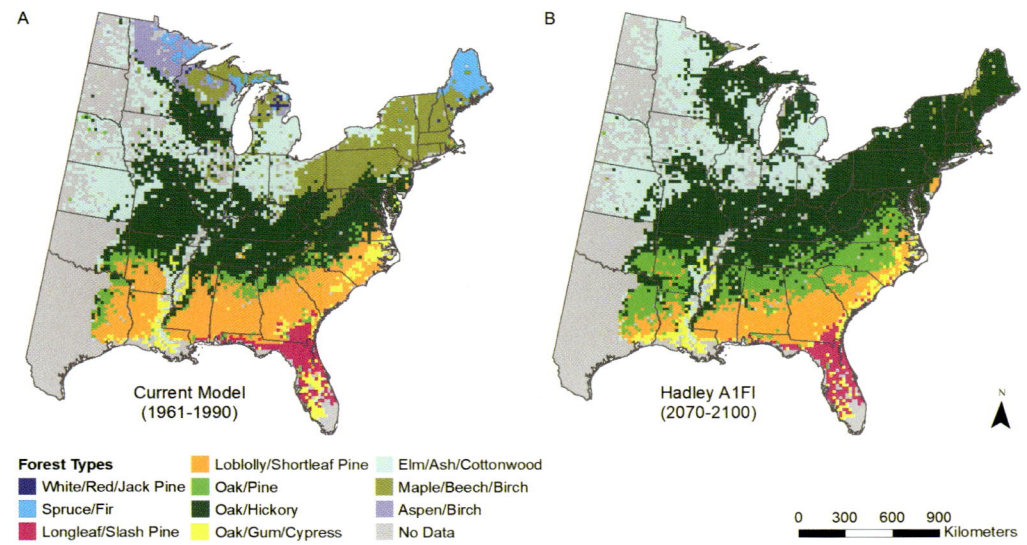

A

B

Current Model
(1961-1990)

Hadley A1FI
(2070-2100)

N

Forest Types
- ■ White/Red/Jack Pine
- ■ Spruce/Fir
- ■ Longleaf/Slash Pine
- ■ Loblolly/Shortleaf Pine
- ■ Oak/Pine
- ■ Oak/Hickory
- ■ Oak/Gum/Cypress
- ■ Elm/Ash/Cottonwood
- ■ Maple/Beech/Birch
- ■ Aspen/Birch
- ■ No Data

0   300   600   900
Kilometers

Fig. 4. Projected changes in dominant forest types in the eastern United States based on (A) current forest cover and (B) changes modeled for 2070–2100. Although environmental conditions are predicted to shift rapidly, tree species will likely migrate at a much slower rate. Data from Prasad et al. (2007), used with permission from the Northern Research Station, USDA Forest Service, Delaware, Ohio.

Problem Plants and Climate Change

Climate change is often perceived to be synonymous with global warming. However, while increased temperatures in many areas are certainly predicted components of climate change, many additional factors are likely to be affected. The Intergovernmental Panel on Climate Change (2013) predicts that the midwestern United States will experience greater variability in climate, including more intense summer heat waves and warmer winter temperatures. Although only a slight increase in precipitation is predicted between 2020 and 2100, the distribution of summer and winter precipitation is likely to change dramatically. The frequency of heavy storms, particularly in the vicinity of Lake Erie, is likely to increase. Precipitation in the form of rain is likely to increase as the amount of snow decreases. These climate changes will result in serious changes in plant hardiness zones and likely changes in types of forest that can grow in the region (Fig. 4).

How will this affect problem plants? The success of invasive and weedy plant species is often ascribed to these species' abilities to grow in a wide range of environmental conditions. However, we also know that many weedy species are capable of rapid evolution, leading to genetic changes that may be best adapted to a particular environment. Invasive species will also respond to climate change, and there is considerable concern that there will be an increase in the number and abundance of problem plant species. For example, if we look at predicted changes in mean annual temperature in Ohio by the year 2100, we see that the entire state is predicted to increase in temperature (Fig. 5).

It is likely that increased temperatures will result in an increase in the number of invasive species. Bradley et al. (2010) predicted that by 2100 *Ligustrum vulgare* (common privet) [pp. 276–277] and *Pueraria montana* var. *lobata*

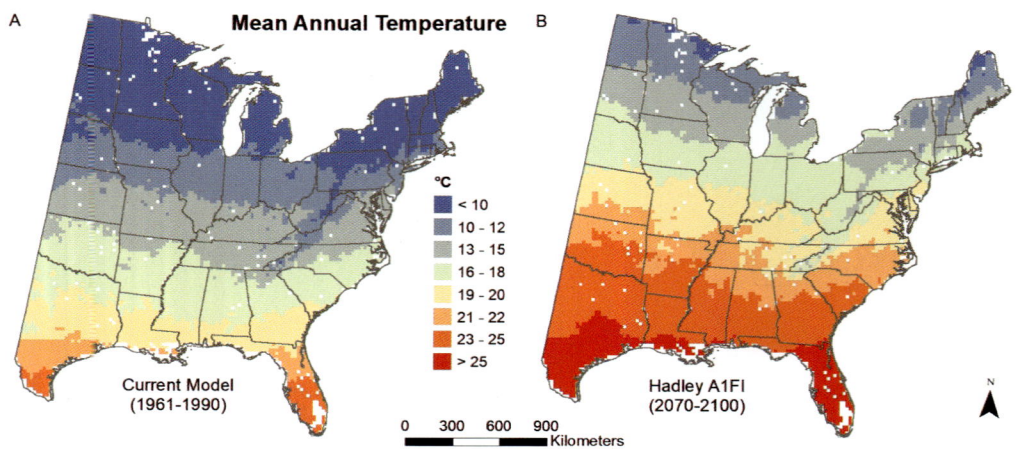

Fig. 5. Projected changes in mean annual temperature for the eastern United States based on (A) current measures and (B) changes modeled for 2070–2100. The majority of Ohio is projected to shift from its current mean annual temperature of 10–12°C (50–54°F) to 16–18°C (60–64°F) by the end of the 21st century. Data from Prasad et al. (2014), used with permission from the Northern Research Station, USDA Forest Service, Delaware, Ohio.

(kudzu) [pp. 233–235] would expand north into Ohio Pennsylvania, New York, and New England states. In fact, both of these species have already reached Ohio. Of course, it is also possible that some of the current invasive species will be driven to local extinction or will be forced farther northward. Regardless of whether some invasive species are extirpated, there will likely be an influx of new invasive species concomitant with rapid evolution of some of the current invasives. Therefore, it is expected that there will be more invasive plant species by 2100, and their effects on the environment will need to be closely monitored.

The Northern Research Station of the USDA Forest Service is particularly concerned about the spread of certain invasive plant species into forests in the northeastern and midwestern United States (see its website: https://www.nrs.fs.fed.us/disturbance/invasive_species/). These include several species that usually occur in warmer temperatures that are predicted to move northward as annual temperatures increase. The emerging threats include *Actinidia arguta* (hardy kiwi vine), *Falcaria*

vulgaris (sickle weed), *Firmiana simplex* (Chinese parasol tree), *Imperata cylindrica* (cogon grass), *Ligustrum lucidum* (glossy privet), *Murdannia keisak* (March dew flower), *Oplismenus hirtellus* ssp. *undulatifolius* (wavy basketgrass), *Rottboellia cochinchininsus* (itch grass), and *Toona sinensis* (Chinese mahogany).

Native Plant Species

Most problematic plant species are nonnative, but some native species occasionally grow out of control. As already mentioned, *Conyza canadensis* (marestail) [pp. 104–105], is a native species that is designated as a prohibited noxious weed in Ohio. Also, species such as the native *Juniperus virginiana* (eastern red cedar), have encroached on grasslands and reduced carrying capacity for livestock over areas of southern Ohio (and the Sandusky peninsula in northwestern Ohio), Kentucky, Oklahoma, Texas, Kansas, Nebraska, and South Dakota. The native *Pinus virginiana* (Virginia pine) is likewise encroaching into grasslands, helped in no small part by the fact that it is often grown on Christmas tree farms and escapes cultivation in regions where it is not native. In some cases, these native plants have clearly negative effects on crops or rangeland quality. Additionally, some native species are undesirable because they are toxic or can cause serious skin irritation, including *Toxicodendron radicans* (poison ivy), *Laportea canadensis* (wood nettle), *Urticaria dioica* (stinging nettle), and *Ageratina altissima* (white snakeroot). Indeed, Abraham Lincoln's mother died of milk sickness after drinking the milk of cows that had eaten white snakeroot. However, in their natural habitats and ranges these native plant species are often important parts of the ecosystem and—especially in the case of *T. radicans*—critical food sources for wildlife.

Many problematic plant species were originally introduced to North America as ornamentals. Despite having documented abilities to escape cultivation and disrupt natural ecosystems, some of these are still widely sold and planted. One of the most important things gardeners and homeowners can do to help prevent further invasion is to remove known problem plants and replace them with native alternatives. There is a growing body of evidence demonstrating the advantages of using native plant species in gardening and landscaping (Tallamy and Shropshire 2009; Narango, Tallamy, and Marra 2018). Nonnative plant species lack evolutionary histories with other native species and support less diverse and abundant insect communities, which in turn support less diverse and abundant wildlife higher up the food chain. Planting carefully selected native species should, therefore, encourage more pollinator and bird species in gardens. Further, native plant species tend to be better adapted to a region's climate and soil, so they are more likely to grow well with minimal input.

Gardeners and landowners are particularly encouraged to start by targeting plant species listed as invasive. We have provided suggestions of some native alternatives to these problem plants. The book *Native Alternatives to Invasive Plants* (Burrell 2006) has many good recommendations, although not all species covered in that publication are appropriate for Ohio. The Ohio Department of Natural Resources (ODNR) has also compiled excellent resources for native plants in the "Go Native" section of its website (http://ohiodnr.gov/gonative). The OIPC also offers several resources at its website (https://www.oipc.info/oipc-and-ohio-material.html). Before purchasing, it is always worth checking the USDA Plants database (plants.sc.egov.usda.gov) to see whether a species is native to Ohio. Details about growing conditions and plant features can be verified using the Missouri Botanical Garden Plant Finder database (https://www.missouribotanicalgarden.org/plantfinder/plantfindersearch.aspx).

There are many very attractive native plants, and their availability in Ohio nurseries is increasing as interest in native landscaping is increasing. Seek out native plant nurseries, which will have the most diverse and appropriate species for a particular area. Never harvest native species from protected land, and never collect endangered or other listed species from the wild. For more about native plants, contact the Midwest Native Plant Society (http://midwestnativeplants.org/), Cincinnati Wildflower Preservation Society (www.cincywildflower.org/), Native Plant Society of Northeastern Ohio (www.nativeplantsocietyneo.org/), or other similar local organizations.

Control and Management

The most successful management plan for any problem plants is prevention. Any plant species included in this book should be considered something to avoid planting or maintaining in a garden. It is also encouraged that you research any new plant species before buying and planting them. Plants may be considered invasive in other states but are not (yet) designated as invasive in Ohio. However, if a plant shows invasive or weedy tendencies elsewhere, it is likely to behave similarly in this state. Even species that grow in different climates, such as in humid parts of the southeast—including *Pueraria montana* var. *lobata* (kudzu) [pp. 233–235] and *Fallopia convolvulus* (black bindweed) [pp. 239–240]—are able to grow in Ohio. The more frequently a species is introduced to a place, and the greater the number of individuals introduced each time (propagule pressure again), the more likely that species is to establish (Lockwood, Cassey, and Blackburn 2005; Davis 2009).

If prevention is no longer possible, it is best to treat infestations of invasive plants or weeds when they are small to prevent them from spreading farther (early detection and rapid response). Controlling a plant before it produces seeds will reduce future problems. Control is generally best applied to the least infested areas before dense infestations are tackled. Consistent monitoring and follow-up are always required for sustainable management.

We provide general control recommendations for each species in this book but these are intended as starting points only. Any management plan should be thoroughly researched before implementation. The OIPC has fact sheets for the most invasive plants in Ohio; these include detailed information on control methods that have been demonstrated to be effective in the state. We strongly recommend readers check the OIPC website for the most up-to-date information.

Biological Control

In many cases, invasive plants have a competitive advantage over native plants because they have escaped their specialist herbivores in their home country. This has led to the growth of biological control studies, where applied entomologists and plant pathologists have sought out the herbivores or diseases controlling invasive plants in their native ranges in the hope of introducing them to control the invasive plants in their new country. In some cases, biological control has been extremely success-

ful. However, sometimes the specialist herbivore switches from the target invasive plant to (usually) closely related native species, resulting in untold damage. Consequently, to test the success of these biological control agents, experts have sought to establish small, isolated areas where specialist herbivores are unlikely to attack native plants. Most biological control agents are insects, but some are pathogens (Myers and Bazely 2003). By and large, such research is done by government and university researchers, and biological control programs typically work at a scale much larger than individual homeowners and farmers are likely to encounter. Biological control is unlikely to be a practical management option for the individual gardener, and it typically involves the introduction of another nonnative species that could become problematic in its own right.

Grazing and Browsing

Grazing and browsing are similar processes that result in the removal of plant material. Strictly speaking, grazing involves removal of herbaceous material (grasses and forbs) and browsing is the removal of woody plants (trees and shrubs). Some herbivore species are considered mixed feeders, which both graze and browse, but these are few. Goats are among the most prominent of these; they readily consume grasses and forbs as well as woody plants. Goats can be introduced to restricted areas to help control invasive plants; however, they are very good at identifying palatable plants and often refuse to eat invasive plant species containing toxic compounds.

Sheep and cattle are almost entirely grazers, so they mostly control the numbers of grasses and forbs. These species can be restricted to a small area, for example, with electric fences, to ensure that they consume target invasive plants. Some ranchers in Ohio do provide gardeners with goats to clear woody plants in an area, but as with biological control, this is not a practical option for most gardeners.

Use of Fire to Control Invasive Plants

Fires can effectively control some invasive plant species, but they can promote the growth or germination of others. Often, fire needs to be used in conjunction with another management technique to ensure that invasive plants (either the same or different species) do not reinvade after fire has cleared an area. The timing of burning is crucial. If burns occur at the beginning of the growing season (before plants produce seeds), fire creates space for seeds to germinate en masse, resulting in heavy invasion. However, if fires are lit at the end of the growing season, those young plants that have started to grow will be killed, and the problem of invasive success is likely to be reduced.

Because of the dangers of a prescribed fire escaping onto nearby properties not intended to be burned, fires should be carefully planned and conducted. Thus, fire is only permitted if conducted by an Ohio-certified prescribed fire manager, who has been trained by the ODNR Division of Forestry. **There are strict regulations on prescribed burning in Ohio; to conduct open burning and prescribed fires during March, April, May, October, and November for the hours of 6:00 A.M.–6:00 P.M., a waiver from the Division of Forestry is required. Notification must be submitted to the Ohio Environmental Protection Agency (OEPA) at least 10 days before the proposed burn.** The ODNR and OEPA websites both provide extensive information about fire rules and regulations. We stress that landowners should not attempt any such burning on their own unless they have the appropriate training, equipment, supplies, and permits. Furthermore, some invasive species may be more flammable than others, exacerbating the potential for escaped fires. Additional information about prescribed fire can be found on the Ohio Prescribed Fire Council's website (https://www.ohioprescribedfire.org/).

Mechanical Techniques

Mechanical techniques are preferred over herbicides; one should always use a less-toxic method before resorting to chemical control.

- **Hand-pulling, digging, and cutting down** problem plants is time and labor intensive but is the most effective and targeted way to clear a small area on private land. Bag and carefully dispose of any reproductive material to ensure that seeds do not continue to develop or accidentally get spread during the removal process. Most home compost piles do not get hot enough to kill seeds, so composting invasive plant waste is not recommended. It is important to minimize disturbance to the soil as much as possible when removing unwanted plants.
- **Solarization** involves the use of a clear or black plastic sheet to cover the invasive plants, allowing the sun to raise the temperature and kill off these plants. To raise the temperature sufficiently to result in significant mortality, this plastic needs to be kept in place for about six to eight weeks. This may be beneficial in cases where only invasive plants occur, but where native plants are present, this process is not feasible. Also, where one is trying to control plants in the forest understory, there is usually insufficient sun to raise the temperatures to the required level. Sunlight can also be useful after invasive plants have been removed; plant material can be placed in the sun to ensure that all viable parts are killed before disposal. This may be particularly useful for killing off clonal (vegetative) plant material.
- **Cutting below the root crown** has been shown to effectively kill off climbers such as *Pueraria montana* var. *lobata* (kudzu) [pp. 233–235] and *Fallopia convolvulus* (black bindweed) [pp. 239–240]. Root crowns look like buds just under ground level, with new sprouts growing from them. Cutting just below the root crown and removing the cut plant from the soil will usually stop new kudzu vines from growing. However, some vines can produce new shoots from tubers underground, and repetition of the

process is often required. Once the root crown's base is exposed, a sharp spade, hatchet, or handsaw can be used to cut the root below the root crown.

- **Additional weeding tools** include root talons, weed wrenches, girdling tools, chainsaws, brush cutters, and brush mowers. A **root talon** has a gripping hook that wraps around a stem, with two forks holding the plant while pressure is applied to the long handle, leveraging it out of the ground. This removes the plant with the root, minimizing the probability that subsequent removal will be required. A root talon is very effectively used with small individuals of *Rhamnus cathartica* (European buckthorn) [pp. 300–301]. A **weed wrench** or **uprooter** is a similar device that uses the physics of a lever to grab and uproot young woody plants. (We have had great success using the Extractigator, produced in Shawnigan Lake, British Columbia, Canada.) It has been shown to be effective on *R. cathartica* and *Ligustrum vulgare* (common privet) [pp. 276–277]. **Girdling tools** cut through the outer bark of woody plants to sever the phloem and cut into the cambium. To be effective, these cuts must extend all the way around the trunk; many trees can recover when there is incomplete bark removal. This method is often used for larger woody trees and shrubs, while the weed wrench and root talon are more suited to younger woody plants. **Chainsaws** and **brush cutters** effectively remove large trees and shrubs, and **brush mowers** or **weed trimmers** work with invasive grasses and forbs. Any of these tools work only in the short term, and repeated treatments will probably be needed.

Less Toxic Chemical Controls

In some cases, plants can be controlled using chemicals that have a lower impact on the local environment and that break down easily. Many homeowners and gardeners want to avoid putting toxic chemicals their yards. There is little formal research on how effectively these methods control individual problem plants, so these approaches are not listed in the control

methods for each species. However, it is worth exploring the less toxic options before resorting to using commercial herbicides.

- **Boiling water** uses a chemical already present in the environment (yes, water is a chemical!). It can be used to scald plants, particularly the root systems. This treatment is most effective on annual plants or perennial herbaceous plants whose aboveground parts have been cut back.
- **Corn gluten meal** can be scattered in an area to prevent seed germination. This is best done early in the spring, before annual plants emerge.
- **Vinegar** is an effective contact herbicide. Household vinegar typically has an acetic acid concentration of 5%. If that is not strong enough, horticultural grades with concentrations of 20–30% are available for purchase. While vinegar is safe, because it breaks down in water, it can also lower the soil pH in an area, so judicious use is recommended.
- **Herbicidal soaps** act as surfactants that break down the outer cuticle on leaves and cause plants to dry out. These are most effective on hot, sunny days.
- **Salt** can also be used to dry out plants, but a build-up in soils can inhibit growth of any plants. Homemade weed-killer solutions often combine dish soap, vinegar, and salt.
- **Plant oils** such as clove, nutmeg, cinnamon, basil, bay leaf, and citrus can also be effective contact herbicides. These are often included in commercially available organic herbicide mixes.

Herbicides

Sometimes, herbicide use cannot be avoided in the control of an invasive plant. Herbicides have both costs and benefits that need to be carefully weighed. Judicious and selective chemical control should be undertaken only if other nonchemical control methods are not effective on their own.

Plants are generally classified into three groups for chemical control: grasses, broadleaf forbs, and broadleaf woody plants. Grasses can be further subcategorized as annuals or perennials. To help identify the correct chemical product to use, it is generally important to know whether a plant is annual or perennial. Likewise, to help determine the correct chemical control program it is essential to know a plant's growth stage (seed, seedling, vegetative, reproductive). To ensure the correct identification of a species and to inhibit any addition to—or propagation from—the seed bank, it is best to perform any management when a plant is actively growing but before seeds form. When undertaking chemical control, special care must be taken to protect nontarget species.

Herbicides are categorized into several different groups based on mode of action, which defines how the chemical works within the plant. Herbicides are also classified according to chemical families and active ingredients that reflect the chemical composition of the product. They are further catalogued as preemergent or postemergent, and they can be selective or nonselective. Preemergent herbicides are applied before a plant germinates, and with this type of treatment the plants die shortly after emergence. It is best to use a preemergent product after an established plant population has been identified and removed or after tillage, which can bring seeds stored in the seed bank back to the surface, where they can germinate. Postemergent herbicides are used on existing plants. Selective herbicides are used to control specific types of plants while not harming others. For instance, triclopyr is a selective herbicide that kills broad-leaved plants but does little or no harm to grasses or conifers. Sethoxydim, in contrast, kills grasses but does not damage broad-leaved plants. Glyphosate is a nonselective herbicide that kills both grasses and broad-leaved plants. Nonselective herbicides kill any plants in a treated area. Contact herbicides work by direct contact, and systemic herbicides are translocated through the plant body. Surfactants are sometimes used to increase mixing ability and activation properties, and tracer

dyes can be used to clearly mark an area that has been treated with a pesticide.

In this book, we refer to herbicides by the active ingredient, because trade names and chemical companies can change. Select trade names are included in appendix D [pp. 363–365]. A responsible chemical control program will rotate products with different modes of action to ensure that the target plants do not become resistant to a specific type of chemical. Refer to appendix D for additional guidelines for the commonly used herbicides. Where chemical control options are provided, this volume does not endorse or recommend a specific product or a chemical manufacturer. Similarly, the exclusion of specific herbicides does not imply that they cannot be used. Always refer to the product label for active ingredients, mixing instructions, application rates, regulations and precautions or contact the manufacturer for any additional product information. **The label is a legal document that must be followed**. Do not use this book as a substitute for instructions given on product labels or to formulate a chemical control program without doing further research on best practices.

As already mentioned, a chemical control program should always be the last course of action. If all other efforts have failed, a chemical control program, if performed correctly, may prove successful. The information provided here is based on the best available resources at the time of writing. Herbicide regulations are constantly changing; always consult herbicide labels for the current guidelines for each active ingredient. County extension officers can also advise on herbicide use—see The Ohio State University Extension's "Locate an Office" page (https://extension.osu.edu/lao) to find contact information for your county extension office. Only licensed and certified commercial applicators are permitted to use herbicides in aquatic habitats. These restrictions are in place to protect human health, drinking water, water bodies, and wildlife.

Fig. 6. Ohio and neighboring states. Plants that are problematic in neighboring states are likely to become problematic in Ohio and vice versa.

sightings to EDDMapS, which can be used to report distributions of invasive and potentially invasive species throughout the United States. The accuracy of the maps is limited by the quality of the data available. If you have a confirmed sighting of a problem plant species in a new locality, we strongly encourage you to register with EDDMapS (https://www.eddmaps.org/report/) and record the observation.

Ohio is part of the Great Lakes Region of the United States, bordering Michigan and Canada (via Lake Erie) to the north; Pennsylvania to the east; West Virginia and Kentucky to the south (separated by the Ohio River); and Indiana to the west (Fig. 6). These neighboring states have similar floras, and many of the problematic species covered in this book may also affect parts of these regions.

Ohio is divided into 88 counties (Fig. 7). Current distribution maps were created for each species in this book, using ESRI Arc-

Maps

The distribution maps reflect actual voucher specimens deposited in Ohio herbaria, namely the Tom S. and Miwako K. Cooperrider Kent State University Herbarium, The Ohio State University Herbarium, and the Willard Sherman Turrell Miami University Herbarium. Additional sources of distribution data include the USDA Natural Resources Conservation Service PLANTS Database and the Early Detection and Distribution Mapping System (EDDMapS), a web-based mapping tool created in collaboration between the National Park Service and the University of Georgia Center for Invasive Species and Ecosystem Health. These maps represent the distributions of potential problem plants in the state of Ohio at the time of this publication and are likely to change as more information is collected. Many species may be more widely distributed than the current maps indicate. The record of distribution depends on the collection and deposit of vouchers submitted to state herbaria and voluntary reporting of

Fig. 7. The 88 counties in Ohio. The distribution maps for the species covered in this book are shaded green for every county in which a species has been recorded.

GIS Desktop V10 software. Green shading represents the occurrence of the species based on at least one record in a county. We do not provide information on species abundance, the map merely reflects presence or absence.

Botanical Information and Scientific Plant Names

The botanical descriptions included in this book are based primarily on Gleason and Cronquist (1991). Several additional print (Chace 2013; Dirr 2009; Kaufman and Kaufman 2007; Czarapata 2005; Weber 2003; Lorenzi and Jeffery 1987) and online sources (listed on pp. 373–374) were used to verify and update the botanical information and control recommendations. We consulted Bojnanský and Fargašová (2007) to fill in missing details for fruits and seeds of European species. Aquatic plant information was supplemented using Crow and Hellquist (1999a, 1999b).

Plant species are identified using Latin scientific names, by genus and species. These names have an advantage because no two species have the same binomial (genus–species). This avoids the confusion that can come with using only common names, which sometimes overlap, even among unrelated species. For example, greater celandine (*Chelidonium majus*) [pp. 161–163] is in the Papaveraceae (poppy family) and lesser celandine (*Ficaria verna*) [pp. 180–182] is in the Ranunculaceae (buttercup family). At a higher taxonomic level, species are grouped into families of related species. As in other areas of science, our understanding of the taxonomic relationships among species, genera, and families is constantly being refined. Taxonomists use morphological features and genetic evidence to inform this classification system. As

more is learned about the evolutionary history and relationships of species, the classification system is sometimes updated with new names at the species, genus, or family level. We have endeavored to use the most current scientific names for the plant species covered in this book by consulting the Royal Botanic Gardens, Kew and Missouri Botanical Garden's Plant List (http://www.theplantlist.org/), the Tropicos database maintained by the Missouri Botanical Garden (http://www.tropicos.org/), and the US Geological Survey and Smithsonian Institution's Integrated Taxonomic Information System (https://www.itis.gov/). Where names have been recently changed or it is likely that readers will identify a species by an older name, we have included the synonyms in the species description. Otherwise, most synonyms are listed in appendix E [pp. 366–370].

Measurements and the Metric System

The metric system has been adopted in every country in the world except for the United States, Liberia, and Myanmar. It is the standard measurement system used in science, even in the United States. With the metric system, it easy to report measurements for small plant features such as or petals or seeds. Consequently, we have chosen to present measurements primarily in metric and provide approximate conversions to imperial units. Most imperial measurements provided in the text have been rounded up or down for practicality. Therefore, the metric measurements are more accurate and precise, and we recommend using them wherever possible. To aid readers less familiar with the metric system, we have provided a more accurate conversion scale from millimeters and centimeters to imperial units and a scale ruler inside the book's back cover.

Explanation of Designations Used in Species Accounts

ODA Invasive = plants designated as invasive by the Ohio Department of Agriculture

OIPC Invasive = plants designated as invasive by the Ohio Invasive Plants Council

Prohibited Noxious Weed = plants designated as prohibited noxious weeds by the Ohio Department of Agriculture

Federal Noxious Weed = plants designated as federal noxious weeds by the United States Department of Agriculture

⚠ **Do Not Touch** = use caution when handling because plants have chemical or mechanical (hairs, prickles, spines, thorns) properties that can cause harm

Grasses

Grasses can be highly successful weeds or invasive species. This is related, in part, to the fact that the seeds of most grasses are wind-dispersed and can be moved long distances. Grasses also often have extensive vegetative reproduction, through structures such as tillers, rhizomes, and stolons (Fig. 8). All of these are primary modifications of the main part of the grass plant, the culm (Gibson 2009). Tillers are aboveground branches that develop at the base of the culm. All grasses produce tillers, but they are particularly associated with a bunch-grass habit. Rhizomes, by contrast, are belowground lateral stems that appear like roots. Stolons, sometimes called runners, are aboveground lateral stems that spread out horizontally from the culm and can start new plants if separated from the culm.

The rate at which tillers, rhizomes, and stolons grow is related to the amount of energy in the culm. Some grass species are considered to have a phalanx life history, which means that they usually have short tillers or rhizomes and move outward in a mass (like the formations used by Greek and Roman armies, described using the same term) (Lovett Doust and Lovett Doust 1982). Phalanx plants tend to develop in low-nutrient environments with lots of light. In contrast, guerilla plants—as the name implies—stealthily spread into areas where there is limited space available, often via long stolons. Guerilla plants tend to grow in moist areas that have high nutrients with little available light. These plants may also occupy disturbed areas that are more nutrient-rich than the surroundings.

Because of their abundant, wind-dispersed seeds and their vegetative growth, grasses tend to be difficult to control. It is important to prevent these plants from setting seed, which is most easily accomplished by repeated removal of flowering stems. Mowing slows the rate of grass spread but is unlikely to control it in most cases. In some habitats, professionally monitored controlled burns are used to maintain certain plant communities. However, fire tends only to retard the growth of invasive grasses, rather than completely remove them. An additional effective control measure is solarization, whereby invasive grasses are covered with plastic sheeting to raise the temperature and kill off the plants. Herbicides that incorporate glyphosate can be applied as nonspecific controls, but there are also grass-specific herbicides, such as sethoxydim.

The literature pertaining to grasses has developed somewhat independently of other plant species, mostly because researchers studying the

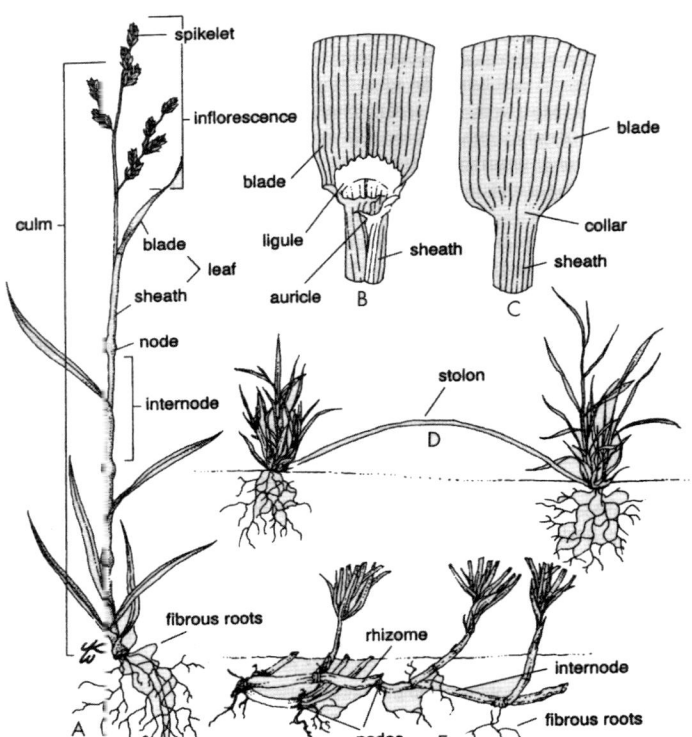

Fig. 8. The basic structure of a grass is based on a culm. Culms may spread aboveground via stolons or belowground via rhizomes. Vegetative structures such as auricles or ligules can be important identifying characteristics for some grass species. Reprinted from Hatch, Schuster, and Drawe (1999), used with permission from Texas A&M University Press.

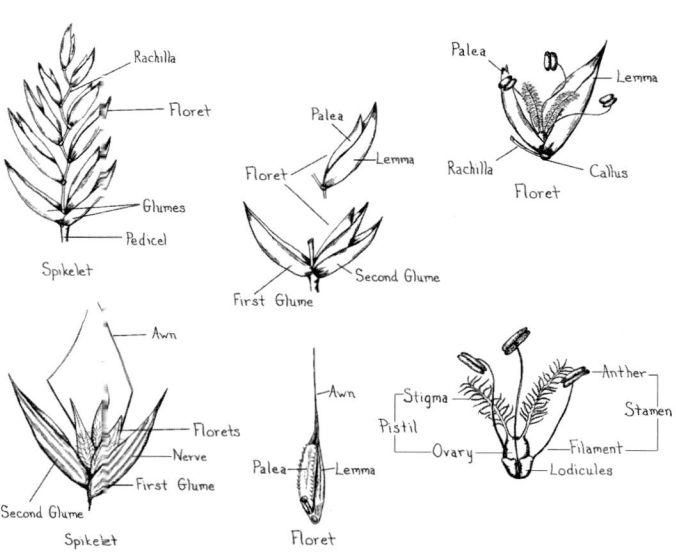

Fig. 9. A grass inflorescence has a number of specialized structures. The floral parts may differ in key identifying features. The spikelets consist of glumes, a palea, and a lemma. The awns, used by many grasses to move their seeds, are also key characteristics. The grass flower has female parts in the pistil, which consists of an ovary and a stigma, and the male parts in the stamen, which consists of a filament and the anthers. Drawing by Susan Kashanski (2002), used with permission from the US Fish and Wildlife Service, National Conservation Training Center, Shepherdstown, West Virginia.

grasses were often rangeland ecologists or pasture scientists. Consequently, terminology has been developed for structures in grasses that is largely distinct from that of other plants. The presence of jointed nodes distinguishes grasses from other grasslike plants. The flowering stems of true grasses are hollow at maturity. There are many terms specific to the identification of grass species, in both the vegetative and the reproductive structures (Figs. 8 and 9).

Cool- and Warm-Season Grasses and Likely Responses to Global Warming

Grasses, which evolved about 80 million years ago, comprise a bewildering array of species. In the course of evolution, grasses initially used one type of photosynthesis (as did most other plants), termed C_3 photosynthesis. This type is more widespread and refers to the way plants obtain carbon dioxide (CO_2) during photosynthesis. All species have the more primitive C_3 pathway, which involves a three-carbon molecule; hence its name. More recently, about 31–34 million years ago, an additional C_4 pathway, which involves a four-carbon molecule, evolved in species that occur in the dry and wet tropics. C_4 plants are typically associated with habitats that have high temperatures and/or frequent droughts. In contrast, C_3 plants tend to grow in cooler climates, often where frost is common. Consequently, grasses with these two photosynthetic pathways are often termed cool- (C_3) and warm-season grasses (C_4.) In Ohio, there are grasses that use both C_3 and C_4 photosynthetic pathways (Teeri and Stowe 1976). It is widely accepted among scientists that with increasing global temperatures, C_4 species will replace C_3 species (see, for example, Edwards and Still 2008). Because cool-season grasses are often critical forage species for cattle, this could have important consequences for economics and food production, as well as for ecosystems.

Poaceae

Arthraxon hispidus

small carpetgrass · small carpgrass · hairy jointgrass · jointhead grass · jointhead arthraxon

Origin: Native to Africa, southern and eastern Asia, and Oceania. This species was first recorded in the United States in the 1870s. Introduction was most likely accidental. It now has a patchy distribution throughout the eastern United States, primarily in the mid-Atlantic region.

Description: A low-growing, warm-season annual grass with broad leaves. *Arthraxon hispidus* resembles the native *Dicanthelium clandestinum* (deertongue grass), but *D. clandestinum* has a pyramid-shaped inflorescence and longer and more pointed leaves, its leaf margins lack hairs, and its stems do not root at the nodes. It is also similar to the nonnative *Microstegium vimineum* (Japanese stiltgrass), but the leaves of *M. vimineum* lack a heart-shaped base.

Habitat: Forest margins, floodplain forest, swamps, lake or pond shorelines, river- and streambanks, wet meadows, pastures, fields, ditches, and roadsides. Prefers full sun and moist to wet soils.

Height: Flowering stems typically 30–50 cm (12–20 in.), occasionally to 1 m (40 in.).

Foliage: Leaves are oval or lance-shaped, 2–7 cm (¾–3 in.) long and 5–20 mm (¼–1 in.) wide, each with a pointed tip and a heart-shaped base that encircles the stem. The leaf blade is hairless, but the margins of the leaves have conspicuous hairs. The leaf has a membranous ligule, 0.5–3.5 mm long, with fine hairs along the margin.

Flowers: Flowers are pale green or purple, in spikes 2–8 cm (¾–3 in.) long, held singly at the ends of stems or clustered in groups of 10 or more diverging from a common point of attachment. Each spikelet is 3–5 mm (⅛–¼ in.) long, containing a single floret with two anthers and an awn that is 6–11 mm (¼–½ in.) long. September–October.

Fruit: Bracts bristly and light brown to yellowish, 4–5 mm (⅛–¼ in.) long and 1 mm (1/32 in.) wide, containing a slender and yellowish caryopsis, 2–3 mm (1/16–⅛ in.) long and 0.4–0.6 mm wide. October–November.

Stems: Stems are slender and hairless except at the nodes. Stems trail along the ground near the base then grow upright.

Root system: Slender, fibrous roots form a shallow root system. Secondary rooting can occur at the nodes where they contact the soil.

Reproduction: By seed, which is dispersed by flooding, animals, machinery, or movement of soil.

Impact: Forms dense monocultures that displace native plant communities, particularly in ephemeral habitats along the margins of water bodies.

Control: Mechanical treatments are most effective when plants are just starting to flower. Pull or mow before seeds are produced. Since this species often grows in wet environments, only aquatic formulations of glyphosate and sethoxydim should be used for control. If this grass is growing in or near a wetland habitat, chemical applications should be made only by a licensed aquatic herbicide applicator.

Stand of *Arthraxon hispidus.* Image by Leslie J. Mehrhoff, University of Connecticut, Bugwood.org.

Leaves and stems of *Arthraxon hispidus.* Image by Leslie J. Mehrhoff, University of Connecticut, Bugwood.org.

Arthraxon hispidus inflorescence. Image by Leslie J. Mehrhoff, University of Connecticut, Bugwood.org.

Arthraxon hispidus infructescence. Image by Leslie J. Mehrhoff, University of Connecticut, Bugwood.org.

Fruits of *Arthraxon hispidus.* Image by Steve Hurst, hosted by the USDA-NRCS PLANTS Database.

Roots and dried stems of *Arthraxon hispidus.* Image by Leslie J. Mehrhoff, University of Connecticut, Bugwood.org.

Poaceae

Bromus inermis

smooth brome · bromegrass · awnless brome · Hungarian brome · Austrian brome · Russian brome

Origin: Native to Europe and eastern Asia. This species, which was intentionally introduced to North America in the late 1800s, has been widely planted as forage for livestock and erosion control along waterways.

Description: A cool-season, perennial rhizomatous grass. This species starts growth in early spring and forms a dense sod.

Habitat: Sunny areas along roadsides, ditches, fields, pastures, and prairies. Can withstand drought, extreme temperature, and periodic flooding.

Height: Stems typically 50–100 cm (20–40 in.).

Foliage: Leaves rolled in bud. Leaves alternate, 10–20 cm (4–8 in.) long and 8–15 mm (¼–⅝ in.) wide, with a raised midvein below and rough edges. Sheaths are fused, except for a notch near the collar. Smooth or with a few small hairs, particularly on the sheaths. A short ligule is present, usually 0.5–1 mm (¹⁄₆₄–¹⁄₃₂ in.). There is often an *M*-shaped constriction between the center and the tip of the leaf that is diagnostic of all bromes but is especially prominent in *Bromus inermis*.

Flowers: Inflorescence is a drooping panicle, 10–20 cm (4–8 in.) long, with 4–10 branching spikelets that are 15–30 mm (⅝–1¼ in.) long and each contain 7–11 florets. Florets have two or three large yellow anthers and white stigmas. May–July.

Fruit: Inflorescence develops a characteristic purple-brown color when mature. Bracts golden or tan, 8–12 mm (¼–½ in.), usually awnless and smooth. Caryopses are brown, flat, 7–9 mm (¼–⅜ in.) long. June–August.

Stems: Smooth or finely haired at nodes.

Root system: Rhizomes with dark brown scales.

Reproduction: Aggressively spreads through tillers and rhizomes. Seeds are dispersed by wind, water, birds, and mammals. Seeds in the soil seed bank remain viable for up to 10 years.

Impact: This species is highly competitive because it starts growth early in the spring and forms a dense sod. This is of particular concern in prairie habitats, where it outcompetes later growing, warm-season prairie species.

Control: Perennial, cool-season grasses like *Bromus inermis* can be controlled by manual, mechanical, and chemical means. Annual spring burning, intensive early spring grazing, or intensive early spring mowing can be used to reduce *B. inermis* cover in prairies dominated by warm-season grasses. Grass-specific herbicides such as sethoxydim can be applied in spring or autumn when native grasses are dormant.

Native Alternatives
Andropogon gerardii (big bluestem)
Calamagrostis canadensis (Canada bluejoint)
Calamovilfa longifolia (prairie sandreed)
Eragrostis spectabilis (purple love grass)
Schizachyrium scoparium (little bluestem)
Sporobolus heterolepis (prairie dropseed)
Tridens flavus (purpletop tridens)
Sorghastrum nutans (Indiangrass)

Stand of *Bromus inermis.* Image by Robert Vidéki, Doronicum Kft., Bugwood.org.

Bromus inermis inflorescence. Image by Patrick J. Alexander, hosted by the USDA-NRCS PLANTS Database.

Bromus inermis infructescences. Image by Matt Lavin, used under a CC BY-SA 2.0 license.

Leaf sheath and stem of *Bromus inermis.* Image by John Cardina, The Ohio State University, Bugwood. org.

Bromus inermis ssp. *inermis* fruit. Image by D. Walters and C. Southwick, Table Grape Weed Disseminule ID, USDA APHIS PPQ, Bugwood.org.

M-shaped constriction on the leaf of *Bromus inermis.* Image courtesy of Fontenelle Nature Association.

Poaceae

Cynodon dactylon

Bermudagrass · Bahama grass · wire grass · devil grass · dog's tooth grass · couch grass · quick grass · star grass · scutch grass · vine grass

Origin: This species likely originated in eastern Africa but now occurs in tropical and subtropical regions worldwide. It is frequently cited as having been introduced to North America as a pasture species in the mid-1800s, but some accounts suggest that it was already established by 1807.

Description: A low-growing, warm-season perennial grass. Spreading stolons and rhizomes form a dense sod.

Habitat: Open woods, disturbed grasslands, riparian areas, plantations, agricultural fields, orchards, pastures, gardens, and roadsides. It grows in almost all soil types. Prefers a warm climate, moderate moisture, and full sun.

Height 5–45 cm (2–18 in.).

Foliage: Leaf blades are grayish green, flat, lance-shaped, 4–15 cm (2–6 in.) long, and 2–5 mm ($^1/_{16}$–¼ in.) wide, alternately arising from the stem. Leaves are finely parallel-ribbed on both surfaces without conspicuous midveins, can be smooth or hairy on the upper surface, and have smooth sheaths. The ligule is a short membrane, 0.2–0.5 mm ($^1/_{64}$ in.), with a conspicuous fringe of white hairs.

Flowers Flowers in spikes, 3–8 cm (1–3 in.) long, clustered at the ends of stems in groups of 4–6 or more, diverging from a common point of attachment. Spikelets are arranged in two rows along the stem of each spike. Each spikelet is 2–3 mm ($^1/_{16}$–⅛ in.) long, containing a single floret with three anthers, awnless. July–October.

Fruit: Brown, ellipsoid caryopses, 1–2 mm long and 0.5 mm wide, slightly compressed. July–October.

Stems: Branching stems creep along the ground and root at the nodes where they con-

tact the soil. Stems are round in cross section or slightly flattened, swollen at the nodes, often slightly purple.

Root system: Deep roots make this species extremely drought tolerant. Mature roots are yellow or brown, while young roots are white. Produces rhizomes that contribute to lateral spread.

Reproduction: By abundant seeds, dispersed by water, soil movement, machinery, as a contaminant of agricultural seed, and in livestock feed and bedding. Seeds remain viable in the soil seed bank for three to four years. Vegetative spread also occurs via creeping stolons and rhizomes.

Impact: Many cultivars—including 'Tifgreen,' 'Riviera,' and 'Wrangler'—have been developed, and this species continues to be planted for erosion control and as a turf grass. Nevertheless, it is considered a noxious weed in several states and is a major weed of many crops worldwide. This species can escape cultivation and outcompete native vegetation in disturbed habitats. Due to its production of allelopathic compounds, it is highly aggressive. *Cynodon dactylon* also produces highly allergenic pollen, making it problematic for human health.

Control: Small patches should be dug out, taking care to remove all rhizomes and stolons. If using a solarization method, mow, irrigate, and cover with a black plastic sheet for six to eight weeks in summer. Grass-specific herbicides such as sethoxydim can be used in early spring, with follow-up applications through the growing season. Nonselective herbicides such as paraquat and glyphosate are most effective in the spring or autumn, when rhizomes are growing.

Inflorescence of *Cynodon dactylon.* Image by Doug Goldman, hosted by the USDA-NRCS PLANTS Database.

Cynodon dactylon plant. Image by Steve Dewey, Utah State University, Bugwood.org.

Cynodon dactylon fruits. Image by Steve Hurst, hosted by the USDA-NRCS PLANTS Database.

Cynodon dactylon leaves and stem. Image by Doug Goldman, hosted by the USDA-NRCS PLANTS Database.

Scrambling growth habit of *Cynodon dactylon.* Image by Doug Goldman, hosted by the USDA-NRCS PLANTS Database.

Cynodon dactylon florets. Image by Doug Goldman, hosted by the USDA-NRCS PLANTS Database.

Poaceae

Dactylis glomerata

orchardgrass · cat grass · cocksfoot grass · cockspur

Origin: Native to Europe, Asia, and northern Africa. This species was introduced to the eastern United States in 1760 as a pasture species and is still widely planted for forage.

Description: A cool-season perennial bunchgrass that grows in dense tussocks with upright flowering stems.

Habitat: Meadows, orchards, pasture, grassland, open woodland, forest, riparian areas, freshwater wetlands, and roadsides. Adapted to well-drained, fertile soils with moderate moisture and temperature. It is shade tolerant and tends to grow in areas with a history of disturbance.

Height: Flowering stems 30–150 cm (12–60 in.).

Foliage: Leaves are folded in the bud. Leaf blades are 20–50 cm (8–20 in.) long and 3–11 mm (⅛–½ in.) wide, strongly keeled with a prominent midvein on the underside, hairless or covered in tiny prickles giving them a rough texture. Leaf margins are minutely toothed. The ligule is a membrane 5–10 mm (¼–½ in.) long, occasionally fringed with short hairs. The leaf sheath is smooth, keeled, with margins fused together to form a closed tube except toward the top.

Flowers: Inflorescence is an upright and stiff panicle, 7–20 cm (3–8 in.) long, with spikelets in a compact cluster. Each spikelet is 5–9 mm (¼–⅜ in.) long and contains two to six florets. Florets are flat, with three anthers, awnless or with a terminal awn to 1 mm ($\frac{1}{32}$ in.). May–September.

Fruit: Ellipsoid caryopses, 2.5–3 mm ($\frac{1}{16}$–⅛ in.) long and 1 mm ($\frac{1}{32}$ in.) wide. June–October.

Stems: Hairless stems, flattened at the base.

Root system: An extensive fibrous root system that produces a dense sod. This species produces very short rhizomes or no rhizomes.

Reproduction: By abundant seed, dispersed by wind, animals, and human activity. It also spreads vegetatively through tillers.

Impact: This species is widespread and is a valuable pasture grass in many places. Its negative impact is relatively small, although it is considered weedy or invasive in parts of the United States and in other countries around the world. *Dactylis glomerata* has been found to suppress native plants in rare habitat types such as savannas, so it is worth monitoring to ensure that it does not spread into high-risk areas.

Control: Plants can be dug out, taking care to remove the root crown to prevent regrowth. So that seed does not set, large stands should be mowed before flowering. Repeated close mowing can eliminate this species entirely.

Dactylis glomerata plant. Image by Ohio State Weed Lab, The Ohio State University, Bugwood.org.

Dactylis glomerata florets. Image by Doug Goldman, hosted by the USDA-NRCS PLANTS Database.

Leaf sheath and ligule on *Dactylis glomerata.* Image by Rob Routledge, Sault College, Bugwood.org.

Dactylis glomerata inflorescence. Image by Catherine Herms, The Ohio State University, Bugwood.org.

Dactylis glomerata stems. Image by Ohio State Weed Lab, The Ohio State University, Bugwood.org.

Dactylis glomerata fruits. Image by Steve Hurst, hosted by the USDA-NRCS PLANTS Database.

Poaceae

Echinochloa crus-galli

barnyardgrass · barn grass · barnyard millet · Japanese millet · wild millet · watergrass · summergrass · cockspur · cocksfoot panicum · chicken panic grass · billion dollar grass · German grass

Origin: The native range of this species is somewhat obscure. This species is thought to have originated in tropical Asia and warmer regions of Europe but is now distributed throughout tropical and temperate regions worldwide. It was likely introduced to North America from Europe as a contaminant in crop seed in the early 1800s. Because it reduces crop yields significantly, *Echinochloa crus-galli* is considered one of the world's worst weeds.

Description: A tall, warm-season annual grass with thick stems that branch at the base. Stems can be solitary or grow in small tufts.

Habitat: Meadows, fields, grasslands, marshes, floodplains, lakeshores, streambanks, ditches, roadsides, and areas under cultivation. Prefers sunny habitats and wet soils.

Height: Flowering stems 30–200 cm (1–6 ft.).

Foliage: Leaves rolled in the bud. Leaf blades 10–50 cm (4–20 in.) long and 5–30 mm (¼–1 in.) wide, flat but with a light or purplish midvein that is raised underneath, typically smooth but occasionally with stiff, rough hairs. Leaves are slightly thickened at the margins. There is no ligule. The sheath is smooth and occasionally reddish.

Flowers: Inflorescence is a panicle, 10–25 cm (4–10 in.) long, with 15–25 branches, each 2–4 cm (¾–1½ in.) long. There are long bristles at the base of each branch that are typically equal to or longer than the spikelets. Each spikelet is 2.5–4 mm (¹⁄₁₆–⅛ in.) long, containing one or two florets. Florets have three anthers, awnless or with an awn to 1 cm (½ in.). Typically August–October, but occasionally earlier.

Fruit: The seed heads often turn purplish as fruits ripen. Ovoid or oblong, brownish caryopses, 1.3–2.2 mm long and 1–1.8 mm wide. September–October.

Stems: Stems grow flat along the ground or in upright tufts. Stem nodes are hairless. Stems and sheaths are often reddish near the base.

Root system: Plants have a fibrous root system with thick, white roots. Lower nodes also develop adventitious roots.

Reproduction: Some individuals produce up to 40,000 seeds per plant. Seeds are persistent in the seed bank for many years. They can be dispersed by water, birds, domestic animals, machinery, and in contaminated crop seed.

Impact: This species is a problematic weed in many crops because it removes almost all available soil nitrogen and its root exudates contain allelopathic compounds. It is also a host for several mosaic virus diseases. In natural habitats, this species is an environmental weed that can outcompete native vegetation, particularly in disturbed wetlands.

Control: Seeds need to be near the surface to germinate, so repeated shallow tilling during the spring or covering with mulch can reduce emergence. Hand pulling is recommended in small populations. Grass-specific herbicides such as sethoxydim, fluazifop, and haloxyfop and nonselective herbicides such as glyphosate, imazapyr, and paraquat offer effective control when applied to young plants.

Echinochloa crus-galli plants. Image by Howard F. Schwartz, Colorado State University, Bugwood.org.

Echinochloa crus-galli florets. Image by Doug Goldman, hosted by the USDA-NRCS PLANTS Database.

Echinochloa crus-galli collar with no ligule. Image by Steve Dewey, Utah State University, Bugwood.org.

Echinochloa crus-galli fruits. Image by D. Walters and C. Southwick, Table Grape Weed Disseminule ID, USDA APHIS PPQ, Bugwood.org.

Echinochloa crus-galli roots. Image by Steve Dewey, Utah State University, Bugwood.org.

Echinochloa crus-galli infructescence. Image by Steve Dewey, Utah State University, Bugwood.org.

Poaceae

Elymus repens

quackgrass · quackgrass rye · quickgrass · couch grass · common couch · dog grass · scutch · quitch · twitch · witchgrass · medusa's head

Origin: *Elymus repens* is native to most of Europe, Asia, and northwest Africa. It was introduced to North America from Eurasia in the 1600s, either accidentally as a seed contaminant or intentionally for use as forage or erosion control.

Description: A cool-season, perennial, rhizomatous grass. It forms a dense sod and can be very aggressive.

Habitat: Grasslands, croplands, pastures, gardens, roadsides, riverbanks, and other open disturbed areas. Grows in a wide variety of soil types and is both drought and salt tolerant.

Height Flowering stems are typically 50–100 cm (20–40 in.).

Foliage: Leaves rolled in the bud. Leaf blades are 15–40 cm long (6–16 in.) and 3–10 mm (⅛–½ in.) wide at the base, relatively flat, tapering to a fine point. Mostly dark green, sometimes with a powdery white bloom. The upper surface and margins are rough and slightly hairy, and the lower surface is smooth. There are fine ribs on both the upper and lower leaf surfaces. Leaves usually have a pair of prominent auricles at the base and a short, membranous ligule.

Flowers: The inflorescence is a slender spike, 8–17 cm (3–7 in.) long. Spikelets are compressed, 1–2 cm (½–¾ in.) long, with 3–8 florets. Each floret has three pale yellow anthers, 4–5.5 mm (⅛–¼ in.). May–September.

Fruit: Fruits are enclosed within the dried yellow-brown bracts, 5–10 mm (¼–½ in.) long, tapering to a point or tipped with a short, straight awn. The caryopses are 4–5 mm (⅛–¼ in.) long.

Stems: Stems are green or occasionally have a whitish bloom. Can be upright or grow along the ground.

Root system: Creeping rhizomes form thick mats. Rhizomes are pale yellow or straw-colored.

Reproduction: Primarily by seed. Seed remains viable in the soil seed bank for up to four years. Spreading rhizomes also give rise to new plants.

Impact: This species is a significant agricultural and horticultural weed throughout large parts of the United States and Canada. *Elymus repens* crowds out native species and cultivated crops.

Control: Annual spring burning and intensive early spring mowing can be used to control this species in prairies dominated by warm-season grasses. A grass-specific herbicide such as sethoxydim can be applied in the early spring or late autumn, when native grasses are dormant.

Elymus repens plants. Image by Ohio State Weed Lab, The Ohio State University, Bugwood.org.

Elymus repens infructescence. Image by Ohio State Weed Lab, The Ohio State University, Bugwood.org.

Inflorescences of *Elymus repens*. Image by Robert Vidéki, Doronicum Kft., Bugwood.org.

Elymus repens fruits. Image by Steve Hurst, hosted by the USDA-NRCS PLANTS Database.

Elymus repens rhizomes and roots. Image by Steve Dewey, Utah State University, Bugwood.org

Elymus repens sheath and auricles. Image by Ohio State Weed Lab, The Ohio State University, Bugwood.org.

Festuca arundinacea

tall fescue · alta fescue · coarse fescue ·
meadow fescue · reed fescue · Kentucky fescue

Origin: Native to Europe, Asia, and northern Africa, this species was introduced to North America in the 1800s as a lawn and pasture grass. It has escaped cultivation and spread into natural areas throughout the United States.

Description: A cool-season, perennial, rhizomatous bunchgrass. *Festuca arundinacea* is long-lived and forms a dense sod. Numerous cultivars have been developed, including 'Kentucky 31' which is still widely planted for lawns, turf, and pasture.

Habitat: Grasslands, old fields, pastures, forest edges, riparian areas, wetlands, roadsides, ditches, lawns, and areas under cultivation. Grows in a range of soil and climatic conditions. Tolerant of drought, poor drainage, low fertility, and salt.

Height: Flowering stems 50–200 cm (20–80 in.).

Foliage: Leaves are rolled in the bud. Leaf blades are dark green, flat, 10–70 cm (4–28 in.) long and 4–10 mm (⅛–½ in.) wide, conspicuously ridged on the upper surface due to prominent veins, smooth on the lower surface. Each leaf has rough margins, a yellowish base, a pair of small auricles fringed with hairs, and a short and membranous ligule up to 2 mm (¹⁄₁₆ in.) long.

Flowers: The inflorescence is an upright or hanging panicle, open or narrowly branched, 10–40 cm (4–16 in.) long, greenish white becoming purplish with age. Each spikelet is 9–14 mm (⅜–½ in.) long and 3 mm (⅛ in.) wide, containing three to eight florets. Florets have three stamens with anthers 3–4 mm (¹⁄₁₆–⅛ in.) long, awnless or with an awn to 4 mm (⅛ in.). May–June.

Fruit: Fruits are enclosed in yellow-brown bracts, 4–9 mm (⅛–⅜ in.) long, with awns up to 4 mm (⅛ in.) long. Each caryopsis is dark brown, 3–5 mm (⅛–¼ in.) long, with a groove running its length. May–July.

Stems: Stems are unbranched, round in cross section, and hairless. Joints are swollen and light green.

Root system: Coarse roots and short rhizomes form dense mats. The root systems can be extensive and quite deep, penetrating to a depth of 150 cm (60 in.) in moist soils.

Reproduction: By seed, dispersed by birds and grazing mammals. Seeds remain viable in the soil seed bank for several years. Vegetative reproduction and spread occur via tillers and rhizomes.

Impact: This species forms dense stands that displace native vegetation. It is most problematic in rare habitats, such as prairie remnants. Although *F. arundinacea* is a popular pasture grass, it should be used with caution. Several varieties are infected with a symbiotic fungus that makes the grass more drought resistant and efficient at nutrient uptake but also produces alkaloids that can be toxic to cattle and horses.

Native Alternatives

Calamagrostis canadensis (Canada bluejoint)
Calamovilfa longifolia (prairie sandreed)
Hesperostipa spartea (porcupine grass)
Panicum virgatum (switchgrass)
Schizachyrium scoparium (little bluestem)
Tridens flavus (purpletop tridens)

Control: Because of the extensive root system, control is difficult. Mowing or disking followed by reseeding with a native grass species is recommended. Mowing or disking can be combined with herbicide. Chemical treatment should occur in spring when plants are actively growing. Effective herbicides include glyphosate, metsulfron, imazapic, imazapyr, fluazifop, and sethoxydim.

Festuca arundinacea plants. Image by James H. Miller and Ted Bodner, Southern Weed Science Society, Bugwood.org.

Festuca arundinacea inflorescence. Image by James H. Miller and Ted Bodner, Southern Weed Science Society, Bugwood.org.

Left: Leaves of *Festuca arundinacea.* Image by Ohio State Weed Lab, The Ohio State University, Bugwood.org. *Right:* Leaf collar and joints of *Festuca arundinacea.* Image by James H. Miller and Ted Bodner, Southern Weed Science Society, Bugwood.org.

Festuca arundinacea fruits. Image by Steve Hurst, hosted by the USDA-NRCS PLANTS Database.

Festuca arundinacea infructescence. Image by Patrick J. Alexander, hosted by the USDA-NRCS PLANTS Database.

Festuca arundinacea root system. Image by Ohio State Weed Lab, The Ohio State University, Bugwood.org.

Poaceae

Festuca pratensis
meadow fescue · meadow ryegrass

Origin: Native to Europe and many parts of Asia, although its precise native range has been obscured by its long use as a forage species. *Festuca pratensis* was introduced to North America in the early 1800s and has been planted extensively for pasture, forage, turf, and soil stabilization. This species occasionally spreads to natural areas

Description: A cool-season, perennial, rhizomatous bunchgrass. It is very similar to *Festuca arundinacea* but lacks hairs on the auricles and does not have conspicuously ridged leaf blades.

Habitat: Fields, meadows, roadsides, old pastures, wetlands, riparian areas, forest margins, and forests. It generally grows on moist and rich soils, especially those that are loamy or heavy.

Height: Flowering stems typically 30–120 cm (12–48 in.).

Foliage: Leaf blades are flat, 30–120 cm (12–48 in.) long and 3–7 mm (⅛–¼ in.) wide. Leaf blades and sheaths are smooth, with conspicuous smooth auricles. Ligules are short and membranous.

Flowers: Inflorescence is a simple to very compound panicle, green to purplish, 10–25 cm (4–10 in.) long, upright or nodding at the tip. The internodes of the panicles are less than twice as long as the spikelets and fold against the stem before and after flowering. Spikelets are 10–15 mm (½–⅝ in.) long, made up of clusters of 4–10 flowers. Each floret has three stamens with pale yellow anthers, 2–4 mm (1/16–⅛ in.). June–August.

Fruit: Fruits are 2.5–4 mm (1/16–⅛ in.) long and narrow, usually lacking awns but occasionally with short awns to 2 mm (1/16 in.). July–September.

Stems: Stems are upright, smooth, hairless, occasionally purplish.

Root system: Tough and coarse roots form dense mats. Tillers and short rhizomes contribute to the persistence and lateral spread of the species.

Reproduction: By seed, most likely dispersed by animals, wind, and water. Seeds remain viable for up to one year and do not accumulate in the soil seed bank. Vegetative reproduction also occurs via tillers and rhizomes.

Impact: The largest impact of this species is related to the conversion of natural areas into pastures, resulting in a dramatic loss of diversity. Like *F. arundinacea*, this species can form associations with fungal symbionts that produce alkaloids toxic to grazing mammals. In natural areas, this species has been found to outcompete native species, such as the endangered *Physaria globosa* (Short's bladderpod).

Control: A combination of disking and summer application of a grass-specific herbicide such as sethoxydim is effective, followed by the reintroduction of native species.

Festuca pratensis stand. Image by Matt Lavin, used under a CC BY-SA 2.0 license.

Developing infructescence of *Festuca pratensis*. Image by T. Voekler, used under a CC BY-SA 3.0 license.

Festuca pratensis leaf collar and auricles. Image by Kristian Peters, used under a CC BY-SA 3.0 license.

Festuca pratensis fruits. Image by Steve Hurst, hosted by the USDA-NRCS PLANTS Database.

Festuca pratensis florets. Image © Marilee Lovit, hosted by the Native Plant Trust.

Joints on the stems of *Festuca pratensis*. Image © Marilee Lovit, hosted by the Native Plant Trust.

Poaceae

Holcus lanatus

common velvetgrass · Yorkshire fog · fog grass
· creeping soft grass · meadow softgrass · soft
meadow grass · woolly soft grass · mesquite grass

Origin: Native to Europe, western Asia, and northwestern Africa. This species was likely introduced to North America numerous times in the 1700s, either as an unintentional contaminant or intentional constituent of forage seed. It was well established by the early 1800s and is now present in much of the United States.

Description: A cool-season, perennial, tufted grass. Its life span varies dramatically with environmental conditions, however, and it grows as an annual or short-lived plant in parts of its range.

Habitat: Grasslands, open woodlands, pastures, orchards, croplands, and roadsides. Tolerates a range of conditions but prefers moist and fertile soils.

Height: 20–100 cm (8–40 in.).

Foliage: Leaves are rolled in the bud. Leaf blades are grayish green, flat, 5–20 cm (2–8 in.) long and 3–10 mm (⅛–½ in.) wide, tapering to a point at the tip. Each leaf has a prominent midrib, and all parts are covered in soft hairs. There is a membranous ligule 2–4 mm ($\frac{1}{16}$–⅛ in.) long, topped with tiny hairs.

Flowers: The inflorescence is a dense panicle, 5–15 cm (2–6 in.) long and up to 5 cm (2 in.) wide, silvery to pinkish or purplish, compressed against the stem early in the growing season and opening slightly later. Each spikelet is 4–6 mm (⅛–¼ in.) long and contains an upper floret with only male parts and a lower floret with both male and female parts. All florets have three stamens with anthers 1.2–2.5 mm ($\frac{1}{32}$–$\frac{1}{16}$ in.) long. The male floret has a 2 mm ($\frac{1}{16}$ in.) awn that becomes hooked when dry. May–August.

Fruit: Fruits are enclosed in yellow-brown bracts, 4–6 mm (⅛–¼ in.) long. Caryopses are pale brown, tapering at both ends, 1.5–3 mm ($\frac{1}{16}$–⅛ in.) long, slightly triangular in cross section, with a groove running the length. June–September.

Stems: Stems grow together in compact tufts, variously creeping or upright. Stems are round in cross section and covered in soft hairs. The bases of the stems are white with pink veins.

Root system: This species lacks rhizomes. Fibrous roots are concentrated at shallow depths.

Reproduction: By prolific seed, dispersed by wind, water, animals, and human activity. Vegetative reproduction also occurs via creeping stolons.

Impact: This species can form dense swards that reduce or exclude native plants. It rapidly colonizes disturbed areas and outcompetes other species for moisture and nutrients. *Holcus lanatus* may produce allelopathic compounds that inhibit the growth of other species.

Control: The most effective control for small populations is hand pulling or hoeing. Intensive mowing before seed production can limit the establishment or spread of this grass. This species is susceptible to fluazifop, sethoxydim, diuron, and hexazinone; spot application of herbicide should be done in the spring when the first seed heads appear.

Holcus lanatus plants. Image by Matt Lavin, used under a CC BY 2.0 license.

Holcus lanatus infructescence. Image by Franz Xaver, used under a CC BY-SA 3.0 license.

Ligule of Holcus lanatus. Image by Harry Rose, used under a CC BY 2.0 license.

Holcus lanatus fruits. Image by Steve Hurst, hosted by the USDA-NRCS PLANTS Database.

Holcus lanatus florets. Image by Matt Lavin, used under a CC BY 2.0 license.

Stand of Holcus lanatus. Image by AnRO0002, used under a CC0 1.0 license.

Poaceae

Microstegium vimineum

Japanese stilt grass · Japanese grass · Mary's grass · Nepalgrass · basketgrass · microstegium · Nepalese browntop · Chinese packing grass

Origin: A native of temperate and tropical Asia, including Japan, Korea, China, Malaysia, and India. This species was accidentally introduced into the United States in the early 1900s, as packing material.

Description: A fast-growing, warm-season annual grass with creeping and scrambling stems that form dense mats. *Microstegium vimineum* resembles a small, delicate bamboo.

Habitat: Moist forests, ditches, streambanks, floodplains, wetlands, trail edges, and roadsides. This species can thrive in both full sun and deep shade, and it prefers moist soils that are high in nitrogen.

Height: Flowering stems are 50–120 cm (20–48 in.).

Foliage: Leaves alternate along the stem. Leaf blades taper at both ends, 5–8 cm (2–3 in.) long and 5–15 mm (¼–⅝ in.) wide, with a distinctive silver midvein on the upper surface. Leaf blades have smooth margins; are slightly hairy on both sides; and are pale green, turning slightly purple in the autumn. Each leaf has a membranous ligule 0.5–2 mm long, usually with tiny hairs.

Flowers: Inflorescence is a cluster of one to six spikes diverging from a common point of attachment, each 2–5 cm (¾–2 in.) long, located terminally or arising from the leaf axils. Spikelets are bristly, arranged in pairs along the stems, with one floret per spikelet. Each floret has from zero to three stamens with anthers 0.7–1 mm (¹⁄₃₂ in.) long. September.

Fruit: Fruits are enclosed in yellow-brown, bristly bracts, 4–8 mm (⅛–¼ in.) long. The caryopses are ellipsoid, 2.8–3 mm (⅛ in.) long, ranging in color from yellow to olive to reddish, and often with a twisted awn, 3–8.5 mm (⅛–¼ in.) long. September–October.

Stems: Slender, hairless stems are branched with a creeping or scrambling growth form that blankets the ground and surrounding vegetation.

Root system: The root system is weak and shallow, making it possible to pull plants out of the ground easily. Adventitious roots grow from nodes along lower sections of the stem that touch the ground.

Reproduction: Each plant can produce 100–1,000 seeds per growing season. Seeds persist in the soil seed bank for one to five years and germinate readily following soil disturbance. Plants also reproduce vegetatively by rooting at the nodes.

Impact: Due to its intense competition for light and space, this aggressive species displaces native vegetation. There is also evidence that it produces allelopathic chemicals that limit growth of other plant species.

Control: Hand pulling and mowing are most effective if they are performed later in the growing season, just before plants produce mature seed. To exhaust the seed bank, multiple removals might be necessary. This species can also be controlled with grass-selective herbicides such as sethoxydim.

Microstegium vimineum plants. Image by Rebekah D. Wallace, University of Georgia, Bugwood.org.

Leaves of *Microstegium vimineum*. Image by Leslie J. Mehrhoff, University of Connecticut, Bugwood.org.

Microstegium vimineum inflorescence. Image by Leslie J. Mehrhoff, University of Connecticut, Bugwood.org.

Microstegium vimineum fruits. Image by Steve Hurst, hosted by the USDA-NRCS PLANTS Database.

Microstegium vimineum roots. Image by Leslie J. Mehrhoff, University of Connecticut, Bugwood.org.

Microstegium vimineum in autumn. Image by Chris Evans, University of Illinois, Bugwood.org.

Microstegium vimineum infestation. Image by Chris Evans, University of Illinois, Bugwood.org.

43

Poaceae

Miscanthus sinensis

Chinese silvergrass · Chinese plume grass ·
Japanese silvergrass · eulalia · zebra grass ·
eulaliagrass · maiden grass · Susuki grass ·
porcupine grass

OIPC Invasive

Origin: Native to eastern Asia, including China, Japan, Taiwan, Korea, and the Pacific coast of Russia. It is unclear when this species was first introduced to North America, but records indicate that it had escaped cultivation and established in natural areas by the 1940s. Despite being classified as weedy or invasive in many states, *Miscanthus sinensis* is still widely used as an ornamental plant. This species has been crossed with *Miscanthus sacchariflorus* to form the hybrid *Miscanthus* x *giganteus,* which is being planted and promoted widely as a biofuel.

Description: A tall, perennial grass that forms dense clumps. This species has an upright growth form and distinctive fan-shaped inflorescences

Habitat: Primarily found in areas with human disturbance, including forest edges, old fields, powerline rights-of-way, and along roadsides and railroads. It is adaptable to a variety of soil types and prefers full sun.

Height: 2–3 m (6–10 ft.) tall.

Foliage. Leaves are folded in the bud. Individual leaves are up to 1 m (40 in.) long and 1–2 cm (½–¾ in.) wide, tapering to a sharp tip. Blades are slightly keeled and have silvery midveins and rough or serrated margins. Each leaf has a fringed membranous ligule, 1–2 mm ($^1/_{32}$–$^1/_{16}$ in.) long. Leaves can be striped green and white, as in the 'Variegatus' and 'Zebrinus' cultivars.

Flowers The inflorescence is a terminal panicle that ranges from silver to light pink in color. The fan-shaped panicle is 20–60 cm (8–24 in.) long and consists of an aggregate of racemes 10–20 cm (4–8 in.) long. Spikelets are

in pairs, 4–6.5 mm (⅛–¼ in.) long, surrounded by a ring of silky hairs 5–8 mm (¼ in.) long. Florets have between zero and three stamens with anthers about 2.5 mm ($^1/_{16}$ in.), and an awn 8–10 mm (¼–½ in.) long that is spirally twisted at its base. September–November.

Fruit: Mature fruits enclosed in bracts are 3–4 mm ($^1/_{16}$–⅛ in.) long, yellowish brown, and encircled at the base with white or purplish hairs. The caryopses are ellipsoid, about 2 mm ($^1/_{16}$ in.) long. Fruits remain on the plants from September–January.

Stems: Stems are upright or arching and unbranched.

Root system: Mature plants have extensive perennial root systems, including rhizomes.

Reproduction: Seeds are dispersed by wind but have low dormancy. Spread within populations is primarily by underground roots or rhizomes.

Impact: The species forms large clumps in disturbed areas, displacing native vegetation. The grass is also extremely flammable, which increases fire risk in areas where it has established.

Control: Mechanical control can be carried out by hand pulling seedlings and shallow-rooted plants. To prevent sprouting from the root system, larger plants must be dug out entirely. Mowing is not recommended, because it can facilitate the spread of this species into new habitats. Periodic spot spraying with glyphosate or imazapyr can control this species.

44

Miscanthus sinensis inflorescence. Image by Leslie J. Mehrhoff, University of Connecticut, Bugwood.org.

Cultivated *Miscanthus sinensis* plant. Image by James H. Miller, USDA Forest Service, Bugwood.org.

Miscanthus sinensis infructescence. Image by Chris Evans, University of Illinois, Bugwood.org.

45

Miscanthus sinensis fruits. Image by Steve Hurst, hosted by the USDA-NRCS PLANTS Database.

Left: Miscanthus sinensis leaves. Image by Britt Slattery, US Fish and Wildlife Service, Bugwood. org. *Right: Miscanthus sinensis* leaves and collar. Image by James H. Miller, USDA Forest Service, Bugwood.org.

Miscanthus sinensis root crowns. Image by Leslie J. Mehrhoff, University of Connecticut, Bugwood.org.

Poaceae

Phalaris arundinacea

reed canary grass · gardener's garters · ribbon grass

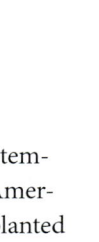

Origin: *Phalaris arundinacea* is native to temperate parts of Europe, Asia, and North America. Nonnative strains, which have been planted in the United States since the 1800s, have become problematic following their repeated introduction for forage, erosion control, bioenergy, fiber production, and phytoremediation.

Description: A cool-season perennial grass that forms large, monospecific stands. Although there is no reliable way to determine the genetic origin of an individual plant, it is generally accepted that aggressive varieties are nonnative.

Habitat: Wetlands, ditches, streambanks, wet meadows, and pastures. It is most successful in sites with disturbed, fertile, and moist soils.

Height: 70–150 cm (28–60 in.).

Foliage: Leaf blades are flat, 10–20 cm (4–8 in.) long and 1–2 cm (½–¾ in.) wide, tapering gradually from the base to the tip. Leaves are hairless but have a rough texture on both surfaces. There is a prominent, membranous ligule, 4–11 mm (⅛–½ in.), and no auricles.

Flowers: A compact panicle 7–25 cm (3–10 in.) long, upright or slightly spreading. Flowers are green, turning purple at maturity. Spikelets are 5 mm (¼ in.) long, typically containing three florets, two of which are smaller and infertile. The fertile florets have three stamens with anthers 2.5–3 mm (¹⁄₁₆–⅛ in.). May–July.

Fruit: The inflorescence color changes to tan as seeds develop. The open branches of the inflorescence draw close to the stem at maturity. Caryopses are shiny brown, 3–5 mm (⅛–¼ in.) long and 1 mm (¹⁄₃₂ in.) wide. June–August.

Stems: Stems are approximately 1 cm (½ in.) in diameter, hairless, and upright.

Root system: Produces short, stout rhizomes that root at the nodes and form a thick, fibrous root system.

Reproduction: Spreads aggressively via rhizomes. Seeds—which can germinate immediately once mature—are dispersed by water, animals, and machinery.

Impact: This species displaces native plants, and the resulting monospecific stands provide poor habitat for wildlife. It is one of the first plants to emerge in spring. After seeds have matured, the stems collapse and form a dense mat that prevents the growth of other plants. Once established, *P. arundinacea* is difficult to eradicate because of the large seed bank and extensive root system.

Control: For the soil seed bank and stored reserves in the rootstock to be exhausted, sites need to be treated for at least three to five years. Hand pulling or digging in late spring before seeds develop can be effective in small populations. To prevent regrowth, complete removal of roots is necessary, and plants must be bagged and removed from the site because seeds can continue to develop after plants have been pulled. Solarization with black plastic can be used on small populations, but plants must be covered for a full growing season. Close mowing is recommended in spring after first

Native Alternatives

Calamagrostis canadensis (Canada bluejoint)
Chasmanthium latifolium (Indian woodoats)
Glyceria grandis (American mannagrass)
Panicum virgatum (switchgrass)
Sorghastrum nutans (Indiangrass)
Spartina pectinata (prairie cordgrass)

growth, at flowering to prevent seed development, and in fall, to reduce growth the following year. In larger populations, plowing, tilling, and disking can reduce the density of stands but may also contribute to the spread of rhizomes. In some cases a combination of mechanical and chemical methods is most successful. Postemergence foliar application of broad-spectrum herbicides such as glyphosate or imazapyr and grass-targeting herbicides such as sethoxydim or fluazifop are effective in this species. It is strongly encouraged that native species be reseeded or planted after control treatments have been used.

Phalaris arundinacea leaves. Image by Leslie J. Mehrhoff, University of Connecticut, Bugwood.org.

Phalaris arundinacea ligule. Image by Caleb Slemmons, National Ecological Observatory Network, Bugwood.org.

Phalaris arundinacea inflorescence. Image by Leslie J. Mehrhoff, University of Connecticut, Bugwood.org.

Phalaris arundinacea infructescence. Image by Leslie J. Mehrhoff, University of Connecticut, Bugwood. org.

Phalaris arundinacea fruits. Image by Steve Hurst, hosted by the USDA-NRCS PLANTS Database.

Phalaris arundinacea infestation. Image by Leslie J. Mehrhoff, University of Connecticut, Bugwood.org.

Poaceae

Phleum pratense

Timothy grass · common cat's tail · meadow
cat's tai · herd grass

Origin: Native to Europe, western and central Asia, and northwestern Africa. *Phleum pratense* is highly palatable and has been cultivated for forage in many parts of the world. Most likely introduced to North America in the early 1700s as a pasture grass, this species has escaped cultivation and is now widespread across most of the United States.

Description: A cool-season, perennial, tufted grass that can grow as single stems or in loose to dense clumps. It starts growing early in the spring, giving it a competitive advantage over other species. The cylindrical inflorescence distinguishes *P. pratense* from most other grasses.

Habitat: Grasslands, meadows, pastures, fencerows, forest margins, woodlands, ditches, roadsides, and other disturbed areas. It can grow in full sun to partial shade and tolerates a range of soil types but does best in nutrient-rich soils.

Height. Flowering stems are typically 50–150 cm (20–60 in.).

Foliage: Leaves are rolled in the bud. Leaf blades are flat, 20–45 cm (8–18 in.) long and 5–10 mm (¼–½ in.) wide, tapering toward the tip. Blades are grayish green and distinctly veined with a prominent midrib. Leaf margins are rough but the blades and sheaths are hairless. There are no auricles and the ligule is elongate, white, membranous, 2–4 mm ($\frac{1}{16}$–⅛ in.) long.

Flowers: The inflorescence is a cylindrical, spikelike panicle, 3–15 cm (1–6 in.) long and 5–10 mm (¼–½ in.) wide. It is grayish or whitish green while immature. Spikelets are densely packed together around the stem along the entire length of the inflorescence. Each spike-

let is 3–4 mm (⅛ in.) long and contains a single white floret, which turns purple with age, with three stamens bearing showy anthers, 2 mm ($\frac{1}{16}$ in.) long. The floret is held within two strongly keeled glumes fringed with soft bristles and tipped with a rigid awn 1–2 mm ($\frac{1}{32}$–$\frac{1}{16}$ in.) long. June–August.

Fruit: The spikelets turn pale tan as they mature. Caryopses are ovoid, 2 mm ($\frac{1}{16}$ in.) long and 1 mm ($\frac{1}{32}$ in.) wide, tapering at both ends. August–September.

Stems: The flowering stems, upright and unbranched, are often swollen at the base forming a small bulblike structure.

Root system: A moderately shallow and fibrous root system with no rhizomes.

Reproduction: Primarily from abundant seed, dispersed by wind, livestock, and other means. Seeds typically remain viable in the soil seed bank for four to five years. Vegetative reproduction also occurs via tillers.

Impact: Although valuable as forage, this species is a concern in natural areas because it spreads quickly and excludes native plants. It also produces large amounts of highly allergenic pollen.

Control: Hand pulling is effective in small populations. Frequent clipping or mowing close to the ground can reduce the overall vigor. Glyphosate can be applied in the early spring to control dense stands of this species.

Phleum pratense plants. Image by Franz Xaver, used under a CC BY-SA 3.0 license.

Roots and swollen stem base of *Phleum pratense*. Ohio State Weed Lab, The Ohio State University, Bugwood.org.

Phleum pratense inflorescences. Image by David Cappaert, Bugwood.org.

Phleum pratense infructescence. Image by Jose Hernandez, hosted by the USDA-NRCS PLANTS Database.

Phleum pratense sheath and ligule. Image by Matt Lavin, used under a CC BY-SA 2.0 license.

Phleum pratense fruits. Image by Steve Hurst, hosted by the USDA-NRCS PLANTS Database.

Poaceae

Phragmites australis ssp. *australis*

common reed · giant reed · ditch reed · reed grass · phragmites · yellow cane

Origin: There are three recognized subspecies of *Phragmites australis* in the United States, all with different origins, which makes this a complicated species to identify and manage. Of these, two subspecies are found in Ohio—a native, *americanus,* and a nonnative, *australis.* The *australis* subspecies was accidentally introduced to the eastern United States in ship ballast during the late 1700s or early 1800s. *Phragmites australis* ssp. *australis* has spread throughout the United States and displaced both the native subspecies and other native plants.

Description: A tall, warm-season perennial grass that forms dense colonies. A number of characteristics can be used to distinguish the native and nonnative subspecies. Many traits overlap, so, for proper identification, it is important to have a clear match on several characteristics. In terms of growth habit, *americanus* grows in low-density stands comingled with native vegetation, whereas *australis* forms dense monocultures. The overall height of *americanus* generally does not exceed 2 m (6 ft.), whereas *australis* may reach 4 m (13 ft.). The leaves of *americanus* are yellowish green, whereas those of *australis* are blue green. The ligules of *americanus* are 1–2 mm ($\frac{1}{32}$–$\frac{1}{16}$ in.) long, while those of *australis* are less than 1 mm ($\frac{1}{32}$ in.) long. Once the leaf dies, the leaf sheaths in *americanus* loosen and typically fall off, while those of *australis* remain tight around the stem. The lower and middle stems of *americanus* are smooth, somewhat shiny and maroon or reddish brown, while those of *australis* are ribbed and rough, dull green or tan. The inflorescence of *americanus* is sparsely flowered, while that of *australis* is densely flowered. Finally, the glumes at the base of the spikelet are longer in *americanus,* which has a lower glume longer than 4 mm ($\frac{1}{8}$ in.) and an upper glume longer than 6 mm ($\frac{1}{4}$ in.), while in *australis* the lower glume is typically less than 4 mm ($\frac{1}{8}$ in.) and the upper is less than 6 mm ($\frac{1}{4}$ in.) (Swearingen and Saltonstall 2010).

Habitat: Tidal and nontidal wetlands, riverbanks, lake and pond shores, ditches, roadsides. Prefers full sun, wet conditions (including shallow water), and fertile soils. Tolerant of salt, partial shade, and highly acidic conditions.

Height: 2–4 m (6–13 ft.).

Foliage: Leaves are rolled in the bud. Blades are blue green, flat, 25–60 cm (10–24 in.) long and 2–6 cm ($\frac{3}{4}$–3 in.) wide, widest in the middle, and gradually tapering to a point at the tip. The margins are rough, and the sheaths are hairless and stay attached to the stem through the winter. There is a short ligule, less than 1 mm ($\frac{1}{32}$ in.) long, and no auricles.

Flowers: The inflorescence is a copiously branched panicle, 20–40 cm (8–16 in.) long, purplish when young, becoming grayish tan and feathery with age. Spikelets are 10–16 mm ($\frac{1}{2}$–$\frac{5}{8}$ in.) long, containing 3–10 florets each, with tufts of soft white hairs up to 1 cm ($\frac{1}{2}$ in.) long along the stalk between florets. At the base of each spikelet is a lower glume less than 4 mm ($\frac{1}{8}$ in.) and an upper glume less than 6 mm ($\frac{1}{4}$ in.). July–August.

Native Alternatives
Calamagrostis canadensis (Canada bluejoint)
Sorghastrum nutans (Indiangrass)
Spartina pectinata (prairie cordgrass)

Phragmites australis ssp. *australis* plants. Image by Leslie J. Mehrhoff, University of Connecticut, Bugwood.org.

Phragmites australis ssp. *australis* leaf collar. Image by Rebekah D. Wallace, University of Georgia, Bugwood.org.

Phragmites australis ssp. *australis* inflorescence. Image by Leslie J. Mehrhoff, University of Connecticut, Bugwood.org.

Phragmites australis ssp. *australis* infructescence. Image by Leslie J. Mehrhoff, University of Connecticut, Bugwood.org.

Phragmites australis ssp. *australis* fruits. Image by Ken Chamberlain, The Ohio State University, Bugwood.org.

Phragmites australis ssp. *australis* rhizomes. Image by Leslie J. Mehrhoff, University of Connecticut, Bugwood.org.

Phragmites australis ssp. *australis* stems. Image by Leslie J. Mehrhoff, University of Connecticut, Bugwood.org.

Fruit: The florets become tan and drop away when mature, leaving behind the glumes and hairy stalks, which persist on plants into winter. Caryopses are 2–3 mm (¹⁄₁₆–⅛ in.) long, enclosed within the floral bracts, which are fringed with long hairs around the base and have a 5–6 mm (¼ in.) awn. August–September.

Stems: Flowering stems are upright and unbranched. The stems are ribbed and rough, the upper parts green while the lower parts are dull green to tan. Stems are slow to deteriorate, and dead stems often remain standing through the next season. Reddish stolons grow along the surface of the ground.

Root system: This species has stout, scaly, creeping rhizomes that form extensive mats. Rhizomes can reach 20 m (65 ft.) in length and 2 m (6 ft.) in depth. Roots are produced at the nodes of the rhizomes.

Reproduction: By seed, primarily dispersed by wind and water. Root fragments can also be moved to new sites via water, animals, and human activities such as boating. Vegetative reproduction by rhizomes and stolons contributes to the persistence and spread of colonies within a site.

Impact: This plant forms dense thickets that outcompete native plants, alter hydrology and wildlife habitat, and block sunlight to the aquatic community.

Control: Early recognition is important because once it is established, this species is difficult to eradicate, due to carbohydrate storage in underground structures. Over several years, repeated cutting in spring and before seeding in autumn can deplete root reserves. In wet areas, mowing in winter minimizes the effects of stem biomass on other plant species and wildlife. Cutting stems below water level can also reduce growth. In small stands, cutting followed by cut stem application of aquatic formulations of glyphosate has been effective. Because this species often grows in or near standing water, it is essential that only licensed aquatic herbicide applicators undertake any chemical management. Whatever control methods are used, careful monitoring and repeated treatment are necessary.

Poaceae

Poa annua

annual bluegrass · speargrass · walkgrass · causeway grass · dwarf meadow grass · six-weeks grass · wintergrass · poa · Michigan bent

Origin: Native to temperate Europe and Asia. Now found worldwide, primarily due to contamination of crop seed.

Description: A cool-season, annual (occasionally short-lived perennial) tufted grass that grows in dense clumps. It is similar in appearance to two related nonnative species that also have weedy tendencies: *Poa pratensis* (Kentucky bluegrass) and *Poa trivialis* (rough bluegrass). *Poa annua* can be distinguished by its small size, spreading growth form, lack of rhizomes, light color, short ligules, and crinkled leaf blades.

Habitat: Fields, margins of streams and lakes, moist or wet forests, areas under cultivation, pastures, gardens, lawns, golf courses, urban parks, and roadsides. Grows in a variety of soil types and can tolerate soil compaction, trampling, mowing, and freezing. Because of its shallow root system, *P. annua* requires relatively high and consistent moisture.

Height: Flowering stems usually 10–20 cm (4–8 in.) but occasionally to 30 cm (12 in.).

Foliage: Leaves folded in the bud. Blades are light green, flat to slightly keeled, 2–15 cm (¾–6 in.) long and 1–5 mm (¹⁄₃₂–¼ in.) wide, tapering abruptly to a canoe-shaped tip that is characteristic of the genus. The leaf blades have a prominent midrib and two parallel veins on either side. Margins are smooth, sheaths and blades are hairless. There is a silvery membranous ligule, 1–5 mm (¹⁄₃₂–¼ in.) long, and no auricles.

Flowers: Inflorescence is a triangular or ovoid panicle, 2–8 cm (¾–3 in.) long, with solitary spikelets branching off the main stem. Spikelets are whitish green, flattened, 3–5 mm (⅛–¼ in.), each containing three to six florets.

Each floret has three anthers, 0.5–1 mm (¹⁄₆₄–¹⁄₃₂ in.) long. April–May.

Fruit: Bracts covering the seed are 2–2.5 mm (¹⁄₁₆ in.) long, with multiple hair-lined ribs. Caryopses are ovoid, 1–1.2 mm (¹⁄₃₂ in.). May–June.

Stems: Flowering stems are smooth and somewhat flattened. Stems are often bent at the base and rooted at the lower nodes.

Root system: Roots are fibrous, spreading horizontally, only 2–3 cm (¾–1 in.) deep.

Reproduction: By seed, spread by animals, water, and wind and as a contaminant in crop or turf seed. An individual plant can produce 1000–2000 seeds in a growing season. Seeds persist in the seed bank for four to six years, but if conditions are suitable they can germinate at any time in the year.

Impact: This species is a major weed of crops such as vegetables, cereals, and sugar beets. It is also problematic in vineyards, orchards, residential and commercial turf, and gardens.

Control: Individual plants can be pulled by hand or dug out before seed set. In lawns and turf, the most effective control is to use deep and infrequent irrigation, which will limit the shallow-rooted *P. annua* while still allowing deeper-rooted grass species to thrive. There is high genetic variability in this species, and many varieties are resistant to common herbicides. Spot application of haloxyfop is effective.

Poa annua plant. Image by Shawn Wright, University of Kentucky, Bugwood.org.

Poa annua inflorescence. Image by AnR00002, used under a CC0 1.0 license.

Poa annua ligule. Image by Doug Goldman, hosted by the USDA-NRCS PLANTS Database.

Poa annua fruits. Image by Steve Hurst, hosted by the USDA-NRCS PLANTS Database.

The canoe-shaped tip on the leaf is an identifying feature in the *Poa* genus. Image by Doug Goldman, hosted by the USDA-NRCS PLANTS Database.

Poa annua florets. Image by Doug Goldman, hosted by the USDA-NRCS PLANTS Database.

Poaceae

Setaria pumila

yellow foxtail · yellow bristle grass · pigeon grass · cattail grass

Origin: The native range of this species is somewhat obscure but is thought to have encompassed parts of Europe, Asia, and Africa. *Setaria pumila* has a long history of cultivation for its starchy grains and is now widespread in tropical and temperate regions worldwide. It was most likely accidentally introduced to North America in the early 1800s.

Description: A warm-season annual grass that grows in large tufts. Several weedy foxtails are common in Ohio. This one can be distinguished from other species by the presence of numerous (5–20) yellow bristles at the base of each spikelet, upper glumes about half as long as the spikelet, leaf sheaths that are smooth on the margins, and long white hairs on the upper surfaces of the leaf blades.

Habitat: Fields, riverbanks, lawns, gardens, areas under cultivation, and roadsides. It prefers full sun but otherwise is adapted to a wide range of soil types and environmental conditions.

Height: Flowering stems 30–130 cm (1–4 ft.).

Foliage: Leaves are rolled in the bud. Blades are green to grayish blue, flat to slightly keeled, 4–30 cm (2–12 in.) long and 4–10 mm (⅛–½ in.) wide, occasionally loosely twisted, with a prominent midrib. The blade's lower surface is smooth and hairless; the upper surface is rough, with long white hairs near the base. The ligule is 1 mm (¹⁄₃₂ in.) long and fringed with short hairs at its base.

Flowers: Inflorescence is an upright, cylindrical, spikelike panicle, 3–15 cm (1–6 in.) long and 15–25 mm (⅝–1 in.) wide, yellowish, with spikelets densely packed together. The spikelets are 3–3.5 mm (⅛ in.) long, each containing one fertile and one sterile floret, green while immature, with 5–20 yellow to golden brown bristles 3–12 mm (⅛–½ in.) long surrounding the base. July–August.

Fruit: Caryopses are ovoid with a blunt point, flattened on one side, 2.5–3.3 mm (¹⁄₁₆–⅛ in.) long and 1.5–2.2 mm (¹⁄₃₂–¹⁄₁₆ in.) wide. The persistent upper glume is about half as long as the seed. August–September.

Stems: Usually upright, with multiple branches at the base. Stems are mostly hairless, green to purplish, and hairless at the joints.

Root system: Roots are short and fibrous.

Reproduction: By seed, dispersed by livestock, birds, and water and as a contaminant in crop seed.

Impact: This species reduces yields in alfalfa, soybeans, and small grain crops. Consumption of large quantities can cause stomatitis (inflammation of the oral cavity) in cattle and horses. It is also considered a nuisance weed in gardens and disturbed areas.

Control: Hand pulling or hoeing is recommended in small populations. Plants should be removed before seed is set. Covering areas with mulch can reduce germination and establishment. Effective herbicides include sethoxydim, dimethyl tetrachloroterephthalate (DCPA), and trifluralin.

55

Setaria pumila fruits. Image by Julia Scher, USDA APHIS PPQ, Bugwood.org.

Setaria pumila plants. Image by Robert Vidéki, Doronicum Kft., Bugwood.org.

Setaria pumila inflorescence. Image by Harry Rose, used under a CC BY 2.0 license.

Setaria pumila stems and collar region. Image by Matt Lavin, used under a CC BY 2.0 license.

Setaria pumila leaves. Image by John Cardina, Ohio State University / OARDC, Bugwood.org.

Setaria pumila spikelets and bristles. Image © Robert L. Carr.

Poaceae

Setaria viridis

green foxtail · green bristle grass · green panicum · green pigeon grass · green millet · wild foxtail millet · bottle grass

Origin: The native range is somewhat obscure but likely included parts of Europe, Asia, and northern Africa. This species, which has a long history of cultivation for its starchy grains, is thought to be the progenitor to *Setaria italica* (foxtail millet). It was most likely introduced to North America in the early 1800s in ship ballast, but multiple deliberate and accidental introductions have occurred since.

Description: A warm-season annual grass that grows in tufts. *Setaria viridis* is the first foxtail to flower in the spring, and development is rapid, so the species can complete several generations within a growing season. It can be differentiated from other foxtails by the presence of one to three light green or purple bristles at the base of each spikelet, an inflorescence that is upright or slightly nodding near the tip, a fringe of hairs along the outer margin of the leaf sheath, and a lack of long hairs on the leaf blade.

Habitat: Fields, lawns, gardens, areas under cultivation, and along roadsides and railroads. It prefers full to partial sun, moist conditions, and highly disturbed sites where soil has been exposed.

Height: Flowering stems 20–70 cm (8–28 in.), occasionally to 100 cm (40 in.).

Foliage: Leaves are rolled in the bud. Leaf blades are light green, flat, 5–40 cm (2–16 in.) long and 4–15 mm (⅛–⅝ in.) wide, drooping, finely but distinctly veined with a prominent midrib on the underside. The lower surface of the leaf blade is smooth, the upper surface is rough and hairless. The ligule is a fringe of hairs 1.5–2 mm long, fused at the base. The leaf sheath is short and fringed with hairs along the outer margin.

Flowers: Inflorescence is a cylindrical, spike-like panicle, 1–15 cm (½–6 in.) long and 4–14 mm (⅛–⅝ in.) wide, green to light brown or purple, upright or slightly nodding near the tip. The spikelets are 1.6–2.5 mm ($^{1}/_{32}$–$^{1}/_{16}$ in.) long, containing one fertile and one sterile floret, green while immature, with one to three light green or purple bristles 2–10 mm ($^{1}/_{16}$–½ in.) long surrounding the base. June–September.

Fruit: Caryopses are ellipsoid with blunt points, 1.8–2.2 mm ($^{1}/_{32}$–$^{1}/_{16}$ in.) long and 1–1.3 mm ($^{1}/_{32}$ in.) wide, enclosed within the floral bracts. The entire spikelet drops off when mature, leaving the bristles behind on the stem. July–October.

Stems: Stems are usually branched, often growing horizontally near the base, then bending to grow upright at the lower joints. Joints are hairless and green or purplish.

Root system: The root system is shallow and fibrous.

Reproduction: Seed is dispersed by livestock, birds, and water and as a contaminant in crop seed. Seeds remain viable in the soil seed bank for 15–20 years.

Impact: This species reduces yields in cereals such as wheat and corn. It commonly occurs in fruit orchards, vegetable crops, ornamentals, and plantation forests. In addition, *S. viridis* is a nuisance weed in gardens and disturbed areas.

Control: Hand pulling or hoeing is recommended in small populations. Because seed develops rapidly, plants should be removed as soon as they appear. Covering areas with mulch can reduce germination and establishment. Effective herbicides include sethoxydim, dimethyl tetrachloroterephthalate (DCPA), and triluralin.

Setaria viridis plant. Image © Katy Chayka, Minnesota Wildflowers.

Setaria viridis spikelets and bristles. Image by Stefan Lefnaer under the CC BY-SA 4.0 license.

Setaria viridis leaf collar and stem. Image by Bruce Ackley, The Ohio State University, Bugwood.org.

Setaria viridis fruits. Image by D. Walters and C. Southwick, Table Grape Weed Disseminule ID, USDA APHIS PPQ, Bugwood.org.

Setaria viridis inflorescences. Image by Matt Lavin, used under a CC BY 2.0 license.

Setaria viridis growing along a roadside. Image by Matt Lavin, used under a CC BY 2.0 license.

Poaceae

Sorghum bicolor

shattercane · wild cane · black amber · broom corn · chicken corn · guinea corn · sorghum · sweet sorghum · forage sorghum · great millet · milo · Sudangrass

Prohibited Noxious Weed

59

Origin: This species is most likely native to Africa, but it has been widely cultivated in tropical and subtropical regions of Africa, Asia, and Europe for thousands of years. *Sorghum bicolor* remains one of the most important cereal crops worldwide, both for human consumption and animal fodder. It was introduced to North America in the early 1800s for agricultural purposes but has escaped cultivation and is considered a noxious weed in many states.

Description: A warm-season annual grass that forms dense clumps. This species bears a superficial resemblance to corn, particularly at the seedling stage. The panicles of shiny, round, red or black seeds break apart easily, hence the common name shattercane. There are many varieties, cultivars, strains, and subspecies.

Habitat: Fields, disturbed soil, and areas under cultivation. Grows best in full sun in fertile, well-drained soils. It is not tolerant of frost, shade, or sustained flooding.

Height: 1–2 m (3–6 ft.), but some varieties can be 3 m (10 ft.) or taller.

Foliage: Leaves are rolled in the bud. Blades are flat, 40–70 cm (16–28 in.) long and 3–10 cm (1–4 in.) wide, with a prominent white midrib. Leaf blades are hairless and covered with a waxy cuticle. The membranous, slightly rounded ligule is 1–4 mm ($\frac{1}{32}$–$\frac{1}{8}$ in.) long. The sheath is smooth or sometimes covered with a white powder.

Flowers: Inflorescence is an upright panicle, 5–60 cm (2–24 in.) long, variably dense or open, with a cylindrical, pyramidal, or obovate shape. Spikelets are 3–7 mm ($\frac{1}{8}$–$\frac{1}{4}$ in.) long, each containing a single floret with three anthers 2–2.8 mm ($\frac{1}{16}$ in.) long. June–September.

Fruit: Caryopses are large, ellipsoid to nearly spherical, 4–5 mm ($\frac{1}{8}$–$\frac{1}{4}$ in.) long and 2–4 mm ($\frac{1}{16}$–$\frac{1}{8}$ in.) wide, often white or reddish brown but covered by the glumes, which range from creamy green to red to dark brown or black at maturity. July–October.

Stems: Stems are round in cross section, rigid and upright, 1–3 cm ($\frac{1}{2}$–1 in.) in diameter, with swollen, typically hairless nodes.

Root system: Roots are fibrous, and the root system is extensive. Can produce tillers but no rhizomes. As in corn, brace roots grow from the lower part of the *S. bicolor* stem to provide stability for tall plants.

Reproduction: By seed, dispersed by wind, water, animals, and agricultural activities. Seeds can remain viable in the soil seed bank for two to three years.

Impact: Although *S. bicolor* is economically important, plants that escape cultivation form dense stands that displace native species. This species competes aggressively with crop plants and can significantly reduce yield. In addition, it serves as an alternate host for pests and pathogens that harm crops. While this species is frequently used for forage, the leaves of plants that are exposed to frost, drought, trampling, or herbicides can produce hydrogen cyanide at levels poisonous to livestock.

Control: For small populations, hand pulling in early summer is recommended. Larger populations should be mowed or tilled before seed set. Spot application of dalapon or glyphosate is effective in this species.

Sorghum bicolor under cultivation. Image by Howard F. Schwartz, Colorado State University, Bugwood.org.

Sorghum bicolor inflorescences. Image by Howard F. Schwartz, Colorado State University, Bugwood.org.

Sorghum bicolor leaves. Image by Bruce Ackley, The Ohio State University, Bugwood.org.

Sorghum bicolor infructescence. Image by Bruce Ackley, The Ohio State University, Bugwood.org.

Sorghum bicolor infructescences. Image by Felicien Amakpe, Cercle nature et développement CENAD NGO, Bugwood.org.

Sorghum bicolor fruits. Image by Steve Hurst, hosted by the USDA-NRCS PLANTS Database.

Poaceae

Sorghum halepense

Johnsongrass · grass sorghum · Arabian millet · Egyptian millet · Morocco millet · evergreen millet · millet-grass · false guinea grass · means-grass · Egyptian grass · Syrian grass · Aleppo grass · Aleppo millet grass

Origin: The precise native range of this species is obscure, but it is thought to have included the Mediterranean region of Europe, southern and western Asia, and northern Africa. *Sorghum halepense* has been cultivated as a fodder species and now has a worldwide distribution. It was introduced to North America multiple times during the 1800s, both deliberately, as a forage crop, and accidentally, as a seed contaminant. This species has escaped cultivation and is considered a noxious weed in at least 19 states.

Description: A warm-season perennial grass that spreads by rhizomes to form dense stands. It is similar in appearance to the nonnative *Sorghum bicolor,* but *S. bicolor* is annual and has a more dense and upright inflorescence. The foliage also resembles the native *Tripsacum dactyloides* (gama grass), but *T. dactyloides* has an inflorescence with spikes of flowers diverging from a common attachment, whereas *S. halepense* has a spreading panicle.

Habitat: Fields, meadows, pastures, areas under cultivation, streambanks, ditches, along roadsides and railroads, and forest margins. Prefers open, disturbed, moist, and fertile soils. Moderately tolerant of salt and drought.

Height: Flowering stalks typically 1–2 m (3–6 ft.), occasionally to 3 m (10 ft.).

Foliage: Leaves are rolled in the bud. Leaf blades are elongate, flat, 20–80 cm (8–32 in.) long and 1–3 cm (½–1 in.) wide, with a prominent white midrib that forms a keel on the underside. Blades are hairless but rough on the underside and margins. There is a prominent, membranous ligule, 3–6 mm ($\frac{1}{16}$–¼ in.) long, often with a fringe of hairs along the top.

Flowers: Inflorescence is an open panicle, 15–50 cm (6–20 in.) long, pale green to purplish, with a pyramidal shape. Spikelets are in groups of two or three; one spikelet of each pair or triplet is fertile with both male and female parts, the others are sterile or only male. Spikelets are hairy, ovoid, 4–7 mm (⅛–¼ in.) long, some with a twisted awn to 15 mm (⅝ in.). Each fertile spikelet contains a single floret with three anthers 1.9–2.7 mm ($\frac{1}{16}$ in.) long and a pair of feathery stigmas. May–July.

Fruit: Caryopses remain enclosed by the glumes, 4–7 mm (⅛–¼ in.) long and 2–3 mm ($\frac{1}{16}$–⅛ in.) wide, reddish brown to shiny black, glossy and finely lined on the surface. The seed within is ovoid, reddish brown, 2–3 mm ($\frac{1}{16}$–⅛ in.) long and 1–2 mm ($\frac{1}{32}$–$\frac{1}{16}$ in.) wide. July–October.

Stems: Stems are round in cross section; upright; light green, becoming reddish near the base; with short hairs at the nodes.

Root system: There are fibrous roots and an extensive system of stout, creeping rhizomes. The rhizomes are fleshy, up to 2 m (6 ft.) long and 1 cm (½ in.) in diameter, whitish, with brown scaly sheaths at the nodes.

Reproduction: By seed, dispersed by wind, water, animals, contaminated seed, and other agricultural activities. Seeds remain viable in the soil seed bank for two to five years. This species can also regenerate from rhizome fragments that are moved within and between sites. Once a population is established, most population growth is through vegetative reproduction by rhizomes.

Impact: This species is considered one of the most problematic agricultural weeds in the United States as well as in many other countries worldwide. Through direct competition and the production of allelopathic compounds that inhibit the growth of other plants, it causes significant reductions in yield for economically important crops such as corn and soybeans. In addition, it is an alternate host for several crop pests and pathogens, and its pollen is extremely allergenic. As with *Sorghum bicolor*, leaves of *S. halepense* plants that have been exposed to stressors such as frost or drought can produce hydro-gen cyanide at levels poisonous to livestock. Where it has spread into natural areas, it forms dense stands that outcompete native species.

Control: Hand hoeing can control the rhizome system. Plants should not be allowed to reach 20 cm in height; repeated cutting at or below this height will eventually exhaust the root reserves. Spot application of glyphosate or dalapon early in the growing season is effective.

Sorghum halepense florets. Image by Steve Dewey, Utah State University, Bugwood.org.

Left: *Sorghum halepense* leaves. Image by James H. Miller and Ted Bodner, Southern Weed Science Society, Bugwood.org. *Right: Sorghum halepense* leaf collar and ligule. Image by Chris Evans, University of Illinois, Bugwood.org.

Sorghum halepense fruits. Image by Steve Hurst, hosted by the USDA-NRCS PLANTS Database.

Sorghum halepense inflorescence. Image by Jil Swearingen, USDI National Park Service, Bugwood.org.

Sorghum halepense rhizomes. Image by Ohio State Weed Lab, The Ohio State University, Bugwood.org.

Forbs

Forbs encompass all herbaceous flowering plants that are not grasses or grass allies (i.e., sedges and rushes). In general, forbs are relatively small, because they do not have woody structures to hold them upright. However, some exceptionally tall forb species, such as *Heracleum mantegazzianum* (giant hogweed) [pp. 77–79], can grow up to 5 m (15 ft.) tall. An important feature of forbs is their life spans, which vary dramatically among species. Some forbs are annuals, and others are biennials. Biennial forbs often grow vegetatively in the first year and send up reproductive (flowering) shoots only in the second year. Some forbs are perennials that live for several years. In most cases, perennials reproduce every year, but some perennial plants are monocarpic. A monocarpic perennial plant grows in a vegetative form for one or more years, flowers and produces seeds once in its lifetime, and dies immediately after the reproductive event.

There are many more forb species than species in other plant growth forms, so there are consequently many more problem plants in this category. Certain plant families are highly represented in this collection, including the Asteraceae and the Apiaceae, both of which have species that produce abundant small seeds. There are also many Brassicaceae problem plants, in part because they contain many secondary chemicals that deter herbivory.

Many forbs possess traits that predispose them to be problematic. For instance, they often reproduce quickly and abundantly. And many forbs are extremely effective at reproducing vegetatively via rhizomes and stolons. In addition, some can regenerate through root or stem fragments. All of these characteristics facilitate the establishment and spread of forbs.

The key to controlling a specific problematic forb is to understand its particular lifecycle and to break it before reproduction can take place. In many cases, one can prevent seeds from germinating initially by using a physical barrier, such as a thick layer of mulch or wood chips. If germination has already occurred, then pulling, hoeing, or mowing seedlings can prevent plants from establishing. Identification of some problem forbs is not possible until the plants are in flower. If plants have already produced flowers, it is important to cut off—and carefully dispose of—flower heads before they have the chance to develop seeds.

Amaranthaceae

Amaranthus palmeri

Palmer's amaranth · Palmer's pigweed ·
carelessweed

Origin: This species is native to the southwestern United States and Mexico, but it has expanded its range, primarily as a cottonseed contaminant. It has become an agricultural weed throughout much of the United States.

Description: An annual with upright, branching stems that end in dense spikes of green flowers. *Amaranthus palmeri* is similar to several other native pigweeds, including *Amaranthus tuberculatus* (tall waterhemp), *Amaranthus retroflexus* (redroot pigweed), and *Amaranthus hybridus* (smooth pigweed). However, *A. retroflexus* and *A. hybridus* have fine hairs on the leaves and stems that are lacking in *A. palmeri*. In addition, the petioles on the basal leaves of this species are as long as or longer than the leaf blades, and it has spikes on its seed head that are longer and spinier than the other native pigweeds.

Habitat: Agricultural fields, pastures, barnyards, streambanks, and along railways and roadsides. Typically grows in full sun and dry soils in areas with a high level of disturbance.

Height: 30–150 cm (1–5 ft.), occasionally taller.

Foliage: Leaves are simple, alternate, ovate or diamond-shaped, 3–10 cm (1–4 in.) long and 1–3.5 cm (½–1½ in.) wide, with a wedge-shaped base and a rounded or sharply pointed tip. The basal leaves and leaves on the lower part of the stem are larger and have petioles that are as long as or longer than the leaf blades, while those higher up on the stem are progressively smaller and have shorter petioles. The margins are smooth, and the leaf surfaces are hairless. There are prominent white veins on the undersides of the leaves. Some leaves have a reddish or white chevron-shaped marking on the blade.

Flowers: Dioecious, with male and female flowers borne on separate plants. The main stem and side branches end in dense spikes of green flowers 15–50 cm (6–20 in.) long and 1–1.5 cm (½–⅝ in.) in diameter. Lateral flowers also occur in small clusters in the leaf axils. The individual flowers are surrounded by spiny bracts, 4–6 mm (⅛–¼ in.) long. Male flowers have five unequal sepals, the outer ones 3.5–4 mm (⅛ in.) long and the inner ones 2.5–3 mm (1/16–⅛ in.) long, and five stamens with yellow anthers. The female flowers have stiffer and spinier bracts; five recurved sepals, the outer ones 3–4 mm (⅛ in.) long and the inner ones 2–2.5 mm (1/16 in.); and a single style with two or three stigmas. June–October.

Fruit: A dry fruit, 1.5–3 mm (1/16–⅛ in.) long, that splits open in the middle to expose a single seed. Seeds are 1–1.3 mm (1/32–1/16 in.) in diameter, glossy and dark reddish brown to black. July–November.

Stems: A single central stem with many upright branches. Stems are hairless.

Root system: A taproot that is often reddish in color.

Reproduction: By seed, which is primarily dispersed by gravity but can also be moved via irrigation; birds; mammals; and farming equipment used for plowing, mowing, and harvesting. A female plant can produce 100,000 to 900,000 seeds or more in a growing season. Seeds remain viable in the soil seed bank for at least three years.

Impact: A major weed in agricultural fields, particularly in cotton, peanut, and soybean crops. It is extremely difficult to control, because it has developed resistance to several common herbicides, including glyphosate. This species can also be toxic to livestock, as its leaves accumulate high levels of nitrates.

Control: Given this species' rapid development of herbicide resistance, chemical control is not recommended. Plants should be pulled by hand and any plant matter carefully disposed of. In agricultural settings, deep tilling is used to bury seeds, followed by several years of no tilling to prevent deep-buried seed from being brought back up to the surface and germinating.

Amaranthus palmeri leaves. Image by Rebekah D. Wallace, University of Georgia, Bugwood.org.

Female flowers of *Amaranthus palmeri*. Image by Gene Sturla, southwestdesertflora.com.

Red petioles in *Amaranthus palmeri*. Image by Gene Sturla, southwestdesertflora.com.

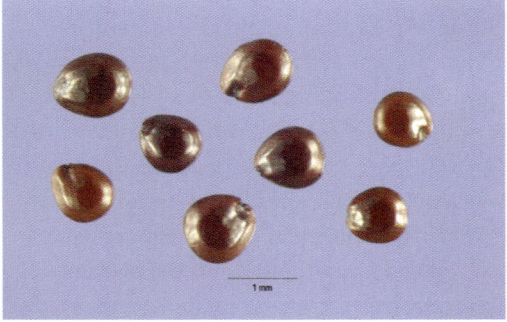

Amaranthus palmeri seeds. Image by Steve Hurst, hosted by the USDA-NRCS PLANTS Database.

Male flowers of *Amaranthus palmeri*. Image by Gene Sturla, southwestdesertflora.com.

Amaranthus palmeri growing in an agricultural field. Image by Rebekah D. Wallace, University of Georgia, Bugwood.org.

Amaranthaceae

Bassia scoparia

kochia · common kochia · burningbush · firebush · common red sage · summer cypress · mock cypress · mirabel · belvadere · poor man's alfalfa · railroad weed · Mexican summer cypress · Mexican fireweed · Mexican firebrush

Synonym: *Kochia scoparia*

Origin: Native to Europe and western Asia. This species was introduced to North America as a forage species in the mid- to late 1800s. It has escaped cultivation and has become a major weed, particularly in the western United States.

Description: An annual with highly branched stems that form a pyramidal or rounded bush, this species is a common and economically important weed in agricultural settings, particularly in arid and semiarid regions. At maturity, the aboveground part of the plant breaks off at the base and forms a tumbleweed.

Habitat: Rangelands, pastures, fields, areas under cultivation, gardens, ditches, and along railways and roadsides. Although particularly adapted to arid and semiarid regions, this species can grow in a wide range of temperatures and climatic conditions. It is tolerant of drought and high salinity.

Height: Typically 30–120 cm (1–4 ft.), occasionally taller.

Foliage: Leaves simple, alternate, 1–6 cm (½–3 in.) long and 4–12 mm (⅛–½ in.) wide, linear or narrowly lance-shaped, tapering to a rounded point. The leaf blades are flat and have three or five prominent veins. Leaves on the main stem are larger than those on the branches, and they become progressively smaller toward the ends of the branches. Leaves are pale green or grayish green, the upper surface is usually smooth, and the lower surface is covered with soft hairs. Margins are unlobed and fringed with long hairs.

Flowers: Inconspicuous greenish flowers, 4–5 mm (⅛–¼ in.) in diameter, solitary or in groups of two to six in the axils of the upper leaves and in 5–10 mm (¼–½ in.) long spikes at the ends of branches. Flowers lack petals but have five fused sepals that form a star shape. Individual flowers can have both male and female parts, with three to five stamens and two stigmas, or they can have female parts only. July–September.

Fruit: Each flower matures into a thin, papery envelope that encloses a single seed. Seeds are flattened, 1.5–2 mm (1/16 in.) long and 1–1.5 mm (1/32–1/16 in.) wide, rounded on one end and tapering to a point on the other, resembling a teardrop. The surface of the seed is somewhat rough, dull brown marked with yellow, with a groove near the base on each side.

Stems: Stems are slender and upright, highly branched, and green, turning reddish in the autumn. During autumn or winter, the central stem can break off at the base and become a tumbleweed.

Root system: Plants have an extensive taproot, 2–5 m (6–15 ft.) deep, with lateral roots that can extend to 7 m (23 ft.) away from the plant.

Reproduction: By seed, which is spread by wind and water. Mature and dried plants become tumbleweeds and scatter seed as the wind moves them. An individual plant can produce up to 50,000 seeds in a growing season. The seeds are short-lived, remaining dormant in the soil for only two or three months, germinating very early the following spring.

Impact: An aggressive competitor for resources that reduces crop yields in agricultural settings. It is also a problem for agriculture because it is an alternate host for beet and tobacco viruses. Furthermore, the leaf litter is allelopathic, inhibiting the growth of other plant species. *Bassia scoparia* is a nutritious forage species—hence the common name poor man's alfalfa—but the presence of oxalates, saponins, alkaloids, and nitrates in the leaves and stems can make it toxic to livestock.

Control: Cutting or mowing just before flowering will reduce seed production. In small populations, stems can also pulled by hand or partially dug out and cut below the root crown to kill the plant entirely. Given that seeds do not last long in the soil, reducing or preventing seed production over a few successive growing seasons can eliminate *B. scoparia* from an area. This species has developed resistance to several common herbicides, including triazines, sulfonylureas, and 2,4-D. Spot treatment with paraquat or glyphosate can still be effective early in the growing season, before leaves develop a thick cuticle.

Bassia scoparia growing in an agricultural field. Image by Jan Samanek, Phytosanitary Administration, Bugwood.org.

Bassia scoparia fruits. Image © Robert L. Carr.

Bassia scoparia stems. Image © Gerald D. Carr.

Bassia scoparia seeds. Image by Steve Hurst, hosted by the USDA-NRCS PLANTS Database.

Bassia scoparia flowers. Image © Gerald D. Carr.

Red autumn color in *Bassia scoparia* plants. Image by Leslie J. Mehrhoff, University of Connecticut, Bugwood.org.

Amaranthaceae

Salsola tragus

Russian thistle · prickly Russian thistle · windwitch · common saltwort · tumbleweed · tumbling thistle

Synonym: *Kali tragus*

Origin: Native to Europe and Asia. This species was accidentally introduced to the western United States in the late 1800s, possibly through contaminated flaxseed.

Description: An annual that is a common weed in disturbed habitats, especially in semi-arid regions. At maturity, the rounded clump of branched, tangled stems breaks off at the base and forms a tumbleweed.

Habitat: Grows in a wide variety of habitat types, commonly in disturbed grassland and desert communities and also in coastal areas. It has a preference for sandy soil and is tolerant of salinity.

Height: 20–100 cm (8–40 in.).

Foliage: Leaves are mostly alternate and narrowly linear, 1–4 cm (½–2 in.) long, tapering to a sharp spine at the tip. Leaves are succulent, dull green or grayish, turning red when mature.

Flowers: Solitary small flowers develop in the leaf axils, each 5–6 mm (¼ in.) in diameter. Flowers lack petals but have five pink to greenish membranous-winged sepals. Each flower is subtended by three leaves. July–September.

Fruit: Fruits, including the wings, are 3–10 mm (⅛–½ in.) in diameter. The coiled embryo is approximately 2 mm (¹⁄₁₆ in.) wide and visible through the nearly transparent seed coat. August–October.

Stems: Stems are smooth or covered with minute bristles. The many branches grow into a tangled, rounded bush.

Root system: This species has a fairly deep taproot and spreading lateral roots.

Reproduction: By seed only. After the fruits develop, the aboveground parts dry out and the stem breaks off near the base. The tumbleweed disperses seed as it is rolled by the wind. A large individual can produce up to 200,000 seeds.

Impact: Mature plants are too woody or spiny for most livestock to browse, so infestation by this species degrades range and pasture lands. Accumulation of nitrates and soluble oxalates in the photosynthetic parts of the plants makes this species poisonous to sheep. Tumbleweeds can pose a road hazard, as rolling plants surprise drivers and cause traffic accidents. A buildup of dry tumbleweeds can also create a fire hazard. In addition, *Salsola tragus* is the primary summer host of *Circulifer tenellus* (beet leafhopper), a vector of curly top virus, which infects several important crops, including beets, tomatoes, beans, melons, and squash.

Control: Any treatment should focus on controlling the immature plants to prevent them from producing seed. Mowing or hand pulling young plants is often effective in areas with limited infestation. Once *S. tragus* is established, it can be controlled with the preemergent herbicides atrazine, bromacil, hexazinone, imazapyr, napropamide, simazine, and sulfometuron. Postemergent herbicides such as glufosinate, glyphosate, and paraquat can also be used, but they must be applied while plants are in the early stages of growth.

Salsola tragus plant. Image by Forest Starr and Kim Starr, Starr Environmental, Bugwood.org.

Salsola tragus seeds. Image by D. Walters and C. Southwick, Table Grape Weed Disseminule ID, USDA APHIS PPQ, Bugwood.org.

Salsola tragus stems. Image © Robert L. Carr.

Salsola tragus incursion. Image by Forest Starr and Kim Starr, Starr Environmental, Bugwood.org.

Salsola tragus flowers. Image by Forest Starr and Kim Starr, Starr Environmental, Bugwood.org.

Salsola tragus tumbleweed. Image by Edmond Meinfelder, used under a CC BY 2.0 license.

Apiaceae

Aegopodium podagraria

goutweed · goutwort · bishop's goutweed · bishop's weed · bishop wort · bull wort · snow-on-the-mountain · dog elder · dwarf elder · ground elder · ground ash · pot ash · ashweed · white-ash-herb · herb Gerard · herb William · English masterwort · wild masterwort · garden plague · farmer's plague · jack jumpabout

Origin: Native to much of Europe and northern Asia. This species has been cultivated for centuries, making its original distribution somewhat unclear. *Aegopodium podagraria* has a long history of use in food, in medicine (namely to treat gout), and as an ornamental plant. It was intentionally introduced to North America and had escaped from cultivation into natural areas by 1863. It is still widely grown as an ornamental plant, particularly horticultural varieties with white leaf margins or variegated leaves.

Description: A creeping perennial that forms a dense groundcover. The flowers are similar to those of *Daucus carota* (Queen Anne's lace) and members of the genus *Angelica*. However, the compound leaves of *A. podagraria* have a biternate arrangement, with three groups of three leaflets. This feature, coupled with the flat-topped inflorescence of white flowers, distinguishes this species from related ones.

Habitat: Forest margins, forest understories, river or stream floodplains, roadsides, and gardens. Grows well in disturbed areas with moist soils and partial sun to full shade.

Height: 30–100 cm (1–3 ft.).

Foliage: Compound leaves with three to nine leaflets, mostly biternate with three groups of three leaflets each but sometimes irregular. Leaflets are 3–8 cm (1–3 in.) long and 1–4 cm (½–2 in.) wide, oblong to ovate, tapering to a point. Leaflets can be deeply cleft into two unequal lobes. The leaflets' margins are sharply toothed, and the leaf surfaces are hair-

less. Leaves alternate along the stem, but long leaf stems also arise directly from the root crown. Upper leaves are reduced, and the base of the petiole has a broad sheath that clasps the stem. Foliage of the wild type is solid green, while many horticultural varieties have pale green leaves with white margins.

Flowers: Small white flowers arranged in dense, flat-topped umbels, 6–12 cm (3–5 in.) wide, with 10–20 smaller umbellets of up to 25 flowers each. Individual flowers are 3 mm (⅛ in.) in diameter, with five notched petals, five stamens, and a single pistil with two long styles. May–June.

Fruit: Small, dry fruits are ovoid, 3–4 mm (⅛ in.) long and 1 mm (1/32 in.) wide, flattened laterally, with longitudinal ribs. As the fruits develop, the long styles persist and curve away from each other. At maturity, the styles fall off and the fruit splits into two parts, each containing a single seed.

Stems: Stems are upright, hollow, and branched. The surfaces of the stems are ridged and hairless.

Native Alternatives
Anemone canadensis (Canada anemone)
Angelica venenosa (hairy angelica)
Asarum canadense (wild ginger)
Aralia nudicaulis (wild sarsaparilla)
Heuchera americana (alumroot or coral bells)
Maianthemum canadense (Canada mayflower)
Osmorhiza longistylis (sweet cicely)
Tiarella cordifolia (heartleaf foamflower)

Typical leaves of *Aegopodium podagraria*. Image by Leslie J. Mehrhoff, University of Connecticut, Bugwood.org.

Developing fruits on *Aegopodium podagraria*. Image by Leslie J. Mehrhoff, University of Connecticut, Bugwood.org.

Variegated leaves of *Aegopodium podagraria* var. *variegatum*. Image by Leslie J. Mehrhoff, University of Connecticut, Bugwood.org.

Aegopodium podagraria roots and rhizomes. Image by Leslie J. Mehrhoff, University of Connecticut, Bugwood.org.

Aegopodium podagraria flowers. Image by Leslie J. Mehrhoff, University of Connecticut, Bugwood.org.

Aegopodium podagraria plants. Image by Leslie J. Mehrhoff, University of Connecticut, Bugwood.org.

Root system: An extensive, fibrous root system with branching rhizomes that spread laterally.

Reproduction: By seed. Vegetative reproduction also occurs via spreading rhizomes and root fragments.

Impact: This species can escape from cultivation and, once established, it is difficult to eradicate. It can form dense colonies that outcompete native understory species and prevent the establishment of shrub and tree seedlings.

Control: Avoid planting this species, particularly in gardens adjacent to natural areas. Although the cultivated varieties with variegated leaves are reported to not be as aggressive, they can revert to the wild form and spread. Small populations can be pulled by hand or dug out, but care must be taken to remove all the roots, as regrowth is possible from root or rhizome fragments. Frequent short mowing can help control the spread of this species. Mowing followed by heavy mulching or solarization can also be effective. A systemic herbicide such as glyphosate can be selectively applied to foliage in early spring.

Apiaceae

Conium maculatum

hemlock · poison hemlock · deadly hemlock · beaver poison · poison parsley · fool's parsley · spotted parsley · spotted corobane · spotted hemlock · carrot fern · herb bennet

**Prohibited
Noxious Weed
OIPC Invasive
⚠ Do Not Touch**

Origin: Native to Europe, western Asia, and northern Africa. This species was introduced to North America in the late 1800s, either accidentally or deliberately. It is a problematic weed in both agricultural areas—affecting pastures and crops—and natural habitats.

Description: A biennial that grows as a large rosette during the first year and develops a flowering stem in the second year. Many relatives in the Apiaceae also have small white flowers in umbels. The following features in combination differentiate *Conium maculatum* from other species: it is taller than most relatives, its stems and leaves are hairless, stems have purple spots, flowers are in compound or branching umbels that have undivided bracts at the base, foliage is multiply pinnately compound, leaflets are pinnately cleft or toothed with leaf veins running to the tips of the teeth, and the foliage gives off a rank odor when damaged.

Habitat: Pastures, areas under cultivation, old fields, fencerows, wet meadows, ditches, river- and streambanks, and roadsides. It prefers frequently disturbed habitats with moist soils and full sun to partial shade.

Height: 1–3 m (3–10 ft.).

Foliage: Leaves alternate, pinnately compound, triangular, 20–40 cm (8–16 in.) long and wide, becoming progressively smaller farther up the stem. Leaves are divided two to four times and have many small leaflets that are cleft or toothed, giving the foliage a fernlike appearance. Leaves are hairless and shiny green. Leaf veins run to the tips of the teeth, distinguishing this species from the native *Cicuta maculata* (water hemlock). The base of the petiole often forms a sheath that clasps the stem. The crushed foliage has a pungent, disagreeable odor.

Flowers: Small white flowers are held in branching, flat-topped umbels about 4–6 cm (2–3 in.) wide. Individual flowers are about 3 mm (⅛ in.) in diameter, with five notched or folded petals, five stamens, and a single pistil with two long styles. June–September.

Fruit: Small, dry fruits are broadly ovoid, 2–4 mm (¹⁄₁₆–⅛ in.) long and 1.5–2.5 mm (¹⁄₁₆ in.) wide, flattened laterally, and covered in wavy ridges. At maturity, the fruit splits into two parts, each containing a single seed.

Stems: Stems are hollow, purple-spotted, ridged, and hairless.

Root system: Each plant has a long, white taproot that is occasionally mistaken for wild parsnip.

Reproduction: Solely by seed. A single plant can produce up to 30,000 seeds. Germination rate is high, and seed viability is approximately three years.

Impact: All parts of *C. maculatum* plants contain highly poisonous alkaloids that are toxic to mammals (including humans) if ingested. In natural areas, this species can displace native plant species. It also reduces the availability of quality forage, contaminates hay, degrades wildlife habitat, and decreases land value.

Control: Wear protective clothing and gloves when undertaking any control measures; the sap of this plant can cause skin irritation. Manual methods such as hoeing, digging, and cutting are effective management options for smaller, isolated populations. Mechanical removal through mowing or tilling are options for large-scale control. Plants should be clipped close to the ground in the spring after bolting but before seed set. Chemical control options include 2,4-D ester or amine formulations, aminopyralid, metsulfuron methyl, glyphosate, and imazapyr.

Conium maculatum leaves. Image by Robert Vidéki, Doronicum Kft., Bugwood.org.

Developing *Conium maculatum* fruits. Image by Doug Goldman, hosted by the USDA-NRCS PLANTS Database.

Conium maculatum stem. Image by Eric Coombs, Oregon Department of Agriculture, Bugwood.org.

Mature *Conium maculatum* fruits. Image by Steve Hurst, hosted by the USDA-NRCS PLANTS Database.

Conium maculatum flowers. Image by Doug Goldman, hosted by the USDA-NRCS PLANTS Database.

Conium maculatum plants. Image by Doug Goldman, hosted by the USDA-NRCS PLANTS Database.

Apiaceae

Daucus carota

Queen Anne's lace · bishop's lace · wild carrot · bird's nest · bee's nest · devil's plague · lace flower · rantipole

⚠ **Do Not Touch**

Origin: Native to Europe, southwestern Asia, and northern Africa. This species was introduced to North America by early colonists in the 1600s, either as a medicinal herb or as a contaminant in crop seed. Most authorities consider it the progenitor—or at least a close relative—of the cultivated carrot. Until recently, this species was listed as a prohibited noxious weed in Ohio.

Description: Typically biennial but occasionally an annual. In the first year, the plant develops a rosette of basal leaves, while in the second year it bolts and produces flowering stalks. Many related species in the Apiaceae also have white flowers and divided leaves. One distinguishing feature of *Daucus carota* is the long bracts at the base of its umbels, which are deeply lobed and have narrow segments. It also blooms later in the summer than most relatives, and there is typically a single reddish-purple flower in the center of the umbel. In addition, *D. carota* occurs in drier soils, rather than wetlands or wet meadows, which differentiates it from *Conium maculatum* (poison hemlock) [pp. 73–74] or *Cicuta maculata* (water hemlock). Finally, the foliage of *D. carota* is more finely divided than that of other related species.

Habitat: Grasslands, meadows, fields, prairies, along railways and roadsides, and gardens. Prefers well-drained soils in full sun to partial shade.

Height: 40–100 cm (16–40 in.).

Foliage: Leaves alternate, pinnately compound, ovate or triangular. Basal leaves are 5–15 cm (2–6 in.) long and 6–10 cm (3–4 in.) wide, with leaves becoming progressively smaller farther up the stem. Leaf blades are divided two or three times and are deeply dissected into narrow segments. Leaves are a bright grayish green and covered in soft hairs, particularly on the veins and leaf margins. Leaf veins run to the tips of the teeth. The base of the petiole often forms a sheath that clasps the stem. The foliage has the bitter, somewhat soapy fragrance typical of carrots.

Flowers: Small white flowers are held in branching, flat-topped umbels about 4–12 cm (2–5 in.) wide. The outer rays are longer than the inner rays, and they arch inward as they age, causing the umbel to become concave when mature. Individual flowers are about 2 mm ($\frac{1}{16}$ in.) in diameter, with five deeply notched petals, five stamens, and a single pistil with two long styles. Most inflorescences contain a few purple-red sterile flowers in the central umbellets. The umbel is subtended by long, narrow bracts that are three-forked or pinnately lobed. May–October.

Fruit: As the umbel matures, the seed head contracts and folds in on itself, causing it to resemble a bird's nest. The small, dry fruits are oblong or ovoid, 2–4 mm ($\frac{1}{16}$–$\frac{1}{8}$ in.) long, and covered with hooked barbs. At maturity, the fruit splits into two parts, each containing a single seed.

Stems: The stems are solid, typically covered with coarse hairs, somewhat grooved, usually striped green but occasionally tinted red.

Root system: Plants have a long, slender, white or yellowish taproot with fibrous secondary roots. The root system becomes tough and woody with age. Roots smell like carrots.

Reproduction: By seed. The hooked barbs on the fruits facilitate dispersal by animals and wind. Seeds remain viable in the soil for one or two years.

Impact: This species invades open habitats, competing with and displacing native plants. It is sometimes a threat to recovering grasslands or prairies, although it tends to decline as native grasses and forbs become established. It is also problematic in agricultural settings. While not toxic to cattle, dairy cows that consume large quantities of *D. carota* produce off-tasting milk, making this plant species undesirable in pastures. This species can also negatively affect commercial carrot cultivation through introgression, which causes genetic contamination in the crops.

Control: Wear protective clothing and gloves while undertaking any control measures, as the sap can cause skin irritation. In small populations, pulling by hand or simply cutting the flowering stems are recommended. In larger stands, mowing close to the ground can be effective. All mechanical control measures should be undertaken in mid- to late summer, before seed set. Young plants are susceptible to 2,4-D and triclopyr.

Daucus carota leaves. Image by Doug Goldman, hosted by the USDA-NRCS PLANTS Database.

Daucus carota flowers. Image by Doug Goldman, hosted by the USDA-NRCS PLANTS Database.

Mature *Daucus carota* fruits. Image by Ken Chamberlain, The Ohio State University, Bugwood.org.

Developing fruits in *Daucus carota*. Image by David Cappaert, Bugwood.org.

Daucus carota umbel. Image by Doug Goldman, hosted by the USDA-NRCS PLANTS Database.

Daucus carota plants. Image by Chris Evans, University of Illinois, Bugwood.org.

Apiaceae

Heracleum mantegazzianum

giant hogweed · giant cow parsnip · giant cow parsley · cartwheel-flower · hogsbane

ODA Invasive
Prohibited
Noxious Weed
⚠ **Do Not Touch**

Origin: Native to the western region of the Greater Caucasus Mountains, located along the border between Russia and Georgia. This species was intentionally introduced to the United States in 1917 as a garden curiosity. It has established populations in the northeastern and Pacific Northwest regions of the United States. This species is considered a noxious weed at the federal level, making it illegal to move throughout the United States.

Description: This species is a very large biennial or monocarpic perennial that grows as a rosette for one or more years, flowers and produces seeds once in its lifetime, and dies immediately after the reproductive event. It is often described as looking like "Queen Anne's lace on steroids" but *Heracleum mantegazzianum* is about five times the size of *Daucus carota* and it has a very different leaf structure. Its large size distinguishes it from related species, as does its deeply lobed, pinnately compound leaves and the presence of over 50 rays in each umbel. This species produces a highly toxic sap that can cause severe blistering of the skin and damage to the eyes.

Habitat: Forest openings and margins, along railways and roadsides, river- and streambanks, abandoned homesteads, and fallow fields. Prefers disturbed areas with rich, moist soils and full sun to partial shade.

Height: 3–5 m (10–15 ft.).

Foliage: Leaves alternate, pinnately compound or divided into three leaflets, up to 3 m (10 ft.) long and 1.5 m (5 ft.) wide. Leaflets are pinnately lobed and deeply incised, with numerous teeth along the margins. The upper surface of the leaf is smooth, but the lower surface is covered in bristles. The petioles are purple spotted, hollow, and also covered with white bristles that have purple blisters at the base. A thick band of white hairs encircles the stem at the base of the petiole. Until the plant becomes reproductive, only basal leaves are produced.

Flowers: Small white flowers are held in umbels up to 80 cm (2 ½ ft.) in diameter. There is a main umbel at the end of the flowering stem and up to eight satellite umbels that attach to the stem at the base of the compound inflorescence. Each umbel has 50–150 unequal hairy rays, each 10–40 cm (4–16 in.) long. Individual flowers have five deeply notched and unequal petals, each 10–12 mm (½ in.) long, five stamens, and a single pistil with two styles. The compound inflorescence is subtended by large, leafy bracts. June–July.

Fruit: Numerous dry fruits, broadly elliptic and flattened, 8–12 mm (¼–½ in.) long and 6–8 mm (¼ in.) wide, winged. Fruits are green, drying to light brown, with three to five dark markings caused by resin canals that extend three-quarters of the length of the fruit and widen at the ends. At maturity, the fruit splits into two parts, each containing a single seed. August–September.

Stems: Stems are hollow, 5–10 cm (2–4 in.) in diameter, green with purple splotches or occasionally continuously purple. The stems are covered in white bristles that have purple blisters at the base.

Root system: There is a thick, yellow, branching taproot, up to 60 cm (2 ft.) long and 15 cm (6 in.) in diameter. The aboveground structures die back over winter, and the plants regrow from the rootstock the following growing season.

Reproduction: By abundant seed, which can

Left: Heracleum mantegazzianum. Right: Heracleum mantegazzianum stem and leaf base. Both images by Robert Vidéki, Doronicum Kft., Bugwood.org.

Heracleum mantegazzianum flowers. Image by Rob Hille, used under a CC BY-SA 3.0 license.

Heracleum mantegazzianum plant. Image by Robert Vidéki, Doronicum Kft., Bugwood.org.

Developing fruits on *Heracleum mantegazzianum*. Image by Jan Samanek, Phytosanitary Administration, Bugwood.

Heracleum mantegazzianum umbel. Image by Robert Vidéki, Doronicum Kft., Bugwood.org.

Mature *Heracleum mantegazzianum* fruits. Image by Steve Hurst, hosted by the USDA-NRCS PLANTS Database.

be dispersed by wind, surface runoff from rain, or animals. An individual plant can produce 5,000–100,000 seeds, which remain viable in the soil for up to 10 years.

Impact: Due to its large size, *H. mantegazzianum* can outcompete many native plant species and can cause erosion of river- and streambanks. However, the major concern related to this species is its threat to human health. The clear, watery sap contains furanocoumarins that sensitize the skin to ultraviolet radiation and can result in severe burns and blistering (phytophotodermatitis) when exposed to the sun. If the sap gets in the eyes it can also cause temporary or permanent loss of vision. If skin contact does occur, wash the affected area with soap and cold water immediately, keep out of sunlight, and seek medical attention immediately.

Control: Protective clothing, gloves, and eyewear are essential for undertaking any control measures. Wash all clothing and equipment immediately after use. Seedlings and young plants can be pulled out. Mature plants can be dug out, but each taproot needs to be cut 8–12 cm (3–5 in.) below ground level. Seeds should be bagged and disposed of carefully; any other plant parts that are removed should be placed in the sun to dry out before they are bagged and disposed of. Repeated cutting will reduce the vigor of a mature plant, but it can regrow from root reserves. It is not recommended to use a mower or weed trimmer because these can spray the sap into the air. Foliar treatments of glyphosate, triclopyr, or imazapyr can control *H. mantegazzianum*. These herbicides should be applied in the spring or early summer, when plants are actively growing.

Apiaceae

Pastinaca sativa

wild parsnip · parsnip · poison parsnip · hart's eye

Origin: Native to most of Europe and western Asia. This species has been cultivated as a vegetable root crop since at least the Middle Ages, and it has subsequently been introduced throughout most of North America. It has escaped from cultivation and is a common weed in disturbed habitats and cropland.

Description: A biennial or monocarpic perennial that grows as a rosette for one or more years, flowers and produces seeds once in its lifetime, and dies immediately after the reproductive event. This species has yellow flowers in flat-topped umbels, which distinguishes it from many related species. *Pastinaca sativa* may be mistaken for the native *Zizia aurea* (golden alexanders), which has brighter yellow flowers and leaflets grouped in threes. The ingestion of *P. sativa* leaves is toxic to livestock and skin contact can cause blistering in humans.

Habitat: Commonly found growing along roadsides, in pastures and abandoned fields, areas under cultivation, and in other highly disturbed habitats. It tolerates a range of soils and moisture levels but prefers full sun.

Height: 50–150 cm (1 ½–5 ft.).

Foliage: Leaves alternate, pinnately compound, 5–10 cm (2–4 in.) long, with leaves becoming progressively smaller farther up the stem. Each leaf has 5–15 oblong or ovate leaflets with variably toothed or lobed margins. Leaflets are yellowish green, shiny, and mostly hairless. When crushed, the foliage gives off a pungent odor of parsnips.

Flowers: The inflorescence is a compound umbel, 10–20 cm (4–8 in.) in diameter, with 15–25 rays, held at the ends of stems or lateral branches. Individual flowers are yellow or greenish yellow, 3–4 mm (⅛ in.) in diameter, with five petals that curve under, five stamens, and a flattened style that forms a prominent greenish-yellow disk in the center with two short stigmas. May–July, although some plants may continue to flower through late summer.

Fruit: The dry fruits are broadly elliptic and flattened, 5–7 mm (¼ in.) long and 4–6 mm (⅛–¼ in.) wide, slightly ribbed, with narrow wings. Fruits are green, ripening to brown. At maturity, the fruit splits into two parts, each containing a single seed. July–September.

Stems: Hairless, angular, deeply grooved, and hollow except at the nodes. Each plant has a single flowering stem that branches at the upper nodes.

Root system: Plants produce a fleshy, white taproot similar to cultivated parsnips.

Reproduction: By seed, primarily dispersed by wind and water. Seeds often remain attached to the dead stalks, and seed dispersal typically takes place from August through November. Seeds remain viable in the soil seed bank for five years.

Impact: This aggressive species spreads rapidly and forms dense stands that outcompete native species. As with *Heracleum mantegazzianum* (giant hogweed) [pp. 77–79], the sap of this species contains furanocoumarins that sensitize the skin to ultraviolet radiation and can result

Native Alternatives
Angelica atropurpurea (purplestem angelica)
Zizia aptera (heartleaf alexanders)
Zizia aurea (golden alenxanders)ss)

80

in severe burns and blistering (phytophotodermatitis) when exposed to the sun. If skin contact does occur, wash the affected area with soap and cold water immediately, keep out of sunlight, and seek immediate medical attention.

Control: Wear protective clothing, gloves, and eyewear when undertaking any control measures. In small populations, seedlings can be pulled by hand or dug out. Cutting through the taproot 2–5 cm (1–2 in.) below the ground level will prevent resprouting. In larger stands, plants can be mown just after flowering but before seed set. Spot application of 2,4-D amine, glyphosate, or triclopyr can be used in early spring.

Pastinaca sativa leaves. Image © Gerald D. Carr.

Pastinaca sativa umbel. Image by Leslie J. Mehrhoff, University of Connecticut, Bugwood.org.

Pastinaca sativa flowers. Image by Patrick J. Alexander, hosted by the USDA-NRCS PLANTS Database.

Developing fruits of *Pastinaca sativa.* Image by Rasbak, used under a CC BY-SA 3.0 license.

Mature *Pastinaca sativa* fruits. Image by Steve Hurst, hosted by the USDA-NRCS PLANTS Database.

Roadside invaded by *Pastinaca sativa.* Image by Leslie J. Mehrhoff, University of Connecticut, Bugwood.org.

Apiaceae

Torilis arvensis

hedgeparsley · spreading hedgeparsley · field hedgeparsley · common hedgeparsley · Canada hedgeparsley

Origin: Native to central and southern Europe, southwestern Asia, and northern Africa. There is little information about the timing or pathway of introduction of this species to North America. It is an increasingly problematic weed in the southern, western, and northwestern regions of the United States.

Description: An annual or occasionally biennial plant with freely branching stems. One distinctive feature of this genus is its fruits, which are covered with prickles. *Torilis arvensis* has earned itself the nickname "tall sock destroyer" because the burlike fruits stick to clothing and are very difficult to remove. This species grows in similar habitats and is somewhat similar in appearance to *Daucus carota* (Queen Anne's lace) [pp. 75–76], but the leaves of the latter are more finely dissected and its floral bracts are long and highly divided. In contrast, there are no floral bracts below the inflorescence in *T. arvensis*. This feature also helps to differentiate it from *Torilis japonica* (Japanese hedgeparsley) [pp. 84–85], another nonnative species that also occurs in Ohio. The confusion between these two species has made it difficult to accurately document their ranges and to assess their overall threat to biodiversity.

Habitat: Pastures, fields, forest margins, river- and streambanks, along railways and roadsides, and urban areas. Tolerates most soil types and prefers areas with disturbance. It can grow in full sun to dense shade.

Height: 30–100 cm (1–3 ft.).

Foliage: Leaves alternate, pinnately compound, 3–16 cm (1–6 in.) long and 2–10 cm (¾–4 in.) wide, triangular in outline. Each leaf has three to seven lance-shaped or ovate leaflets with variably toothed or lobed margins.

The leaves are covered with short, white hairs and have a prominent midrib.

Flowers: The inflorescence is a compound umbel, 5–8 cm (2–3 in.) in diameter, with 2–12 long and hairless rays, each supporting 2–18 flowers. The individual flowers are white or pink, 2–3 mm (1/16–1/8 in.) in diameter, with five unequal and deeply notched petals, five stamens, and a single style with two short stigmas. There are no bracts at the base of the umbel. June–August.

Fruit: The small, dry fruits are oblong or ovoid, somewhat flattened, 3–5 mm (1/8–1/4 in.) long and 2–3 mm (1/16–1/8 in.) wide, covered with straight prickles about 1 mm (1/32 in.) long. Fruits are green or pinkish while developing, turning brown with age. At maturity, the fruit splits into two parts, each containing a single seed. July–September.

Stems: Stems branch freely and are covered in short, white hairs, giving the stem a rough feel.

Root system: A slender taproot.

Reproduction: By seed. The prickles on the fruits facilitate dispersal by animals and humans.

Impact: This species is spreading quickly and has the potential to be increasingly problematic. The fruits are particularly a nuisance to animals and humans. In pasture, the fruits can cause severe problems for livestock by getting lodged in the eyes, ears, and nose.

Control: In small populations, pulling by hand is an effective control measure. Larger stands can be mowed soon after the onset of flowering but before seed set. Herbicides such as 2,4-D, triclopyr, glyphosate, and chlorsulfron are effective in controlling this species.

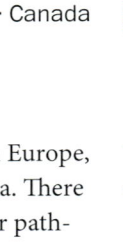

Torilis arvensis leaves. Image by Michael Becker, used under a CC BY-SA 3.0 license.

Torilis arvensis flowers. Image by Stefan Lefnaer, used under a CC BY-SA 4.0 license.

Torilis arvensis stems. Note that there are no bracts at the base of the inflorescence. Image by Stefan Lefnaer, used under a CC BY-SA 4.0 license.

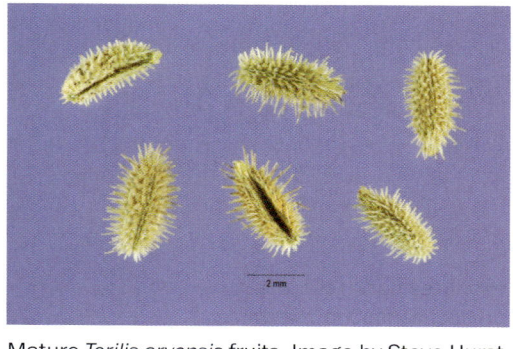

Mature *Torilis arvensis* fruits. Image by Steve Hurst, hosted by the USDA-NRCS PLANTS Database.

Torilis arvensis umbel. Image by Stickpen, used under a CC0 1.0 license.

Developing fruits on *Torilis arvensis*. Image © Robert L. Carr.

Apiaceae

Torilis japonica

Japanese hedgeparsley · erect hedgeparsley ·
upright hedgeparsley

Origin: Native to western Europe, central Asia, northern Japan, and northern Africa. The first known record of this species in the United States is from 1917. It occurs in parts of the northeastern, midwestern, and northwestern United States and is an increasingly problematic weed within this range. It is now considered an invasive species in Wisconsin.

Description: An annual or biennial plant with upright stems. This species first grows as a rosette of parsleylike leaves, then sends up a flowering stalk in the first or second year. *Torilis japonica* can be distinguished from the closely related *Torilis arvensis* (hedgeparsley) [pp. 82–83] by the presence of two to eight linear bracts at the base of the inflorescence and hooked tips on the prickles of its fruits.

Habitat: Pastures, fields, forest margins, forests, along railways and roadsides, and urban areas. Often found in partial to full shade, but it can tolerate a wide range of light availabilities.

Height: 30–120 cm (1–4 ft.).

Foliage: Leaves alternate, pinnately compound, 5–16 cm (2–6 in.) long and 3–10 cm (1–4 in.) wide, triangular in outline. Each leaf has three to seven lance-shaped or ovate leaflets with variably toothed or lobed margins. The leaves are covered with short, white hairs and have a prominent midrib.

Flowers: The inflorescence is a compound umbel, 4–5 cm (2 in.) in diameter, with 4–12 long rays each supporting 10–20 flowers. The individual flowers are white or pink, 2–3 mm (¹⁄₁₆–⅛ in.) in diameter, with five unequal and deeply notched petals, five stamens, and a single style with two short stigmas. There are two

to eight linear bracts at the base of each umbel. June–August.

Fruit: The small, dry fruits are oblong or ovoid, somewhat flattened, 3–5 mm (⅛–¼ in.) long and 2–3 mm (¹⁄₁₆–⅛ in.) wide, covered with hooked prickles. Fruits are green or pinkish while developing, turning brown with age. At maturity, the fruit splits into two parts, each containing a single seed. July–September.

Stems: Stems are upright and branched. The stem's surface is grooved and covered in short, white hairs, giving it a rough feel. The stem may be reddish purple at the base and the nodes.

Root system: A slender taproot.

Reproduction: By seed. The hooked prickles on the fruits facilitate dispersal by animals and humans.

Impact: This species spreads rapidly and creates dense stands that outcompete native vegetation. As with *T. arvensis,* its fruits are a nuisance for humans and they can cause problems for livestock by getting lodged in the eyes, ears, and nose.

Control: In small populations, pulling by hand is an effective control measure. Larger stands can be mowed soon after the onset of flowering but before seed set. Herbicides such as 2,4-D and triclopyr are effective in controlling this species.

Torilis japonica leaves. Image © Katy Chayka, Minnesota Wildflowers.

Bracts at the base of Torilis japonica inflorescence. Image © Gerald D. Carr.

Torilis japonica flowers. Image © Gerald D. Carr.

Torilis japonica fruits. Image by Dan Tenaglia, Missouriplants.com.

Torilis japonica stem. Image by Dan Tenaglia, Missouriplants.com.

mm

Torilis japonica fruit. Image © Gerald D. Carr.

Torilis japonica plant. Image © Gerald D. Carr.

Asparagaceae

Ornithogalum umbellatum

star-of-Bethehem · Pyrenees star-of-Bethehem · grass lily · sleepydick · nap-at-noon · eleven-o'clock lady

Origin: Native to the Mediterranean regions of Europe, western Asia, and northern Africa. This horticultural species is grown for its attractive flowers. It is not clear when *Ornithogalum umbellatum* was first introduced to North America, but records show that it was established in natural areas by the 1940s. This species is classified as a noxious weed in Alabama and a potentially invasive species in Connecticut.

Description: A perennial that grows from bulbs. This species has narrow, grasslike foliage and white, star-shaped flowers. It is similar in appearance to plants in the genus *Allium* and the native *Nothoscordum bivalve* (false garlic or crowpoison), although these species all have flat-topped umbels of flowers and the filaments supporting the anthers in the flowers are threadlike. In contrast, *O. umbellatum* has spreading racemes of upright flowers and triangular filaments supporting the anthers. The closely related *Ornithogalum nutans* is also cultivated widely but has drooping flowers and a more spikelike inflorescence, and its filaments develop a pair of teeth at the top that surround the anthers. In addition, *O. nutans* rarely escapes into natural areas.

Habitat: Forests, forest margins, river- and streambanks, grasslands, fields, roadsides, and gardens. It grows in moist, fertile, well-drained soils in full sun to partial shade.

Height: Flowering stems 10–40 cm (4–16 in.).

Foliage: Narrow, linear, grasslike leaves, 15–30 cm (6–12 in.) long and 2–8 mm (1/16–1/4 in.) wide, with a white stripe along the length of the upper surface. Leaves are basal and form dense clumps. The margins of leaves are smooth, and venation is parallel. The foliage begins to droop and fade once the flowering stems develop.

Flowers: Inflorescence is a spreading raceme of 4–20 flowers. Individual flowers are 1.5–2 cm (1/2–3/4 in.) in diameter, with six tepals (three petals and three modified sepals that look like petals), six stamens, and a single pistil. The tepals are lance-shaped and white, with a broad green stripe on the outer side. The stamens have a structure distinctive to this species, with triangular filaments holding the anthers. Flowers open around midday and close at sunset or during cloudy weather. May–June.

Fruit: Fruits are oblong-ovoid capsules, 8–18 mm (1/4–3/4 in.) long and 6–15 mm (1/4–5/8 in.) wide. Each capsule has three cells that contain 1–15 nearly spherical black seeds, 1.5–1.7 mm (1/16 in.) long and 1.3–1.5 mm (1/16 in.) in diameter.

Stems: Flowering stems are hairless and lack leaves.

Root system: An ovoid bulb that is typically 1–2.5 cm (1/2–1 in.) long and wide. New bulbs and bulblets are produced each year.

Reproduction: Plants are capable of producing seed but most reproduction is vegetative. Bulblets allow the species to spread within a site and can be dispersed by water.

Native Alternatives
Anemone acutiloba (sharp-lobed hepatica)
Anemone canadensis (Canadian anemone)
Thalictrum thalictroides (rue anemone)
Sanguinaria canadensis (bloodroot)

Impact: This species is still widely cultivated and therefore has the potential to increase its current range. It can be a nuisance in gardens because it spreads aggressively and can escape into neighboring natural areas. In addition, it forms locally dense stands and can be a particular problem along river- and streambanks, where it crowds out native riparian plants. This species is also problematic because it contains cardiac glycosides that are toxic to humans and livestock.

Control: Small populations can be eradicated by digging up the entire plant, taking care to remove all of the bulb material. Bulbs should be dried out thoroughly before disposal to ensure that they do not regrow. Larger populations can be controlled using herbicides such as 2,4-D, paraquat, or glyphosate.

Ornithogalum umbellatum flower. Image by Leslie J. Mehrhoff, University of Connecticut, Bugwood.org.

Early spring foliage of *Ornithogalum umbellatum*. Image by Leslie J. Mehrhoff, University of Connecticut, Bugwood.org.

White stripe on leaves of *Ornithogalum umbellatum*. Image by Leslie J. Mehrhoff, University of Connecticut, Bugwood.org.

Ornithogalum umbellatum fruits. Image by Leslie J. Mehrhoff, University of Connecticut, Bugwood.org.

Ornithogalum umbellatum flowering plants. Image by Leslie J. Mehrhoff, University of Connecticut, Bugwood.org.

Ornithogalum umbellatum bulbs. Image by Leslie J. Mehrhoff, University of Connecticut, Bugwood.org.

Asteraceae

Arctium minus

common burdock · lesser burdock · small burdock · wild burdock · wild rhubarb · hardock · hurr bur · louse-bur · button-bur · clotbur · cocklebur · thorny bur · bardane · petite bardane · beggar's lice · beggar's button · cockle button · cuckoo-button · stick button · cuckold dock · love leaves

Origin: Native to most of Europe and parts of western Asia. It was most likely accidentally introduced to North America by early European colonists and was already well established by the mid-1600s. It is widespread across most of the United States.

Description: A biennial or short-lived perennial with a strong taproot and very large, heart-shaped leaves that somewhat resemble rhubarb. Two other nonnative species in this genus occur in the United States but have not yet been recorded in Ohio. *Arctium tomentosum* (woolly burdock) can be differentiated by the abundant cobwebby hairs in its floral bracts; *Arctium lappa* (great burdock) is taller, has larger flowers arranged in flat-headed clusters, and has solid petioles. *Arctium minus* can also be confused with thistles, but it has much larger, spineless leaves, and there are hooks at the ends of its floral bracts.

Habitat: Forest openings, forest margins, streambanks, roadsides, fencerows, fields, pastures, and areas under cultivation. Grows best in areas with moist, well-drained, nitrogen-rich soils. It prefers full sun but can grow in partial shade.

Height: 50–150 cm (1–5 ft.).

Foliage: Basal rosette leaves are large and heart-shaped, 30–60 cm (1–2 ft.) long and 20–40 cm (8–16 in.) wide, with long and hollow petioles that are sometimes tinged purple. Leaves on the flowering stems are alternate, each leaf ovate with a rounded tip, becoming progressively smaller farther up the stem. All leaves have an upper surface that is dull green and coarsely veined, and a lower surface that is pale green to whitish and covered in woolly hairs.

Flowers: Composite flower heads are borne singly or in small clusters at the tips of branches or in leaf axils. Each flower head is nearly spherical, 1.5–2.5 cm (½–1 in.) in diameter, containing 30 or more pink to violet (rarely white) disc florets. The florets have five purple-tipped stamens that are fused to the floral tube; these surround a slender white pistil with a divided style that extends above the rest of the flower. The base of the flower head is surrounded by spiny green phyllaries with hooked tips. July–September.

Fruit: At maturity, the phyllaries dry to brown and stiffen, turning the seed head into a thistlelike bur. The flower heads produce many small, dry fruits, each containing a single seed. The fruits are brown, oblong, angled in cross section, 4–8 mm (⅛–¼ in.) long, broader at the top and crowned with a row of bristles 1–3.5 mm (1/32–⅛ in.) long. September–October.

Stems: Flowering stems typically arise from the rosette in the second year of growth, but this species may not bloom until the third or fourth year. The stems are hollow, grooved, much branched, covered with white woolly hairs when young but becoming hairless with age.

Root system: A thick, fleshy taproot grows up to 30 cm (1 ft.) deep.

Reproduction: By seed. The hooked bristles of the seed head attach to fur or clothing, facilitating dispersal. A single plant typically produces 10,000–15,000 seeds. These remain viable in the soil seed bank for 10–20 years.

Impact: An aggressive species that can form dense stands. The large leaves shade out most neighboring species. It is also a secondary host for several plant pathogens—including *Erysiphe cichoracearum* (powdery mildew) and *Phymatotrichum omnivorum* (root rot)—that affect economically important plants.

Control: For any control method, it is important to remove the flowering stalks before seed set. Young plants can be dug out, taking care to remove as much of the taproot as possible. For larger plants, use a spade to slice through the taproot about 10 cm (4 in.) below the soil surface. In large stands, mowing has been shown to dramatically reduce seed production. If herbicide use is deemed necessary, 2,4-D, glyphosate, and clopyralid are effective against this species.

Arctium minus leaves. Image by Matt Lavin, used under a CC BY-SA 2.0 license.

Phyllaries on an *Arctium minus* seed head. Image © Gerald D. Carr.

Arctium minus stem. Image by Enrico Blasutto, used under a CC BY-SA 3.0 license.

Arctium minus fruits. Image by Steve Hurst, hosted by the USDA-NRCS PLANTS Database.

Arctium minus flowers. Image © Gerald D. Carr.

Hollow petiole of an *Arctium minus* leaf. Image by Kris Maes, used under a CC BY-SA 4.0 license.

Asteraceae

Artemisia vulgaris

mugwort · common mugwort · common wormwood · riverside wormwood · wild wormwood
· felon herb · green-ginger · chrysanthemum
weed · old Uncle Henry · sailor's tobacco ·
naughty man · old man · Saint John's plant

Origin: Native to Europe, Asia, northern Africa, and northwestern Alaska. Records suggest that this species was introduced to eastern North America as early as the 1600s for use as a medicinal herb and as a bittering agent in beer brewing. It has spread from cultivation and is most problematic in the northeastern United States.

Description: A rhizomatous perennial with a bushy form. Its leaves are highly variable, but some are multiply lobed and toothed, resembling those of cultivated chrysanthemums. Several species of *Artemisia* are native to the United States. Its wider and strongly bicolored leaves—with a green and hairless upper surface and a finely hairy and powdery white underside—and hairless stems except around the flower clusters differentiate *Artemisia vulgaris* from these other species.

Habitat: Forest margins, pastures, along railways and roadsides, and areas under cultivation. Prefers dry to slightly moist, well-drained soils in full sun to partial shade.

Height: 60–200 cm (2–7 ft.).

Foliage: Leaves alternate, deeply lobed, 3–12 cm (1–5 in.) long and 2–8 cm (¾–3 in.) wide, becoming progressively smaller farther up the stem. Leaves are highly variable; those around the flowers are linear but other leaves are deeply divided into wedge- or spatula-shaped lobes along the central vein. The lobes can be further divided and have toothed margins, but they are more often toothless. Leaves are strongly bicolored, with green and hairless upper surfaces and densely matted, woolly white hairs on the undersides. Foliage has a pungent odor when crushed.

Flowers: Composite flower heads are borne in pyramidal, branching panicles, 5–20 cm (2–8 in.) long, at the tips of branches and in the upper leaf axils. Each flower head is small and indistinct, 5 mm (¼ in.) long and 3 mm (⅛ in.) wide, ovoid, upright or drooping, containing 7–10 reddish-brown disc florets. The florets have pale yellow pistils with a divided style that extends slightly above the rest of the flower. The base of the flower head is surrounded with phyllaries that are whitish or silvery green, due to a covering of fine hairs. July–September.

Fruit: The flower heads produce many small, dry fruits, each containing a single seed. The yellowish-brown fruits are oblong, 1–1.5 mm (¹⁄₃₂–¹⁄₁₆ in.) long and 0.5 mm (¹⁄₆₄ in.) wide, glossy, finely striate. August–October.

Stems: Multiple upright stems arise from the ground, mostly smooth and unbranched in the lower part of the plant, becoming more branched and softly hairy around the flower clusters. Stems are angular in cross section, grooved, and often purple or reddish in color.

Root system: An extensive rhizome system develops in the upper 20 cm (8 in.) of soil. Rhizomes are light brown and grow to 2 cm (¾ in.) in diameter.

Reproduction: Primarily by spreading rhizomes and rhizome fragments. Seeds contribute to the establishment of new populations.

Impact: Forms dense colonies that displace native species. There is evidence that *A. vulgaris* is allelopathic, further limiting the growth of other species. It is a problematic weed in orchards and plant nurseries.

Control: Small populations can be pulled by hand or dug out, but be aware that plants will resprout from rhizome fragments. Repeated mowing over several years can control spread.

If mowing is not effective by itself, it can be combined with application of glyphosate, picloram, triclopyr, or clopyralid.

Variation in leaf shape of *Artemisia vulgaris.* Image by the Ohio State Weed Lab, The Ohio State University, Bugwood.org.

Artemisia vulgaris stem and leaves. Image © Gerald. D. Carr.

Artemisia vulgaris fruits. Image by Steve Hurst, hosted by the USDA-NRCS PLANTS Database.

Artemisia vulgaris seed heads. Image by AnR00002, used under a CC0 1.0 license.

Artemisia vulgaris flower heads. Image by Doug Goldman, hosted by the USDA-NRCS PLANTS Database.

Artemisia vulgaris seedlings. Image by Doug Goldman, hosted by the USDA-NRCS PLANTS Database.

Asteraceae

Carduus nutans

musk thistle · nodding thistle · nodding
plumeless thistle · bristle thistle

Origin: Native to Europe, western Asia, and northern Africa. This species was accidentally introduced to North America in the mid-1800s. It has spread throughout the United States and is a major weed in pastures because it is unpalatable to livestock.

Description: A biennial plant with extremely spiky leaves and stems. The seedling emerges in early spring, develops into a rosette in the first year, and overwinters in the rosette stage. In the midspring of the the second year, plants send up a multibranched flowering stem. The spines on its stems distinguishes *Carduus nutans* from native thistles. The large flower heads and broad, triangular, purplish bracts differentiate this species from all other thistles.

Habitat: Meadows, grasslands, pastures, old fields, and disturbed sites such as roadsides, building sites, and ditches. It grows in a wide range of soil conditions but is particularly common in areas with calcareous sand and soils derived from limestone. Prefers full sun but can tolerate partial shade.

Height: Stems 30–200 cm (1–7 ft.).

Foliage: Basal rosette leaves are 10–40 cm (4–16 in.) long and 3–10 cm (1–4 in.) wide. Leaves on the flowering stems are alternate, becoming progressively smaller farther up the stem. Each leaf is deeply lobed and has spine-pointed tips, the upper surface dark green and waxy with a light green midrib, the lower surface with soft hairs along the main veins.

Flowers: Large, globose flower heads are solitary or clustered at the ends of the branches, drooping or nodding to a 90°–120° angle when mature. Individual flower heads are 4–8 cm (2–3 in.) in diameter, each containing 100–1,000 disc florets with reddish-purple floral tubes and long reddish-purple pistils that extend above the rest of the flower. Triangular phyllaries at the base of flower head are broad, often purplish, spine-tipped, and overlapping for several rows. June–October.

Fruit: The flower heads produce many small, dry fruits, each containing a single seed. The light brown fruits are 4 mm (⅛ in.) long and 1 mm (1/32 in.) in diameter. White, plumelike bristles up to 4 cm (2 in.) long attach to the tips of the fruits and aid in wind dispersal. July–November.

Stems: The upper stem is covered with soft hairs. In lower parts, the leaf bases extend down the stem, forming winglike structures covered with sharp spines.

Root system: Each plant forms a deep taproot.

Reproduction: By seed. Plants have an average production of about 10,000 seeds, but some individuals produce up to 120,000 seeds. Seed is dispersed over long distances by wind, water, livestock, and wildlife. Some seeds can remain viable in the soil for up to 10 years.

Impact: This species is unpalatable to livestock, so it frequently becomes established in pastures. It spreads rapidly and forms extensive stands, displacing native vegetation and crowding out forage plants.

Control: Because *C. nutans* reproduces only by seed, it is essential to destroy the plant before flowering. The stems can be closely mowed or cut after they have begun growing but before flower formation occurs. Regrowth will often occur, so follow-up treatments will most likely be necessary. Mowing after flowering should be avoided, as it spreads seed. Chemical control should be focused on young rosettes; they are easier to kill than more mature rosettes or plants with flower stalks. Early spring application of 2,4-D, clopyralid, MCPA, triclopyr, and picloram is effective for this species.

Carduus nutans flowering plant. Image by Norman E. Rees, USDA Agricultural Research Service, Bugwood.org.

Carduus nutans leaves. Image by Leslie J. Mehrhoff, University of Connecticut, Bugwood.org.

Carduus nutans stem. Image by Bruce Ackley, The Ohio State University, Bugwood.org.

Carduus nutans seed head and flower head. Image by Leslie J. Mehrhoff, University of Connecticut, Bugwood.org.

Carduus nutans fruits. Image by Ken Chamberlain, The Ohio State University, Bugwood.org.

Carduus nutans flower head. Image © Gerald D. Carr.

Asteraceae

Centaurea stoebe
ssp. *micranthos*

spotted knapweed

Synonym: *Centaurea maculosa*

ODA Invasive
OIPC Invasive
⚠ Do Not Touch

Origin: Native to eastern Europe and western Asia. This species was most likely introduced to North America as a contaminant in alfalfa seed or in soil used as ship ballast in the late 1800s. It is a significant invasive plant in natural areas and rangelands, particularly in the western United States.

Description: A biennial or short-lived perennial that typically survives three to five years. Plants may flower once or up to three years in succession. This species, which has a bushy form, produces numerous flowers that somewhat resemble thistles or the cultivated *Centaurea cyanus* (bachelor's buttons). The common name is derived from the coloration of the phyllaries, which give the base of the flower head a spotted appearance. These phyllaries distinguish *Centaurea stoebe* from related species, by the black marking at the tip of the bract and the fringe of white or brownish-black teeth.

Habitat: Pastures, rangeland, meadows, prairies, old fields, edges of cultivated fields, along railways and roadsides, oak and pine barrens, and lake dunes. This species is most commonly found in highly disturbed habitats with full sun but it is increasing in undisturbed natural areas.

Height: 30–120 cm (1–4 ft.).

Foliage: The basal rosette leaves are 15–20 cm (6–8 in.) long and 3–5 cm (1–2 in.) wide, oblong, pinnately divided and deeply lobed into narrow segments. Leaves on the flowering stems are alternate; those in the lower part of the stem are similar in shape to the rosette leaves but smaller, 3–9 cm (1–4 in.) long, while the upper leaves are linear, becoming progressively smaller farther up the stem. All leaves are a pale grayish green, with fine but rough hairs.

Flowers: Flower heads are borne singly at the ends of branching stems. Each flower head is 1.5–2.5 cm (½–1 in.) in diameter and contains 20–50 pink to purplish-pink tubular disc florets. The florets around the outer edge are sterile and have five long and widely spreading lobes. The florets in the center are shorter, and each has a column of stamens with white tips surrounding a divided style that extends above the rest of the flower. The base of the flower head is covered with overlapping phyllaries. Each floral bract is ovate, grayish green with dark green parallel veins toward the base, tapering to a black tip with a fringe of 1–2 mm (¹⁄₃₂–¹⁄₁₆ in.) long, white or brownish-black teeth. June–October.

Fruit: Seed heads are about 1 cm (½ in.) long with numerous small, dry fruits, each containing a single seed. The light brown fruits are 3–4 mm (⅛ in.) long and 1 mm (¹⁄₃₂ in.) wide, with a tuft of white hairs at the tip. July–November.

Stems: Multiple highly branched upright stems arise from the base. Stems are slender, stiff, ridged, and covered with rough hairs.

Root system: A deep taproot.

Reproduction: Primarily by seed, dispersed as a contaminant in crop seed or hay, by animals, and over short distances by wind and gravity. Individual plants produce an average of 1,000 seeds in a growing season but can produce up to 25,000 seeds in sites with high water availability. Seeds remain viable in the soil for five to eight years. Plants also extend lateral roots to form new rosettes adjacent to the parent plant.

Impact: Can spread rapidly and form a dense cover that displaces native vegetation. In addition to direct competition, *C. stoebe* produces chemicals called catechins, which inhibit the

First-year rosette of *Centaurea stoebe.* Image by Leslie J. Mehrhoff, University of Connecticut, Bugwood.org.

Centaurea stoebe stem and leaves. Image © Robert L. Carr.

Centaurea stoebe flower head. Image © Robert L. Carr.

Centaurea stoebe fruits. Image by Steve Hurst, hosted by the USDA-NRCS PLANTS Database.

Centaurea stoebe phyllaries. Image © Robert L. Carr.

Centaurea stoebe infestation. Image © Robert L. Carr.

growth of other plant species. A bitter compound called cnicin makes this plant unpalatable to many mammalian herbivores; thus it is less likely to be eaten than other species. As its cover increases, there is a reduction in forage potential and an overall degradation of rangelands. Areas where native vegetation has been replaced with *C. stoebe* also have increased surface runoff and erosion, because the taproots do not stabilize the soil as effectively as the shallow fibrous roots of native grasses.

Control: Wear protective clothing and gloves while undertaking any control measures; the stems of this plant are quite rough, and some of the chemicals it contains can cause skin irritation. Small numbers of isolated plants can be pulled by hand, dug out, or clipped repeatedly. Cutting or mowing plants before flowering will prevent seed production. Use caution later in the growing season, as mowing mature seed heads can facilitate seed dispersal. Cut plants should be bagged and removed from the site. Chemical control with 2,4-D, picloram, and clopyralid can be effective for this species. With any control method, it is essential to revisit the site for several years to eliminate any new plants that grow from root fragments or the seed bank.

Asteraceae

Cichorium intybus

chicory · common chicory · chickory · succory · wild endive · ragged sailors · blue sailors · blue daisy · blue dandelion · blue weed · coffeeweed · wild bachelor's buttons

Origin: Native to Europe, Asia, and northern Africa. This species has been cultivated as a food and medicinal plant for thousands of years. It was introduced to North America in the 1700s and primarily grown for its edible leaves and roots. *Cichorium intybus* has escaped cultivation and spread throughout the United States and Canada, where it is a common weed of roadsides and other highly disturbed areas.

Description: A perennial plant with showy light blue to lilac flower heads. It initially grows as a rosette of irregularly toothed leaves, and later in the season sends up flowering stems. The light blue flower heads differentiate *C. intybus* from most other closely related species. Some native and nonnative *Lactuca* and *Mulgedium* species (wild lettuce) have similar but much smaller blue flower heads.

Habitat: Pastures, old fields, meadows, along railways and roadsides, fencerows, and gardens. This species usually grows in areas with high levels of disturbance but it occasionally encroaches in open natural areas. It prefers full sun and moderate levels of moisture but can withstand moderate drought once established.

Height: 30–100 cm (1–3 ft.).

Foliage: Basal leaves are lance-shaped, 5–35 cm (2–14 in.) long and 1–8 cm (½–3 in.) wide, variously toothed or lobed such that they resemble dandelion foliage. The leaves on the flowering stem are alternate, similar in form to the basal leaves but smaller and the bases clasp the stem, becoming progressively smaller and less toothed farther up the stem. Leaves are dark green, with a prominent white midrib. Both the upper and lower surfaces of the leaf blade are roughly hairy.

Flowers: Showy light blue to lilac flower heads are borne in clusters of one to four, widely spaced along the stems and ends of branches. Each flower head is 3–4 cm (1–2 in.) wide, containing 12–25 strap-shaped ray florets that have squared ends notched with five teeth. The individual ray florets each have several dark blue anthers that are fused to form a tube through which a single style extends. The tip of the style is split into two curling stigmas. The base of each flower head is surrounded by six to eight long inner phyllaries and five or six shorter outer phyllaries that are green to purplish and sparsely covered with glandular hairs. Each flower head opens for a single day only. In hot and sunny weather, the flower heads open in the morning and close around noon, while in cooler weather they will stay open for the whole day. June–September.

Fruit: The flower heads produce many small, dry fruits, each containing a single seed. The fruits are dark brown, 2–3 mm (1/16–⅛ in.) long and 1–1.5 mm (1/32–1/16 in.) wide, wedge-shaped, often pentagonal in cross section, with a ribbed surface and a fringe of short, white bristles on the top. July–October.

Stems: Flowering stems are upright, sparsely branched, roughly hairy in the lower portion,

Native Alternatives

Campanula rotundifolia (harebell)
Delphinium exaltatum (tall larkspur)
Symphyotrichum cordifolium (common blue wood aster)
Symphyotrichum laeve (smooth aster)
Symphyotrichum novae-angliae (New England aster)

becoming nearly hairless higher up, green to reddish brown, often woody near the base. The stems contain latex and exude a white milky sap when cut.

Root system: The root system is a thick, often divided, and somewhat woody taproot.

Reproduction: By seed, dispersed by animals, machines, and human activities. An individual plant can produce 3,000 seeds in a growing season.

Impact: A persistent weed in disturbed areas such as roadsides, also problematic in cropland and pastures. Plants grow quickly and aggressively and are difficult to remove once established due to their taproots.

Control: In small populations, plants should be cut using a shovel as far below the root crown as possible. If herbicide is the only option, 2,4-D can be applied to the rosettes before flowering stalks are formed.

Left: Flowering *Cichorium intybus* plants. *Right: Cichorium intybus* leaves and stem. Both images by Robert Vidéki, Doronicum Kft., Bugwood.org.

Cichorium intybus flower head. Image by David Cappaert, Bugwood.org.

Cichorium intybus leaves and roots. Image © Gerald D. Carr.

Cichorium intybus seed heads. Image by Gianni Careddu, used under a CC BY-SA 4.0 license.

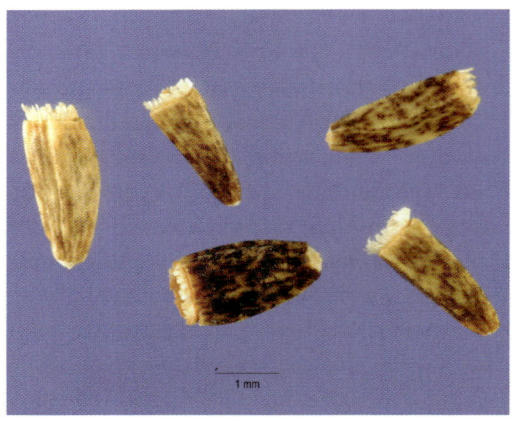

Cichorium intybus fruits. Image by Steve Hurst, hosted by the USDA-NRCS PLANTS Database.

Asteraceae

Cirsium arvense

Canada thistle · creeping thistle · California thistle · field thistle · corn thistle · cursed thistle · green thistle · hard thistle · perennial thistle · prickly thistle · small-flowered thistle · way thistle

OIPC Invasive
Prohibited
Noxious Weed
⚠ **Do Not Touch**

Origin: This species is most likely native to Europe and western Asia, but it has been associated with human settlement for so long that it is difficult to establish the precise native range. *Cirsium arvense* was accidentally introduced to North America in contaminated crop seed sometime in the 1600s. It was long thought to have been introduced first to Canada (hence the common name), from where it spread to the United States. However, more recent studies have determined that this species was probably introduced to both Canada and the New England region of the United States at about the same time. It is considered one of the most significant agricultural weeds in the world and is also problematic in natural areas.

Description: An upright perennial that spreads rapidly via lateral roots and forms extensive clones. Both the creeping roots and the clonal nature of this species differentiate it from other thistles in the region. In addition, *C. arvense* is the only thistle that has separate male and female plants (it is dioecious, or nearly so). It can also be identified by its slender stalks; numerous small flower heads with pale pink or lavender florets; and phyllaries at the base of each flower head, which have very short pointed tips rather than the well-developed spines in other thistles.

Habitat: Forests, wet to moist grasslands, river- and streambanks, lakeshores, marshes, roadsides, drainage ditches, and pastures. Prefers full sun in moist, fertile soils.

Height: 30–150 cm (1–5 ft.).

Foliage: Leaves are alternate, with variable appearance and size, generally oblong to lance-shaped, 3–30 cm (1–12 in.) long and 1–6 cm (½–3 in.) wide. Leaves unlobed to deeply lobed, with wavy and spine-toothed margins. Spines are typically yellow, 1–7 mm ($\frac{1}{32}$–¼ in.) long. Leaves on the upper parts of the plant clasp the stem. The leaf surfaces are mostly hairless above and can be hairless or slightly hairy on the underside.

Flowers: The flowers of this species are usually dioecious but not always so, with some plants bearing hermaphrodite flowers. Flower heads are numerous, singly or in clusters of two to five at the tips of the main stems and branches. Individual flower heads are 1–2 cm (½–¾ in.) tall and 1–1.5 cm (½–⅝ in.) in diameter, the male heads slightly smaller and more spherical than the flask-shaped female heads. The flower heads contain 20–50 tubular disc florets, pale pink to lavender. Each female floret has a lavender pistil that extends above the rest of the flower. All flower heads are surrounded at the base by six to eight rows of overlapping phyllaries that are often tinged purple and have a very short pointed tip. Florets with female parts produce a sweet fragrance. June–August.

Fruit: The flower heads produce many small, dry fruits, each containing a single seed. The fruits are light brown, 2–3 mm ($\frac{1}{16}$–⅛ in.) long and about 1 mm ($\frac{1}{32}$ in.) wide, smooth and finely grooved, with a tuft of white to light brown hairs at the top. July–September.

Stems: Multiple stems arise from the lateral roots and then branch near the top. The stems are light green, ridged, slightly hairy when young, and hairier with age. Unlike other non-native thistle species, the stems of *C. arvense* are not typically spiny.

Cirsium arvense leaves. Image by Jan Samanek, Phytosanitary Administration, Bugwood.org.

Cirsium arvense seed heads. Image by Leslie J. Mehrhoff, University of Connecticut, Bugwood.org.

Cirsium arvense stems. Image by Leslie J. Mehrhoff, University of Connecticut, Bugwood.org.

Cirsium arvense fruits. Image by Julia Scher, Federal Noxious Weeds Disseminules, USDA APHIS PPQ, Bugwood.org.

Cirsium arvense infestation. Image by Leslie J. Mehrhoff, University of Connecticut, Bugwood.org.

Cirsium arvense flower heads. Image by Leslie J. Mehrhoff, University of Connecticut, Bugwood.org.

Root system: A slender taproot and creeping lateral roots that typically spread 1–2 m (3–6 ft.) in a year but can reach up to 6 m (20 ft.). Aerial shoots arising from the lateral roots die back over winter, and new stems grow from the rootstock in the spring.

Reproduction: Vegetative reproduction from the lateral roots is the main means of reproduction and spread within a site. Also, root fragments are capable of forming new plants. An individual plant can produce 1,500–5,000 seeds in a growing season. Most seeds germinate within 1 year, but seeds can remain viable in the soil seed bank for 20 years. Seeds are dispersed by wind, water, and farm machinery.

Impact: A significant agricultural weed worldwide. This species has been problematic in the United States since the late 1700s. It reduces crop yield through competition, interferes with harvest, and is a host for insects and pathogens that damage crops. In addition, because this species is highly unpalatable, it reduces forage potential in pasture and rangeland. It also invades natural areas, where it displaces native plants and reduces plant and animal species diversity.

Control: In small stands, repeated cutting and mowing several times a year over several growing seasons can kill clones. Mowing followed by solarization is also effective. In areas with extensive cover, tilling followed by autumn application of clopyralid, 2,4-D, glyphosate, or picloram can be successful in controlling this species.

Cirsium vulgare

bull thistle · common thistle · spear thistle · plume thistle

⚠ Do Not Touch

Origin: Native to Europe, western Asia, and northern Africa. This species was first introduced to eastern North America during colonial times, with introductions to western North America during the late 1800s. It spread from these initial centers and is now reported in all 50 states and all but two Canadian provinces.

Description: Typically biennial but can occasionally grow as an annual or a short-lived monocarpic perennial. It first forms a rosette and later develops a flowering stalk with spiny, spreading, winged stems. The prickly phyllaries at the base of the flower heads distinguish this species from native thistles. In *Cirsium vulgare*, each bract curls outward and narrows into a sharp spine, while native thistles have flattened, overlapping bracts. A species with similarly spiny bracts is the nonnative *Carduus acanthoides* (plumeless thistle), which generally has smaller flowers (less than 3 cm [1 in.] in diameter) and leaves that are mostly hairless on the upper surface.

Habitat: Pastures, rangelands, roadsides, ditches, and fencerows. This species is most commonly associated with areas of cultivation that have repeated disturbance. It does best in full sun and high-fertility soils.

Height: 30–200 cm (1–7 ft.).

Foliage: Leaf form varies with developmental stage. The first leaves of the rosette are elliptic, coarsely toothed, with spines along the margin. Mature basal leaves are lance-shaped, 15–30 cm (6–12 in.) long and 5–10 cm (2–4 in.) wide, deeply pinnately lobed with secondarily lobed segments, the lobes and teeth tipped with stout spines about 8 mm (¼ in.) long. The leaves on the flowering stems are alternate, becoming pro-gressively smaller toward the ends of branches. Leaves are covered with coarse hairs above and woolly hairs on the underside.

Flowers: Flower heads are solitary or in small clusters at the ends of branches. Individual flower heads are 5 cm (2 in.) tall and 4–5 cm (1 ½–2 in.) in diameter, containing up to 200 tubular disc florets, pinkish red to purple, divided into five long lobes that give the flower head a feathery appearance. The flower heads are surrounded at the base by phyllaries that curl outward and narrow into a sharp spine. June–September.

Fruit: The flower heads produce many small, dry fruits, each containing a single seed. The fruits are light brown, 3.5–5 mm (⅛–¼ in.) long and about 1.5 mm (¹⁄₁₆ in.) wide, yellow streaked with dark brown, with a tuft of white hairs at the top, 2–3 cm (¾–1 in.) long. July–October.

Stems: The stems are stout, hairy, and covered with spiny wings. They branch to varying degrees, but the overall form remains upright.

Root system: Plants form a deep and branching taproot.

Reproduction: By seed, primarily dispersed by wind but also can be moved by water, animals, and human activities. Some plants produce up to 120,000 seeds in a growing season, of which there is a high germination rate. Most seeds only remain viable in the soil seed bank for one or two years.

Impact: This species is generally not considered as problematic as *Carduus nutans* (musk thistle) [pp. 92–93] and *Cirsium arvense* (Canada thistle) [pp. 99–101], but it is one of the spiniest thistles. As such, it can injure livestock and get caught in the wool of sheep, causing a decrease in the value of the final product. The rosettes of this species can cover large areas and outcompete both crops and native plant species.

Control: For small populations, cut rosettes below the root crown about 5–10 cm (2–4 in.) below ground level, using a hoe or spade. In larger populations, mowing late in the growing season before seed maturation can help reduce population size. The herbicides 2,4-D and picloram are effective at controlling this species, but they must be applied at the rosette stage because plants become resistant as the flower stalks develop.

Left: Cirsium vulgare rosette. Image by Ohio State Weed Lab, The Ohio State University, Bugwood.org. *Right: Cirsium vulgare* stems. Image by Rob Routledge, Sault College, Bugwood.org.

Cirsium vulgare flower head. Image by Loke T. Kok, Virginia Polytechnic Institute and State University, Bugwood.org.

Cirsium vulgare fruits. Image by Julia Scher, Federal Noxious Weeds Disseminules, USDA APHIS PPQ, Bugwood.org.

Cirsium vulgare seed heads. Image by Forest Starr and Kim Starr, Starr Environmental, Bugwood.org.

Cirsium vulgare plants. Image by Forest Starr and Kim Starr, Starr Environmental, Bugwood.org.

Asteraceae

Conyza canadensis

marestail · horseweed · Canadian horseweed · Canadian fleabane · horseweed fleabane · coltstail · butterweed

Synonym: *Erigeron canadensis*

Origin: This species is thought to be native to North America, but it is classified as a prohibited noxious weed in Ohio. *Conyza canadensis,* widespread throughout the United States, was most likely transported to Europe in the 1600s and is now a weed in most tropical and temperate habitats worldwide.

Description: An annual plant that typically germinates in autumn and persists as a rosette of leaves over the winter. In the spring, it develops a single, upright, leafy stem that ends in a highly branched inflorescence of insignificant white and yellow flower heads. This common, widespread weed may be mistaken for *Solidago* spp. (goldenrods) during the early stages of growth. However, the flowers are much closer in appearance to *Erigeron* spp. (fleabanes).

Habitat: Fields, meadows, fencerows, gardens, vacant lots, along railways and roadsides, pastures, orchards, plantations, and cultivated land with low or no tillage. This species prefers dry, disturbed soils but will grow in moist habitats.

Height: To 180 cm (7 ft.).

Foliage: Plants form a basal rosette of lance-shaped, dark green, sparsely hairy leaves, to 10 cm (4 in.) long and 1 cm (½ in.) wide, with coarsely toothed margins. The rosette deteriorates as the flowering stem begins to elongate. Stem leaves are alternate but densely arranged so that they almost appear whorled. Leaves become progressively smaller farther up the stem, but only slightly so, 2–10 cm (¾–4 in.) long and 4–15 mm (⅛–⅝ in.) wide. Leaves are often toothed along their outer margins and are edged with stiff hairs. The upper and lower surfaces of the blades are mostly smooth, with sparse hairs along the outer length of the central vein.

Flowers: The flowering stem ends in a highly branched panicle of 50–300 small flower heads. Individual flower heads are 4–5 mm (⅛–¼ in.) long and 3–5 mm (⅛–¼ in.) in diameter, cylindrical or somewhat urn-shaped, each containing 25–45 white, strap-shaped ray florets that surround 8–30 tubular, yellow disc florets. The base of each flower head is surrounded by tiny, linear phyllaries in three or four overlapping rows. July–August.

Fruit: Each flower head produces an average of 60–70 small, dry fruits, each containing a single seed. The fruits are light brown, 1–2 mm (¹⁄₃₂–¹⁄₁₆ in.) long and 0.5 mm (¹⁄₆₄ in.) wide, with a tuft of light brown hairs at the top, 3–5 mm (⅛–¼ in.) long. August–September.

Stems: Stems are stout, covered in stiff hairs, and unbranched except near the flowers. The stems grow upright, although they may start to lean over once the flowers develop and the plant becomes top-heavy.

Root system: A short taproot with fibrous lateral roots.

Reproduction: By seed, dispersed by wind. An individual plant can produce up to 250,000 seeds. The seeds remain viable in the soil seed bank for two to three years.

Impact: Primarily a weed of perennial crops and untilled areas under cultivation, including fruit orchards, vineyards, and ornamental nurseries. It is becoming increasingly problematic in untilled annual crops, such as cotton and soybean. This species may be allelopathic, and it causes a reduction in crop yield. In addition, it is a secondary host for numerous plant pathogens—including tomato spotted wilt virus and cucumber mosaic virus—as well as a range of nematodes and insect pests. It is also a frequent weed of gardens.

Control: In small populations, pulling by hand is effective, provided that the entire taproot is removed. The seeds require light for germination, so shallow tilling or the use of mulch to cover the soil layer can provide control. This species has developed widespread resistance to herbicides, making chemical control difficult. However, spot treatment with the herbicides 2,4-D and picloram can be used.

Conyza canadensis rosette. Image by Robert Vidéki, Doronicum Kft., Bugwood.org.

Conyza canadensis seed heads. Image by AnRO0002, used under a CC0 1.0 license.

Conyza canadensis stem and leaves. Image by Rob Routledge, Sault College, Bugwood.org.

Conyza canadensis fruits. Image by D. Walters and C. Southwick, Table Grape Weed Disseminule ID, USDA APHIS PPQ, Bugwood.org.

Conyza canadensis flower heads. Image © Gerald D. Carr.

Conyza canadensis infestation. Image by AnRO0002, used under a CC0 1.0 license.

Asteraceae

Hieracium caespitosum

meadow hawkweed · yellow hawkweed · field hawkweed · king devil · devil's paintbrush · yellow paintbrush · yellow king devil · yellow devil · yellow fox-and-cubs

Origin: Native to Europe and some northern and western portions of Asia. It was introduced to the northeastern United States as an ornamental in the 1800s. Several subsequent introductions to the northwestern United States occurred, and this species was first recorded in Washington in 1969. This species, which has spread inland from these centers of introduction on the two coasts, is now a declared weed in Washington, Oregon, Montana, and Idaho.

Description: A creeping perennial that forms a low rosette of basal leaves from which an upright and leafless flowering stem develops. It has yellow flower heads similar to dandelions, but these are smaller and borne in clusters. The leaves and stem are hairy and exude a milky sap when crushed. There are several native hawkweeds in Ohio, which can be distinguished from this nonnative species by the presence of numerous alternate leaves along the flowering stems, hairs on the lower surface of the leaves only, and a lack of black glandular hairs on the branches of the inflorescence.

Habitat: Fields, pastures, meadows, forest openings, river- and streambanks, lakeshores, roadsides, gardens, and lawns. This species is often found in areas with sandy or gravelly soils and slightly acidic conditions, but it can tolerate a range of soil types and can grow in full sun to partial shade.

Height: 30–80 cm (1–3 ft.).

Foliage: Plants form a low rosette of basal leaves that are simple, lance- or spatula-shaped, 8–20 cm (3–8 in.) long and 1–4 cm (½–2 in.) wide, with rounded tips. With the exception of one or two very small leaves on the lower part of the stem, the leaves are almost exclusively basal.

In some plants, the basal leaves lie flat on the ground, while in others they grow more upright. The smooth leaf margins have a dense fringe of hairs along the edges. Both the upper and lower surfaces of the leaf blade are dull green and covered with long, stiff hairs, while the lower leaf surfaces also have short, star-shaped hairs.

Flowers: The flowering stem ends in a flat-topped cluster of dandelion-like flower heads. There are typically 5–25 flower heads per cluster but occasionally up to 50. Individual flower heads are 1–2 cm (½–¾ in.) in diameter, containing strap-shaped ray florets that have a squared end bearing several teeth. The base of each flower head is surrounded by linear phyllaries that have black glandular hairs along their midveins. The branches of the inflorescence are also covered with black glandular hairs. June–August.

Fruit: The flower heads produce an average of 12–50 small, dry fruits, each containing a single seed. The fruits are shiny and black, 1–2 mm (1/32–1/16 in.) long and less than 0.5 mm (1/64 in.) wide, with a spreading plume of fine white bristles at the top, 4–6 mm (1/8–¼ in.) long. July–September.

Stems: Flower stems are upright, unbranched, leafless except for one or two small leaves on the lower part of the stem, and covered in hairs. The hairs near the base of the stem are long and white, while those on the upper part of the stem are black and glandular.

Root system: A shallow, fibrous root system with slender rhizomes.

Reproduction: By seed, dispersed by wind, animals, and human activities. Seeds germinate as soon as they are released from the plant and do not form a persistent seed bank. Vegetative reproduction is also possible through stolons, which creep along the soil surface and root at the nodes; through adventitious root buds that develop on the fibrous roots; or from rhizomes.

Impact: Forms dense colonies that outcompete other species, including important range and pasture species, as well as native plant species. Some sources suggest that this species may be allelopathic.

Control: *Hieracium caespitosum* is difficult to control because it can reproduce by both seed and vegetative means. In small populations, plants can be dug out by hand, taking care to ensure that shallow roots, rhizomes, and stolons are removed. In more extensive stands, close mowing can prevent seed production. This species is not tolerant of heavy shade, so management practices that favor more competitive species will ultimately reduce *H. caespitosum*. Research has shown that clopyralid or 2,4-D will control this species.

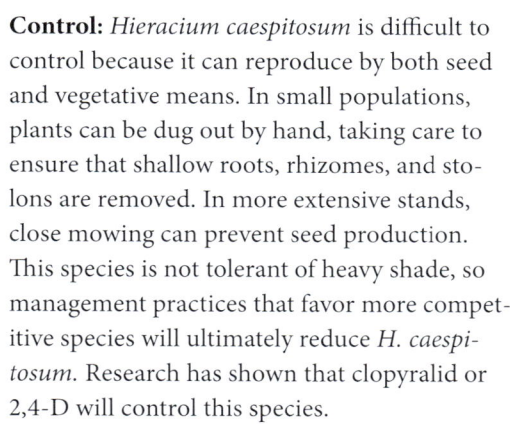

Hieracium caespitosum flower heads. Image © Robert L. Carr.

Hieracium caespitosum leaves. Image © Robert L. Carr.

Hieracium caespitosum stem. Image by Julia Kruse, used under a CC BY-SA 3.0 license.

Hieracium caespitosum seed heads. Image used with permission from www.FloraFinder.org.

Glandular hairs on the stems on *Hieracium caespitosum*. Image © Robert L. Carr.

Hieracium caespitosum infestation. Image by Tom Heutte, USDA Forest Service, Bugwood.org.

Asteraceae

Lactuca serriola

prickly lettuce · wild lettuce · China lettuce · compass plant · horse thistle · milk thistle · wild opium · scarole

Origin: This species is native to the Mediterranean region of Europe, western and central Asia, and northern Africa. It was most likely accidentally introduced to North America in the late 1800s, as a contaminant in crop seed. *Lactuca serriola* is now a common weed found throughout the United States.

Description: This annual or biennial plant forms a low rosette of rounded leaves that gives rise to a single, tall flowering stem. The light yellow flower heads are similar to dandelions, but they are smaller and borne in spreading panicles. When cut, the leaves and stem exude a milky sap that contains slightly analgesic and sedative latex compounds, which is the reason for the common name wild opium. The presence of prickles along the midrib on the underside of the leaves distinguishes *L. serriola* from native *Lactuca* and *Mulgedium* species and from nonnative species in the genus *Sonchus* (sow thistles).

Habitat: Pastures, orchards, cultivated fields, old fields, along railways and roadsides, fencerows, and gardens. This species prefers areas with high levels of disturbance. It often grows in dry conditions with sandy or gravelly soils.

Height: 30–200 cm (1–7 ft.).

Foliage: The leaves are highly variable, typically oblong to lance-shaped, up to 30 cm (1 ft.) long and 10 cm (4 in.) wide. Leaves on the flowering stem are alternate and become progressively smaller farther up the stem. The margins of the leaves are often deeply pinnately lobed and edged with short teeth and tiny prickles. The bluish-green leaves have a prominent white midrib; this vein is lined with prickles on the underside of the leaf. Angular lobes clasp the stem on either side of the leaf base. The foliage gives off a rank odor when crushed.

Flowers: Small, pale yellow, dandelion-like flower heads develop in an airy, branching panicle at the end of the stem. Individual flower heads are 11–13 mm (½ in.) in diameter, containing 12–20 strap-shaped ray florets that have a squared end notched with five teeth. The base of each flower head is surrounded by bluish-green phyllaries that can be tinged purple. July–September.

Fruit: The flower heads produce many small, dry fruits, each containing a single seed. The fruits are brown, 3–4 mm (⅛ in.) long and 1 mm (1/32 in.) wide, oblong with a tapering tip, with five to seven longitudinal ribs on each side. The tip of the fruit attaches to a threadlike beak that ends in a spreading plume of fine white bristles, similar to that of a dandelion. August–October.

Stems: Flower stems are upright, often prickly on the lower portion, becoming smooth higher up, light green or dull white, occasionally with a reddish tinge. When cut, stems exude a bitter white sap.

Root system: A long, stout, white taproot.

Reproduction: By seed, primarily dispersed by wind. A single individual can produce 45,000 seeds that germinate soon after they are released. Seeds remain viable in the soil seed bank for three years.

Impact: This species is primarily a nuisance in agriculture, particularly in orchards, pastures, and crops with minimal tilling. It does not encroach on high-quality natural areas, but it is an aggressive colonizer in any disturbed habitats.

Control: In small populations, plants should be pulled by hand or cut using a shovel below the root crown before seed set. Covering the area with mulch can help prevent subsequent seedling establishment. The herbicides 2,4-D, isoxaben, and glyphosate are all effective at controlling *L. serriola* to some degree. There is evidence of herbicide resistance developing in this species, so it is critical that the same herbicide not be used every year to prevent further resistance from developing.

Lactuca serriola leaves with prickles along the lower midrib. Image by Bruce Ackley, The Ohio State University, Bugwood.org.

Lactuca serriola leaves clasping the stem. Image by Bruce Ackley, The Ohio State University, Bugwood.org.

Lactuca serriola plants. Image by Forest Starr and Kim Starr, Starr Environmental, Bugwood.org.

Lactuca serriola flower head. Image © Gerald D. Carr.

Lactuca serriola seed head. Image © Gerald D. Carr.

Lactuca serriola fruits. Image by Steve Hurst, hosted by the USDA-NRCS PLANTS Database.

Asteraceae

Leucanthemum vulgare

oxeye daisy · common daisy · moon daisy · dog daisy · horse daisy · dun daisy · butter daisy · field daisy · white daisy · marguerite daisy · Maudlin daisy · Maudlinwort

Origin: This species is native to Europe and western Asia. It was introduced to the United States as an ornamental plant in the 1800s. *Leucanthemum vulgare* is still commonly sold as a wildflower and in meadow seed mixes. Until recently, this species was listed as a prohibited noxious weed in Ohio.

Description: A perennial plant with characteristically daisy-like flower heads that have white petals surrounding a yellow center. This species is one of the parent plants of the cultivated *Leucanthemum* x *superbum* (Shasta daisy) and, as such, is similar in appearance. However, the Shasta daisy tends to have larger flowers, typically over 5 cm in diameter. The pinnately lobed leaves of *L. vulgare* differentiate it from other similar species.

Habitat: Pastures, fields, prairies, meadows, forest margins and openings, river- and streambanks, lakeshores, along railways and roadsides, lawns, and gardens. Prefers mesic to dry soils with a high level of disturbance, in full sun to partial shade.

Height: Typically 30–90 cm (1–3 ft.), but occasionally taller.

Foliage: Plants first form a rosette of basal leaves that are stalked, spatula-shaped to obovate, 10–25 cm (4–10 in.) long and 3–7 cm (1–3 in.) wide, pinnately lobed, with irregular rounded teeth along the margins. Stem leaves are smaller, 3–8 cm (1–3 in.) long and 2–15 mm (1/16–5/8 in.) wide, alternate, clasping the stem, spatula-shaped to narrowly lance-shaped, with toothed or deeply lobed margins. Leaves are often dark green, glossy, fleshy, and sparsely haired.

Flowers: Flower heads are solitary at the ends of stems and branches, with 1–15 flower heads per plant. Individual flower heads are 3–5 cm (1–2 in.) in diameter, containing 15–35 white ray florets surrounding a center of 400–500 yellow disc florets. The ray florets are strap-shaped, rounded at the ends, and indistinctly three-toothed. The base of each flower head is surrounded by green phyllaries that can be edged with brown. May–September.

Fruit: Flower heads produce up to 200 dry fruits, each containing a single seed. The fruits are brown, 2–3 mm (1/16–1/8 in.) long and up to 1 mm (1/32 in.) wide, ridged with ten lighter brown nerves. Unlike many other species in the family, the fruits lack tufts of hair to aid in wind dispersal. June–October.

Stems: Each rosette gives rise to 1–40 flowering stems. Stems are slender, upright, and mostly hairless.

Root system: Shallow, unbranched roots and creeping rhizomes.

Reproduction: By seed, primarily dispersed by gravity in the area surrounding the parent plant. Seeds remain viable in the soil seed bank for 20 years or more. Plants also spread

Native Alternatives

Echinacea purpurea 'Alba' (purple coneflower)
Erigeron pulchellus (robin's plantain)
Helianthus giganteus (giant sunflower)
Helianthus occidentalis (oxeye sunflower)
Helianthus tuberosus (Jerusalem artichoke)
Parthenium integrifolium (American feverfew)
Symphyotrichum novi-belgii 'White Opal' (Michaelmas daisy)

via creeping rhizomes and can regenerate from rhizome fragments.

Impact: This species is problematic in pastures because cows that eat it produce off-tasting milk. It is also a concern in agricultural settings because it is a secondary host for several crop diseases and pests. In addition, it can form dense stands that displace native vegetation in highly disturbed natural areas.

Control: In small populations, plants can be pulled or dug out, taking care to remove all shallow roots and rhizomes. Heavy mulching over the winter can cause the rootstocks to rot. Herbicides such as picloram, imazapyr, and 2,4-D have been used for this species, usually just as plants begin to flower.

Leucanthemum vulgare fruits. Image by Steve Hurst, hosted by the USDA-NRCS PLANTS Database.

Leucanthemum vulgare rosette. Image by Ohio State Weed Lab, The Ohio State University, Bugwood.org.

Leucanthemum vulgare stem and leaves. Image © Katy Chayka, Minnesota Wildflowers.

Leucanthemum vulgare roots. Image by Steve Dewey, Utah State University, Bugwood.org.

Leucanthemum vulgare flower heads. Image by David Stephens, Bugwood.org.

Leucanthemum vulgare plants. Image by David Stephens, Bugwood.org.

Asteraceae

Packera glabella
cressleaf groundsel · butterweed · yellowtop

Synonym: *Senecio glabellus*

Origin: This species is native to central and southeastern North America, including southern portions of Ohio, but its range and coverage have been steadily expanding over the past two decades. It is extremely successful in areas with high disturbance, and it has become a major weed in pastures and cropland. Despite being native to parts of the state, *Packera glabella* is a prohibited noxious weed in Ohio.

Description: An annual or winter annual that germinates in the autumn, stays green throughout the winter, and flowers and sets seed the following growing season. It is taller than most other native *Packera* species, and its hollow, red-lined stem is distinctive. The shape of its leaves is reminiscent of rocket or cress species from the Brassicaceae (mustard family), hence the common name cressleaf groundsel.

Habitat: Open floodplain forests, forest margins, river- and streambanks, edges of ponds, meadows, marshes, roadsides, ditches, pastures, and agricultural fields. Prefers wet to moist areas and can grow in full sun to partial shade.

Height: 15–80 cm (½–3 ft.).

Foliage: Plants first form a low rosette of basal leaves, 5–25 cm (2–10 in.) long and 1–6 cm (½–3 in.) wide, deeply pinnately lobed with obovate lateral lobes and a larger, more circular terminal lobe. The leaf margins have irregular, pointed or rounded teeth. The stem leaves are alternate, becoming progressively smaller farther up the stem. Leaves are hairless and mostly light green, although the midribs and petioles are often reddish.

Flowers: Flat-topped inflorescences of 8–30 small, yellow flower heads develop at the ends of stems and branches. Individual flower heads are 1.5–2.5 cm (½–1 in.) in diameter, consisting of 5–15 yellow ray florets surrounding a center of 35–50 yellow disc florets. The base of each flower head is surrounded by a single row of hairless, light green, linear phyllaries. Flowers fragrant. March–May.

Fruit: Flower heads produce dry fruits, each containing a single seed. The fruits are light brown, 1.5–2 mm (¹⁄₁₆ in.) long and 0.5 mm (¹⁄₁₆ in.) wide, with sparsely hairy or hairless ribs, each topped with a plume of white bristles. May–July.

Stems: Plants produce single stems that are hollow, stout, and unbranched except for short flowering stems that develop in the axils of some upper leaves. The stem is hairless and light green with conspicuous, red, longitudinal veins.

Root system: Shallow, fibrous roots.

Reproduction: By seed, dispersed by wind.

Impact: The leaves and stems contain pyrrolizidine alkaloids, which are toxic to mammals, making *P. glabella* an undesirable species to have in pastures. This species is becoming increasingly common in agricultural settings. It can also form dense stands in highly disturbed natural areas.

Control: Because *P. glabella* is a native species and not considered a weed in many situations, control methods for it have not been researched extensively. Isolated plants in areas of concern should be pulled by hand. Spring mowing before seed set can reduce the population. Autumn or spring applications of glyphosate or a commercial premix of glyphosate plus 2,4-D have been found to control this species.

Packera glabella rosette. Image by Robert Thomas, Loyola University.

Packera glabella flower heads. Image by Gerald J. Lenhard, Louisiana State University, Bugwood.org.

Packera glabella plants. Image by Gerald J. Lenhard, Louisiana State University, Bugwood.org.

Packera glabella fruits. Image by Bruce Ackley, The Ohio State University, Bugwood.org.

Packera glabella stems. Image by Dan Tenaglia, Alabama plants.com.

Roadside stand of *Packera glabella* plants. Image by Robert Thomas, Loyola University.

Asteraceae

Sonchus arvensis

perennial sowthistle · field sowthistle · creeping sowthistle · corn sow thistle · tree sow thistle · milk thistle · swine thistle · dindle · gutweed · hare's colewort · hare's lettuce · hare's palace

114

Origin: Native to Europe and western Asia. This species was most likely accidentally introduced to North America as a contaminant of crop seed. First recorded in Pennsylvania in 1814, it has spread throughout most of the United States, except for the southeastern and central regions.

Description: A creeping perennial that has large, dandelion-like flowers borne in clusters at the tops of tall stems. All *Sonchus* spp. in Ohio are nonnative but *Sonchus arvensis* is the only perennial species of this genus found in the state. In addition, it has larger flower heads than in the other species, which are all less than 2 cm (¾ in.) in diameter. It is similar in appearance to *Lactuca serriola* (prickly lettuce), but *L. serriola* has prickles along the midrib on the underside of its leaves.

Habitat: Cultivated fields, meadows, river- and streambanks, lakeshores, along railways and roadsides, and gardens. This species prefers areas with high levels of disturbance. It typically grows in moderately moist soils in areas with full sun to partial shade.

Height: 40–200 cm (1–7 ft.).

Foliage: Leaves are alternate, typically oblong to lance-shaped, 6–40 cm (3–16 in.) long and 2–15 cm (¾–6 in.) wide, with extremely prickly margins. The leaves are concentrated near the base of the stem, where they are larger and more deeply pinnately lobed, becoming less numerous and progressively smaller and less lobed farther up the stem. All leaves have rounded lobes at the base that clasp the stem. The leaf surfaces are hairless, and the midrib on the underside of the leaf lacks prickles.

Flowers: Yellow, dandelion-like flower heads

develop in branching clusters at the ends of the stems. Typically only a few flower heads are open at a time. Individual flower heads are 3–5 cm (1–2 in.) in diameter, containing hundreds of strap-shaped ray florets. The base of each flower head is surrounded by overlapping green phyllaries that in the subspecies *arvensis* are covered with tiny glandular hairs. June–October.

Fruit: The flower heads produce numerous small, dry fruits, each containing a single seed. The fruits are reddish brown, oblong, 2.5–3.5 mm (⅛ in.) long and 1 mm (¹⁄₃₂ in.) wide, with five to seven longitudinal ribs on each side. The tip of the fruit attaches to a plume of long, white hairs, 8–14 mm (¼–½ in.) long. July–November.

Stems: Stems are upright, hollow, unbranched except in the inflorescence. The cut stems exude a milky sap that contains latex. Plants of the subspecies *glabrescens* have hairless stems, while those of the subspecies *arvensis* have tiny glandular hairs on the upper parts of the stems and on the flower stalks.

Root system: A deep taproot and creeping lateral roots that thicken as the plant matures. The lateral roots are fleshy and somewhat fragile, but they can extend several meters from the parent plant, allowing an individual to spread over a wide area.

Reproduction: By seed, primarily dispersed by wind. Vegetative reproduction is also important in this species, through creeping roots and regeneration from root fragments.

Impact: This species is primarily a nuisance in agriculture, particularly in cereal crops. It does not encroach on high-quality natural areas, but it does colonize open habitats that have been disturbed.

Control: Deep tilling has been shown to help control this species in cultivated fields. The herbicides 2,4-D amine and glyphosate have shown partial control, particularly when applied while the plants are actively growing.

Sonchus arvensis flower head. Image © Robert L. Carr.

Basal leaves of *Sonchus arvensis.* Image by New York State IPM Program at Cornell University, used under a CC BY 2.0 license.

Sonchus arvensis seed head. Image © Robert L. Carr.

Sonchus arvensis roots. Image by Steve Dewey, Utah State University, Bugwood.org.

Sonchus arvensis fruits. Image by Julia Scher, USDA APHIS PPQ, Bugwood.org.

Sonchus arvensis leaves clasping the stem. Image by Michael Shephard, USDA Forest Service, Bugwood.org.

Asteraceae

Taraxacum officinale
dandelion · common dandelion · faceclock

Origin: This species is most likely native to southern Europe or western Asia, but it has been associated with human settlement for so long that it is difficult to establish the precise native range. It has a long history of use as a medicinal herb and in food preparation. It has been introduced to virtually every region in the world. The earliest record of *Taraxacum officinale* in North America is from the 1600s, but it may have been introduced earlier via the Bering land bridge or with Viking explorers.

Description: A perennial species that forms a rosette of leaves with distinctly lobed and toothed margins. Stout, hollow stems develop from the rosette, each bearing a single yellow flower head. When damaged, the leaves, roots, and flowering stems of *T. officinale* exude a milky sap containing latex. Another closely related nonnative species—*Taraxacum erythrospermum* (red-seeded dandelion)—also occurs in Ohio. These two species can be distinguished from other Asteraceae by the presence of only a single flower head on each stem, outer phyllaries that curve downward from the flower heads, and leaves with terminal lobes that are larger than the other lobes.

Habitat: Fields, roadsides, open shorelines, forest margins and clearings, gardens, and lawns. Prefers moist soil and full sun to partial shade.

Height: 5–30 cm (2–12 in.); stem height is highly variable in response to mowing.

Foliage: Each plant forms a low rosette of basal leaves 6–40 cm (3–16 in.) long and 1–10 cm (½–4 in.) wide, broader toward the outer tip than at the base, and variably pinnately lobed. The lobes are distinctly triangular, with a terminal lobe larger than the others. The leaf margins are slightly wavy and coarsely toothed. Each leaf has a prominent, hollow midrib that is green or sometimes reddish near the base. Leaves are typically hairless, although some are hairy on the undersides, particularly along the midrib.

Flowers: Each stem ends in a single flower head 3–5 cm (1–2 in.) in diameter, containing 150–200 yellow, strap-shaped ray florets that spread outward from the center. The base of the flower head is surrounded by two rows of green phyllaries. The inner bracts overlap to form a tube that encloses the base of the flower head, while the outer bracts curve downward from the flower head. Most flowers form April–May, but sporadic flowering occurs as early as February and as late as June. In some locations there is secondary flowering in the autumn.

Fruit: The flower heads produce many small, dry fruits, each containing a single seed. The fruits are light brown, 3–4 mm (⅛ in.) long and 1–1.5 mm (¹⁄₃₂–¹⁄₁₆ in.) wide, tapering at both ends, with 4–12 longitudinal ribs and tiny teeth near the top. The fruit attaches to a threadlike beak, 7–9 mm (¼–⅜ in.) long, that ends in a spreading plume of fine white bristles.

Stems: One or more hollow stems arise from each rosette. The stems are light green or tinged with red or purple, particularly toward the base. The surface of the stem can be smooth or sparsely covered with cobwebby hairs.

Root system: A deep, thick, branched taproot.

Reproduction: By seed, dispersed by wind. A single plant can produce 3,000–23,000 seeds in a growing season.

Impact: One of the most common and recognizable weeds, it is considered by many to be a nuisance in lawns and gardens. It is also becoming increasingly problematic as a weed in agricultural fields with reduced tilling.

Control: Do not let this plant go to seed. Plants can be dug out or cut below the root crown using a spade or hoe. Mowing during flowering can reduce seed production, but plants often respond by flowering closer to ground level. Spot treatment with 2,4-D in the spring can control *T. officinale.*

Taraxacum officinale leaves. Image by Robert Vidéki, Doronicum Kft., Bugwood.org.

Taraxacum officinale seed head. Image by Joseph Berger, Bugwood.org.

Taraxacum officinale plant. Image by Robert Vidéki, Doronicum Kft., Bugwood.org.

Taraxacum officinale fruits. Image by Steve Hurst, hosted by the USDA-NRCS PLANTS Database.

Taraxacum officinale flower head. Image © Gerald D. Carr.

Taraxacum officinale infestation. Image by Benjamin Gimmel, used under a CC BY-SA 3.0 license.

Asteraceae

Tussilago farfara

coltsfoot · horsefoot · assfoot · sowfoot ·
bullsfoot · foalfoot · foalswort · horsehoof ·
hoofs · coughwort · clayweed · cleats · colt-
herb · donnhove · dovedock · dummyweed ·
fieldhove · English tobacco

Origin: Native to Europe, western Asia, and northwestern Africa. This species was most likely introduced to North America by European colonists for use as a medicinal herb and was first recorded growing in natural areas in 1840. *Tussilago farfara* has spread extensively through the northeastern United States and is also present in disjunct populations in Washington.

Description: A rhizomatous perennial with flowering stalks that emerge in spring before the leaves. The yellow flower heads superficially resemble dandelions, but this species has leaflike bracts along the flowering stems, flower heads containing both ray and disc florets, and large, heart-shaped leaves with woolly undersides. The leaves are very similar in shape to the nonnative *Petasites hybridus* (common butterbur), which is locally established in floodplains in northeastern Ohio and may become increasingly problematic in these habitats. However, *P. hybridus* can be distinguished by its pale pink, tubular flowers, borne in a dense spike.

Habitat: Forests, grasslands, wetlands, river- and streambanks, edges of lakes and ponds, fields, ditches, roadsides, and gardens. Tolerant of a wide range of soil types and conditions but requires consistent soil moisture. Can grow in full sun or partial shade and prefers areas with a high level of disturbance.

Height: 5–50 cm (2–20 in.).

Foliage: A rosette of basal leaves emerges in the spring after the flowering stems. Leaves are large and heart-shaped, 5–20 cm (2–8 in.) long and 5–20 cm (2–8 in.) wide, on long, upright petioles. The leaves are palmately veined and have shallow lobes and angular teeth along

their edges. The upper surfaces of mature leaves are smooth and bright green to bluish green, while the lower surfaces retain a dense cover of fuzzy white hairs.

Flowers: Flowering stems emerge in spring before the leaves. Each stem ends in a single flower head, 2.5–3.5 cm (1–1 ½ in.) in diameter, containing 3–80 yellow disc florets surrounded by 150–500 yellow ray florets. The tubular disc florets have five triangular petals and contain only male parts, while the strap-shaped ray florets have female parts. The base of the flower head is surrounded by overlapping phyllaries with a reddish or purplish tinge. During development, the immature flower heads often droop toward the ground. Flower heads often close on cloudy days and overnight. March–April.

Fruit: The flower heads produce many small, dry fruits, each containing a single seed. The fruits are light brown, 4–5 mm (⅛–¼ in.) long and 0.5 mm (1/64 in.) wide, with 5–10 longitudinal ribs. The fruit attaches to a plume of long white bristles, 8–12 mm (¼–½ in.) long. April–May.

Stems: Multiple flowering stems arise from a single root crown. There are alternate, leaflike bracts along the stems, each 1–2 cm (½–¾ in.) long. These bracts often have a reddish or purplish tinge, and both the bracts and stems are covered in woolly hairs.

Root system: Each plant has a primary taproot and an extensive system of thick, white rhizomes.

Reproduction: By seed, dispersed by wind. Most seeds germinate soon after dispersal. Seeds do not remain viable in the soil seed bank. Vegetative spread also occurs via rhizomes.

Impact: This species aggressively colonizes moist, open areas with a high level of disturbance. It forms dense stands that can interfere with crop production or displace native species.

Control: Young plants can be pulled by hand, but the deep rhizomes make it difficult to control more established individuals by pulling or tilling. However, some studies have suggested that *T. farfara* is a poor competitor and that limiting disturbance and establishing taller native vegetation may help control this species. Cutting flowering stems before seed set will also reduce spread. Spot application of glyphosate in midsummer, once all the leaves have developed, can kill rhizomes.

Tussilago farfara leaves. Image by Leslie J. Mehrhoff, University of Connecticut, Bugwood.org.

Tussilago farfara seed heads. Image by Leslie J. Mehrhoff, University of Connecticut, Bugwood.org.

Tussilago farfara stem. Image by Rob Routledge, Sault College, Bugwood.org.

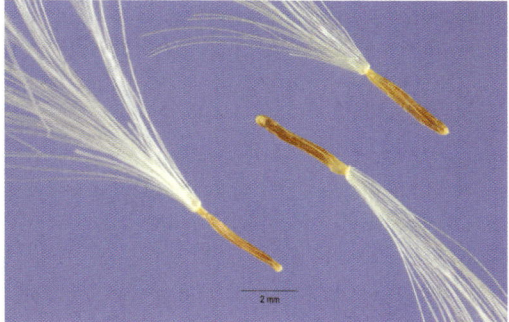

Tussilago farfara fruits. Image by Steve Hurst, hosted by the USDA-NRCS PLANTS Database.

Tussilago farfara flower heads. Image by Jan Samanek, Phytosanitary Administration, Bugwood.org.

Tussilago farfara plants growing along stream edge. Image by Leslie J. Mehrhoff, University of Connecticut, Bugwood.org.

Brassicaceae

Alliaria petiolata

garlic mustard · mustard root · poor man's
mustard · garlic root · garlicwort · hedge garlic
· penny hedge · jack-by-the hedge · jack-in-the-
bush · sauce-alone

Origin: Native to Europe, western and central Asia, and northwestern Africa. This species was introduced to North America in the 1800s by European settlers, for use in cooking and as a medicinal herb.

Description: This biennial is currently considered one of the worst invasive plant species in Ohio. First-year plants form basal rosettes in the early spring, and these remain green through the following winter. Plants bolt and flower in the second year. All plant parts give off a strong odor of garlic when crushed.

Habitat: This species invades both high-quality forests and disturbed habitats such as roadsides, trails, and gardens. It does not tolerate acidic soils and can grow in both sun and shade.

Height: Rosettes grow close to the ground, reaching a height of 3–15 cm (1–6 in.). Flowering stems are typically 30–100 cm (1–3 ft.).

Foliage: The first-year rosette has dark green, kidney-shaped leaves with scalloped edges, 3–6 cm (1–3 in.) in diameter. Leaves on the flowering stalks are heart-shaped or triangular, 5–8 cm (2–3 in.) long and wide, coarsely toothed.

Flowers: Plants have numerous small, white flowers in clusters at the top of the flowering stems. Each flower has four petals, approximately 5–6 mm (¼ in.) long and 2–3 mm (¹⁄₁₆–⅛ in.) wide, arranged in a cross shape. In the center, there is a short cylindrical style and six green stamens (two of which are shorter than the others) with pale yellow anthers. May–June.

Fruit: Fruit is an erect, slender, green silique. The siliques appear soon after flowering commences and grow to be 4–6 cm (2–3 in.) long.

Seeds are shiny and black, ellipsoid, 2.5–3 mm (¹⁄₁₆–⅛ in.) long, developing in a single row within the fruit. Siliques turn light brown and burst open when mature, dispersing the seeds explosively. June–July.

Stems: Most individual plants have only one or two stems, but they may grow more if the primary stem has been cut or damaged.

Root system: Each plant has a slender, white taproot. The root system has the potential to grow multiple flowering stalks if only the top of the plant is removed without damaging the roots.

Reproduction: By seed. A single plant can produce up to 900 seeds, which scatter several meters from the parent plant. Seeds can remain viable in the soil for up to seven years. In addition to explosive dispersal, seeds are spread by human and animal vectors or by water in riparian areas.

Impact: Begins growing in the early spring and tends to form dense stands that shade out native wildflowers and tree seedlings. In addition, plants produce allelochemicals that inhibit seed germination of other species and suppress the symbiotic mycorrhizal fungi that are beneficial to many native forest forbs, shrubs, and trees.

Control: With all control methods, treatments should be continued annually until the soil seed bank is exhausted. Be aware that removing only the second-year plants enhances the survival and growth of smaller first-year plants, so both developmental stages should be targeted. Hand pulling, which can be effective for small populations of garlic mustard, should be done before seed capsules are formed on second-year plants. It is important to pull out the taproot to prevent any regrowth. Cutting can be effective for larger populations. Once plants have begun to flower, stems should be cut at ground level. Glyphosate foliar spray is sometimes used on populations where mechanical control measures are impractical, but extreme care must be taken to avoid nontarget species. Any herbicide treatment should be done in the early spring, when most other nontarget vegetation is dormant.

Alliaria petiolata leaves in basal rosette. Image by Rob Routledge, Sault College, Bugwood.org.

Alliaria petiolata flowers. Image by Chris Evans, University of Illinois, Bugwood.org.

Stem leaf of *Alliaria petiolata.* Image by Chris Evans, University of Illinois, Bugwood.org.

Alliaria petiolata seeds. Image by Steve Hurst, hosted by the USDA-NRCS PLANTS Database.

Alliaria petiolata fruit with seeds. Image by Leslie J. Mehrhoff, University of Connecticut, Bugwood.org.

Alliaria petiolata infestation. Image by Leslie J. Mehrhoff, University of Connecticut, Bugwood.org.

Brassicaceae

Barbarea vulgaris

yellow rocket · garden yellow rocket · common yellow rocket · wound rocket · winter rocket · wintercress· rocketcress · bittercress · yellow weed · herb barbara

Origin: Native to Europe, Asia, and northern Africa. This species was most likely introduced to North America around 1800, perhaps as a contaminant of hay or crop seed, but the precise timing and pathway of its introduction is not clear. It has spread throughout much of the continental United States and is absent or infrequent only in the most southern states.

Description: This species is typically biennial, although it is occasionally a winter annual or short-lived perennial. It produces a rosette of leaves that remain green through the winter. One to several flowering stalks develop from the rosette, bearing rounded clusters of yellow flowers. It is most likely to be confused with the native species *Barbarea orthoceras* (American yellowrocket)—which is not known to occur in Ohio but is present to the north and west—and the nonnative species *Barbarea verna* (early winter cress). *Barbarea vulgaris* is most easily distinguished from *B. orthoceras* by the lack of hairs on the auricles of the stem leaves, and it can be differentiated from *B. verna* by the basal leaves, which have one to four pairs of lateral lobes in *B. vulgaris* and 4–10 pairs of lateral lobes in *B. verna*.

Habitat: Agricultural fields, pastures, wet meadows, construction sites, along railways and roadsides, and gardens. Prefers highly disturbed areas with full sun to partial shade and moist to mesic conditions.

Height: 20–80 cm (8–32 in.).

Foliage: Basal leaves are up to 15 cm (6 in.) long and 6 cm (3 in.) wide, oddly pinnately lobed, with one to four pairs of ovate lateral lobes and a larger, rounded terminal lobe. The leaves on the flowering stems are alternate and vary in shape based on location. Leaves on the lower to middle part of the stems are similar to the basal leaves, but they are progressively smaller, have shorter petioles, and have fewer lateral lobes farther up the stem. Each leaf on the upper part of the stem is ovate or wedge-shaped and has a pair of auricles that clasp the stem. All leaves are dark green and hairless, with wavy or bluntly toothed margins.

Flowers: Abundant small, yellow flowers develop in rounded clusters at the tops of the flowering stems and branches. Each flower has four yellow petals, 6–8 mm (¼–⅜ in.) long, four yellowish-green sepals, a single pistil with a cylindrical style, and six stamens (two of which are shorter than the others) with pale yellow anthers. Mildly fragrant. April–June.

Fruit: Fruit is a slender green silique 1–4.5 cm (½–2 in.) long and 2 mm (¹⁄₁₆ in.) wide, ending in a short beak, 2–3 mm (¹⁄₁₆–⅛ in.) long. The siliques are square in cross section and curve upward. Seeds are light brown, ovate, 1.2–1.5 mm (¹⁄₁₆ in.) long, flattened, developing in a single row within the fruit. The siliques turn light brown when mature and split open to release the seeds. May–July.

Stems: Multiple stems arise from the base, and the stems branch in the upper portions. Stems are hairless, green to purplish, ridged with eight darker purple veins.

Root system: Each plant has a taproot that can reach to 50 cm (20 in.) and a highly branched, fibrous secondary root system.

Reproduction: Primarily by seed, which can remain viable in the soil seed bank for five years. New rosettes can also develop from vegetative buds on the root.

Impact: This species is primarily a weed in agricultural settings, particularly cereal crops and hay. It can be a secondary host for pests of cruciferous vegetable crops and for diseases such as the cucumber mosaic virus.

Barbarea vulgaris stem and leaves. Image by Salicyna, used under a CC BY-SA 4.0 license.

Barbarea vulgaris flowers. Image by Hectonichus, used under a CC BY-SA 3.0 license.

Barbarea vulgaris fruits. Image © Katy Chayka, Minnesota Wildflowers.

Control: In small populations, pulling by hand or hoeing are recommended. Fruits and seeds can continue to mature if plants are cut or pulled when flowering, so all plant parts should be bagged and disposed of carefully. In more extensive stands, populations can be controlled by applying 2,4-D before seed develops.

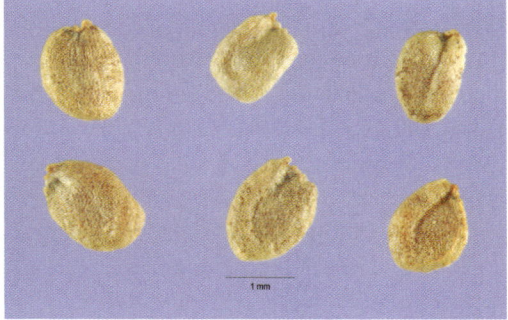

Barbarea vulgaris seeds. Image by Steve Hurst, hosted by the USDA-NRCS PLANTS Database.

Barbarea vulgaris plants. Image © Gerald D. Carr.

Field of *Barbarea vulgaris.* Image by Tero Laakso, used under a CC BY-SA 2.0 license.

Brassicaceae

Hesperis matronalis

dame's rocket · dame's gillyflower · dame's violet · damask violet · dames-wort · dame's gilliflower · night scented gilliflower · queen's gilliflower · rogue's gilliflower · winter gilliflower · summer lilac · sweet rocket · mother-of-the-evening

ODA Invasive
OIPC Invasive

Origin: Most likely native to southern Europe and western Asia, but it has a long history of use as a garden ornamental, food and forage plant, and medicinal herb. This species, which was probably introduced to North America in the 1600s, has spread throughout much of the United States. *Hesperis matronalis* has been present on the continent for so long that it is often incorrectly assumed to be native. It is sometimes included in meadow and wildflower seed mixes, so contents should be checked carefully before purchase or use.

Description: This biennial or short-lived perennial has upright stems and showy white, pink, lavender, or deep purple flowers. It is often mistaken for native or cultivated *Phlox* species. However, *Phlox* have opposite leaves and five petals on each flower. Another species that could cause confusion is the nonnative *Lunaria annua* (silver dollar plant), which is in the same plant family and also has large purple flowers, but its leaves are heart-shaped and it produces distinctive coin-shaped seedpods.

Habitat: Open woodlands, forest margins, fields, river- and streambanks, ditches, and roadsides. Prefers damp soils in full sun to partial shade.

Height: 40–100 cm (1 ½–3 ft.), occasionally taller.

Foliage: This species first produces a rosette of basal leaves that are petioled, hairy, ovate or lance-shaped, with smooth or slightly toothed margins. The basal leaves remain green through the winter but die back as the flowering stems develop. The stem leaves are alternate, hairy, lance-shaped with a rounded or wedge-shaped base, 4–15 cm (2–6 in.) long and 1–4 cm (½–2 in.) wide, tapering to a sharp tip. Leaves are progressively smaller farther up the stem. Stem leaves have fine teeth along the margins.

Flowers: The upper stems end in large, branching racemes of flowers, 30 cm (12 in.) long. Flower color can be white, pink, lavender, or deep purple. Each flower is 1.5 cm (⅝ in.) in diameter, with four rounded petals, a single pistil with a cylindrical style, and six stamens (two of which are shorter than the others) with pale yellow anthers. The base of the flower is enclosed with four hairy, oblong sepals that form a slender tube. Fragrant, particularly at night. May–June.

Fruit: Fruit is a slender green silique 6–10 cm (3–4 in.) long and 2–2.5 mm (¹⁄₁₆–⅛ in.) wide. The siliques are round in cross section and slightly constricted between the seeds. Seeds are brown, oblong, 3–4 mm (⅛ in.) long and 1–1.5 mm (¹⁄₃₂–¹⁄₁₆ in.) wide. The siliques turn light

Native Alternatives

Chamerion angustifolium (fireweed)
Clarkia pulchella (pinkfairies)
Geranium maculatum (wild geranium)
Lobelia siphilitica (great blue lobelia)
Penstemon digitalis (smooth penstemon)
Penstemon hirsutus (hairy beard-tongue)
Phlox divaricata (wild blue phlox)
Phlox glaberrima (smooth phlox)
Phlox paniculata (garden phlox)
Phlox pilosa (prairie phlox)
Physostegia virginiana (obedient plant)

Basal rosette of *Hesperis matronalis* leaves. Image by Leslie J. Mehrhoff, University of Connecticut, Bugwood.

Hesperis matronalis stems and leaves. Image by Leslie J. Mehrhoff, University of Connecticut, Bugwood.org.

Hesperis matronalis flowers. Image by David Cappaert, Bugwood.org.

Hesperis matronalis fruits. Image by Leslie J. Mehrhoff, University of Connecticut, Bugwood.org.

2 mm

Hesperis matronalis seeds. Image by Steve Hurst, hosted by the USDA-NRCS PLANTS Database.

Hesperis matronalis stand. Image by Leslie J. Mehrhoff, University of Connecticut, Bugwood.org.

brown when mature and split open from the bottom to the top to release the seeds. June–July.

Stems: A single upright stem typically grows from the base, but occasionally there are multiple stems. Stems are covered with short hairs, and they branch sparingly toward the top.

Root system: Shallow, with a knobby root crown. There are several vertical roots and fibrous lateral roots.

Reproduction: By seed, mostly dispersed close to the parent plant. A single plant can produce up to 20,000 seeds. Seeds remain viable in the soil seed bank for one to three years.

Impact: Forms dense stands that displace native species, particularly in the forest understory. In areas with agriculture, it is also a concern as a secondary host for diseases of cruciferous vegetable crops. The inclusion of *H. matronalis* in meadow and wildflower seed mixes is a serious concern because it both facilitates the spread and perpetuates the incorrect assumption that this is a desired native species. Consequently, this species can become well established before it is recognized as a problem.

Control: Do not purchase or plant meadow and wildflower seed mixes that include this species. Plants can be pulled by hand, taking care to remove the root crown. If plants are cut or pulled when they are in flower, the fruits and seeds can continue to mature, so all plant parts should be bagged and disposed of carefully. Large populations can be treated with glyphosate or triclopyr, applied to foliage in the early spring or late autumn when other plants are dormant.

Brassicaceae

Sinapis arvensis

wild mustard · corn mustard · charlock mustard · charlock

Synonyms: *Brassica arvensis, Brassica kaber*

Origin: Generally considered native to Europe, southwestern Asia to the Himalayas, and northern Africa. There is archeological evidence that pre-Columbian cultures in North America used this species, but it was probably reintroduced by European settlers in the 1600s. It is most likely the nonnative strain that has become a widespread and serious weed in agriculture. Until recently, this species was listed as a prohibited noxious weed in Ohio.

Description: An annual or winter annual plant with yellow flowers. *Sinapis arvensis* is easily confused with a number of other yellow-flowered mustards, including *Sinapis alba* (white mustard), *Brassica nigra* (black mustard), *Brassica juncea* (brown mustard), *Brassica napus* (cultivated rape), and *Raphanus raphanistrum* (wild radish). It can be distinguished from these relatives by the following characteristics in combination: there are stiff downward-pointing white hairs on its stem, there is reddish or purplish pigmentation at the junction between branches and stems, it has comparatively large flowers, and a seed often develops in the beak at the top of the seedpod.

Habitat: Agricultural fields, pastures, meadows, river- and streambanks, along railways and roadsides, and gardens. Prefers full sun and calcareous soils.

Height: 20–80 cm (8 in.–3 ft.).

Foliage: Basal leaves are 4–18 cm (2–7 in.) long and 2–5 cm (¾–2 in.) wide, with a petiole 1–4 cm (½–2 in.) long, variable in shape but mostly obovate to lance-shaped, often oddly pinnately lobed with 1–4 pairs of lateral lobes and a much larger terminal lobe. The leaves on

the flowering stems are alternate, broadly triangular, becoming progressively smaller, shorter petioled or with no petiole, and with fewer or no lateral lobes farther up the stem. The leaf surfaces are sparsely hairy, and the leaf margins are coarsely toothed and slightly wavy.

Flowers: Bright yellow flowers develop in rounded racemes at the tops of the flowering stems and branches. Each flower is about 1.5 cm (⅝ in.) in diameter, with four petals, a single green pistil with a cylindrical style, and six stamens (two of which are shorter than the others) with bright yellow anthers. At the base of the flower, there are four short, spreading, yellow sepals. May–July, with winter annuals blooming earlier in the growing season than summer annuals.

Fruit: Fruit is a green silique 3–5 cm (1–2 in.) long and 2.5–3.5 mm (1/16–⅛ in.) wide, tapering to a flattened, 1–1.5 cm (½–⅝ in.) long beak at the top. A feature of this species is that a seed often develops in the beak. The siliques are mostly hairless or are rarely bristly, have prominent veins along their length, and are slightly angular in cross section. Seeds are dark brown to black, spherical, 1–1.5 mm (1/32–1/16 in.) in diameter. The siliques turn light brown when mature and split open to release the seeds. June–August.

Stems: The stem is simple or branched, the branches often with a reddish or purplish pigmentation at their junction with the main stem. The surface of the stem is covered with stiff, downward-pointing white hairs.

Root system: A taproot.

Reproduction: By seed, dispersed in the area surrounding the parent plant. A single individual can produce 2,000–3,500 seeds. This species has a strongly persistent seed bank, with seed remaining viable for 75 years or more.

Impact: This species is primarily a weed in agricultural settings, particularly cereal crops and canola. It can be a secondary host for pests and pathogens of cruciferous vegetable crops.

Control: In small populations, pulling by hand or hoeing are recommended. In more extensive stands, it can be successfully controlled with 2,4-D or chlorsulfuron.

Sinapis arvensis leaves and stem junctions. Image © Peter M. Dziuk, Minnesota Wildflowers.

Sinapis arvensis fruits. Image by Teun Spaans, used under a CC BY-SA 3.0 license.

Sinapis arvensis stems. Image © Peter M. Dziuk, Minnesota Wildflowers.

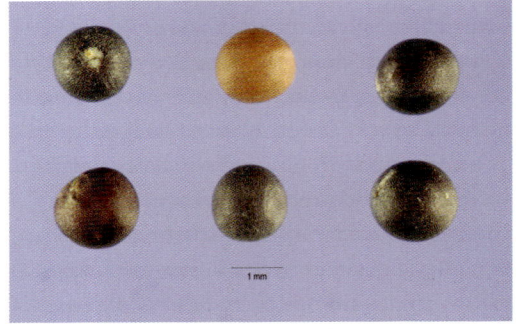

Sinapis arvensis seeds. Image by Steve Hurst, hosted by the USDA-NRCS PLANTS Database.

Sinapis arvensis flowers. Image by Hectonichus, used under a CC BY-SA 3.0 license.

Sinapis arvensis plants. Image by Zampel, used under a CC BY-SA 3.0 license.

Brassicaceae

Thlaspi arvense

field pennycress · Frenchweed · stinkweed ·
fanweed · bastard cress · Mithridate mustard

Origin: Native to Europe and Asia. This species, first recorded in 1701 near Detroit, Michigan, was most likely introduced to North America by French colonists. It is now a common and widespread weed found throughout the United States.

Description: An annual or winter annual with white flowers, flattened oval seedpods that are notched at the top, and a fetid odor when crushed. There are two closely related nonnative species similar in most features. *Microthlaspi perfoliatum* (perfoliate pennycress)is less common and it is much shorter and has smaller seedpods. *Thlaspi alliaceum* (roadside pennycress) is a newly spreading and potentially invasive species with smaller and more convex seedpods, a hairy lower stem, and a strong garlic odor when crushed.

Habitat: Agricultural fields, pastures, meadows, river- and streambanks, along railways and roadsides, and gardens. This species grows in a wide range of environmental conditions but prefers highly disturbed habitats.

Height: 10–80 cm (4 in.–3 ft.).

Foliage: Basal leaves are lance-shaped or obovate, 1–5 cm (½–2 in.) long and 4–23 mm (⅛–1 in.) wide, with a petiole 0.5–3 cm (¹⁄₆₄–1 in.) long, a wedge-shaped base and rounded tip, and a smooth or wavy margin. The basal leaves die back as the flowering stems develop. Leaves on the stem are alternate, oblong, 2–10 cm (¾–4 in.) long and 5–25 mm (¼–1 in.) wide, with a rounded or blunt point at the tip and bluntly toothed or wavy margins. Leaves are progressively smaller farther up the stem. The leaves on the lower to middle part of the stems have short petioles or attach directly to the stem, and those on the upper part of the stem have pairs of auricles that clasp the stem and taper to a blunt point. Leaves are hairless.

Flowers: Small white flowers develop in rounded racemes at the tops of the flowering stems and branches. The racemes elongate as the growing season progresses, and there are typically open flowers at the top of the inflorescence and fruits developing below. Each flower is about 3–6 mm (⅛–¼ in.) in diameter, with four white petals, a single green pistil with a cylindrical style, and six green stamens (two of which are shorter than the others) with bright yellow anthers. At the base of the flower there are four green sepals. April–June.

Fruit: Fruit is a flattened green silicle, circular or oval, 1–2 cm (½–¾ in.) long and 8–20 mm (¼–1 in.) wide, with a membranous wing 4 mm (⅛ in.) wide and a pronounced notch at the tip 5 mm (¼ in.) deep. The silicle has two cells, each containing five to eight seeds. The seeds are dark brown to black, ovoid and slightly flattened, 1.2–2.3 mm (¹⁄₃₂–¹⁄₁₆ in.) long and 1–1.5 mm (¹⁄₃₂ in.) wide, with curved concentric ridges on the surface. The silicles turn yellow then brown as they mature. May–July.

Stems: The stem is upright, hairless, branched in the upper parts, and sharply ridged.

Root system: A taproot with fibrous lateral roots.

Reproduction: By seed, often dispersed by farm machinery. A single plant can produce 1,600 to 20,000 seeds. Seeds can remain viable in the soil seed bank for 20 years.

Impact: This species is primarily a weed in agricultural settings. It reduces crop yields and is a contaminant of commercial seed stock.

Thlaspi arvense is a secondary host for a wide range of crop pests and pathogens. In addition, the seeds contain a large quantity of a mustard oil, allyl isothiocyanate, which can cause gastric distress or other complications in cattle and gives their milk and meat an off-flavor.

Control: Mulching can help prevent germination of *T. arvense.* In small populations, pulling by hand or hoeing are recommended. This species is susceptible to 2,4-D and MCPA when applied in spring.

Thlaspi arvense basal leaves. Image by Robert Vidéki, Doronicum Kft., Bugwood.org.

Thlaspi arvense fruits. Image by Matt Lavin, used under a CC BY-SA 2.0 license.

Thlaspi arvense seeds. Image by Steve Hurst, hosted by the USDA-NRCS PLANTS Database.

Left: Thlaspi arvense stem and leaves. *Right: Thlaspi arvense* flowers. Both images © Katy Chayka, Minnesota Wildflowers.

Thlaspi arvense plants. Image by Matt Lavin, used under a CC BY-SA 2.0 license.

Cerastium fontanum

common mouse-ear chickweed · common
chickweed · common mouse-ear · big
chickweed · starweed

Origin: The exact native range of this wide-spread and weedy species is somewhat obscure, but most records suggest that it is native to Europe, western Asia, and northern Africa. It was accidentally introduced to North America and is now established throughout the United States and Canada.

Description: A short-lived perennial (occasionally an annual or biennial) with spreading stems that form mats on the ground. Short hairs cover the stems and leaves, and its flowers have five styles. The nonnative species *Stellaria media* (common chickweed) is similar in appearance but has hairless leaves, only a single row of hairs along its stems, and flowers with three styles.

Habitat: Forest margins, fields, pastures, roadsides, lawns, and gardens. Prefers disturbed soils with full sun to partial shade.

Height: 15–50 cm (½–2 ft.).

Foliage: Leaves simple, opposite, narrowly ovate to lance-shaped, 1–4 cm (½–2 in.) long and 3–15 mm (⅛–⅝ in.) wide. The leaves on the prostrate stems are closely spaced together, while those on the flowering stems are farther apart. Leaves have a prominent midvein on the upper surface. The margins are smooth, and both the upper and lower surfaces of the leaf blade are covered with hairs.

Flowers: Upright flowering stems end in small, flat-headed cymes. At the base of each inflorescence, there is a pair of hair-covered leaflike bracts. Individual flowers are 1–1.5 cm (½–⅝ in.) in diameter, with five deeply notched white petals, five green sepals, five styles, and ten stamens with pale yellow or greenish anthers. The sepals are lance-shaped,

hairy, and almost as long as the petals. Flowering is intermittent from April to September.

Fruit: The fruit is a narrowly cylindric capsule 8–12 mm (¼–½ in.) long, slightly curved, with an opening at the top surrounded by ten small teeth. The capsule contains several reddish-brown seeds, 0.4–1.2 mm ($\frac{1}{64}$–$\frac{1}{32}$ in.) long, irregularly shaped and slightly flattened. May–October.

Stems: Multiple stems arise from the base, typically spreading along the ground and becoming upright when flowering. Stems can be green or purplish, and they are covered in hairs.

Root system: The root system is shallow. Each plant forms a small taproot with fibrous secondary roots.

Reproduction: Primarily by seed. Vegetative reproduction also occurs through the formation of roots where nodes come into contact with the soil surface.

Impact: This common weed is particularly a nuisance in lawns and gardens. It can also be problematic in pastures and croplands.

Control: Individual plants can be pulled by hand. In lawns, proper mowing height and frequency will allow grass to outcompete this species. A commercial premix of 2,4-D plus mecoprop is recommended for selective control of this species in lawns.

Cerastium fontanum leaves. Image by Ohio State Weed Lab, The Ohio State University, Bugwood.

Cerastium fontanum fruit. Image by Rasbak, used under a CC BY-SA 3.0 license.

Cerastium fontanum stems. Image © Gerald D. Carr.

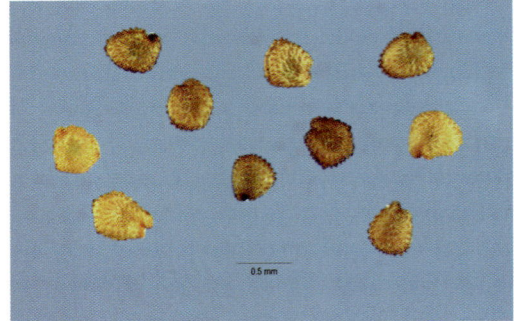

Cerastium fontanum seeds. Image by Steve Hurst, hosted by the USDA-NRCS PLANTS Database.

Cerastium fontanum flowers. Image by Peter O'Connor, used under a CC BY-SA 2.0 license.

Cerastium fontanum plants. Image by Theodore Webster, Ohio State Weed Lab Archive, The Ohio State University, Bugwood.org.

Caryophyllaceae

Saponaria officinalis

bouncing bet · common soapwort · soapweed · soaproot · crowsoap · scourwort · latherwort · chimney pink · hedge pink · wild sweet William · sweet Betty

OIPC Invasive

Origin: Native to Europe and western Asia. This species was introduced to North America for ornamental and medicinal purposes, as well as for its soaplike properties. It has escaped cultivation and spread throughout most of the United States. Despite its weedy tendencies, *Saponaria officinalis* is still grown in flower and herb gardens.

Description: A rhizomatous perennial with showy pink or white flowers. The roots, stems, and leaves of this plant contain high concentrations of saponins, which produce a soapy lather when mixed with water. *S. officinalis* flowers are similar to those of native *Phlox* species, but its leaves have three conspicuous parallel veins that distinguish it from *Phlox*. The nonnative species *Silene latifolia* (white campion) and *Silene vulgaris* (bladder campion) are related to *S. officinalis,* but they have more deeply lobed flower petals and a distinctive swollen calyx around the base of the flower.

Habitat: Forest margins, meadows, fields, river- and streambanks, sand bars, ditches, along railways and roadsides, and fencerows. Prefers disturbed soils with full sun to partial shade.

Height: 40–80 cm (1 ½–3 ft.).

Foliage: Leaves simple, opposite, narrowly elliptic to lance-shaped, 4–12 cm (2–5 in.) long and 2–4 cm (¾–2 in.) wide. Most leaves attach directly to the stem, otherwise they have short, broad petioles. There are three conspicuous parallel veins on the upper surface, which are particularly prominent on the larger leaves. The margins are smooth, and the leaf surfaces are hairless.

Flowers: The main stem and side branches end in dense clusters of three to seven white or light pink flowers. Individual flowers are 1.5–2.5 cm (⅝–1 in.) in diameter and have a tube 1.5–2 cm (⅝–¾ in.) long, with five slightly notched petals, two styles, and ten stamens with pale yellow anthers that extend out of the flower. The floral tube is enclosed by a hairless green to reddish calyx. The flowers open in the evening and remain open for three days. As the flowers mature, the spreading petals curve backward. Fragrant, particularly at night. June–September.

Fruit: An oblong capsule, 1.5–2 cm (⅝–¾ in.) long, that develops within the calyx of each flower. When mature, the capsule splits open at the top, forming four teeth. The capsule contains numerous flattened and round or kidney-shaped black seeds, 1.6–2 mm (1/16 in.) long and wide, with a granular surface. July–October.

Stems: One to several upright stems arise from the base, mostly unbranched except near the top. The stems are stout and hairless, slightly swollen at the nodes.

Root system: Short orange rhizomes that are fleshy, becoming woody with age.

Reproduction: By seed, primarily dispersed to the area surrounding the parent plant. This plant also spreads vegetatively through rhizomes.

Impact: Can invade natural areas and form dense colonies that displace native species. Because of the high concentration of saponins, all parts of *S. officinalis* plants, particularly the seeds and roots, are toxic to mammals.

Control: For small populations, pull the plants out by hand and dig out more extensive rhizomes with a shovel. In large populations, repeated mowing just before flowering is recommended. For **Native Alternatives**, see *Hesperis matronalis* [pp. 124–126].

Saponaria officinalis leaves. Image by Patrick J. Alexander, hosted by the USDA-NRCS PLANTS Database.

Developing fruits on *Saponaria officinalis*. Image by AnR00002, used under a CC0 1.0 license.

Saponaria officinalis stem. Image by Muscari, used under a CC BY-SA 3.0 license.

Saponaria officinalis seeds. Image by Steve Hurst, hosted by the USDA-NRCS PLANTS Database.

Saponaria officinalis flowers. Image by Joan Simon, used under a CC BY-SA 2.0 license.

Roadside stand of *Saponaria officinalis*. Image by Michael Shephard, USDA Forest Service, Bugwood.org.

Dipsacaceae

Dipsacus fullonum

common teasel · Fuller's teasel · wild teasel · venuscup teasel · Venus' cup · Venus' basin · barber's brush · church broom · brushes and combs · gypsy-combs · card teasel · card-thistle

ODA Invasive
OIPC Invasive
⚠ **Do Not Touch**

Origin: Native to Europe, western Asia, and northern Africa. This species was introduced to North America in the 1700s by European settlers who used the spiny seed heads to comb wool, as a dye for wool, and as an ornamental in gardens and flower arrangements. It has escaped cultivation and established populations in the northeastern and northwestern United States.

Description: A biennial or short-lived monocarpic perennial that grows and overwinters as a rosette for one or more years, flowers and produces seeds once in its lifetime, and dies after the reproductive event. The genus is distinguished by the leaves on the flowering stem, which are fused together and form a cup that collects rainwater. *Dipsacus fullonum* is more widespread than the closely related nonnative species *Dipsacus laciniatus* (cutleaf teasel) [pp. 137–138], which also occurs in Ohio. These two species can be differentiated because the former has purple flowers and unlobed leaves, while the latter has white flowers and deeply pinnately lobed leaves.

Habitat: Most commonly found in highly disturbed habitats such as pastures and abandoned fields and along railways and roadsides. However, it also invades high-quality habitats such as prairies, savannas, and sedge meadows. This species prefers open, sunny habitats with moist and fertile soils.

Height: 50–200 cm (1 ½–7 ft.), occasionally to 3 m (10 ft.).

Foliage: Basal leaves are simple, obovate to lance-shaped, 10–40 cm (4–16 in.) long and 4–10 cm (2–4 in.) wide. Each basal leaf has a petiole, rounded teeth along the edges, scattered stiff hairs on the upper surface, and prickles along the veins and midrib on the underside. Leaves on the flowering stem are opposite, lance-shaped, smaller than the basal leaves, the margins becoming less toothed higher up the stem. The paired stem leaves are fused together around the stem, forming a cup-like structure that collects rainwater.

Flowers: Each stem ends in a single ovoid flower head, 4–12 cm (2–5 in.) long and 3–5 cm (1–2 in.) wide, densely crowded with spiky bractlets, and surrounded at the base by a whorl of long and narrow prickly bracts that curve upward around the flower head. Purple, lavender, or dark pink flowers begin opening in a band around the middle of the flower head, and flowering progresses sequentially toward the top and bottom, forming two narrow bands of open flowers. Each flower has a spiky bractlet at its base, typically about 2 cm (¾ in.) long but often longer toward the top of the flower head, straight but gradually tapering to a sharp point. The individual flowers are tubular, about 1 cm (½ in.) long, with four rounded lobes, four purple stamens that extend above the rest of the flower, and a single pistil. July–September.

Fruit: The fruits are yellowish or grayish-brown achenes, 4–5 mm (⅛–¼ in.) long and 1–1.5 mm (¹⁄₃₂–¹⁄₁₆ in.) wide, ridged, and square in cross section. Each fruit contains a single seed. September–November.

Stems: Stems are upright, hollow, and branching. They are green, occasionally striped with red, and are deeply furrowed, with curved prickles along the ridges.

Root system: A taproot that grows to 75 cm (30 in.) deep and 2.5 cm (1 in.) in diameter

135

at the root crown. In addition to the taproot, there are fibrous secondary roots.

Reproduction: By seed, typically dispersed in the area around the parent plant but can be moved by flooding, mowing equipment, and the transport and disposal of seed heads in dried floral arrangements. An individual plant can produce 3,000 seeds or more. Seeds remain viable in the soil seed bank for two years.

Impact: This species develops dense monocultures that displace native plant species and may restrict wildlife movement. It is a particular concern in riparian, prairie, and savanna habitats.

Control: Rosettes can be dug out by hand, making sure to remove as much of the taproot as possible to prevent resprouting. Repeated cutting or mowing before flowering will exhaust root reserves. Immature seeds from cut flower heads may still germinate. Rosettes can be controlled by applying glyphosate, triclopyr, or 2,4-D in early spring.

Rosettes of basal leaves of *Dipsacus fullonum*. Image by Steve Dewey, Utah State University, Bugwood.org.

Dipsacus fullonum fruits. Image by Steve Hurst, hosted by the USDA-NRCS PLANTS Database.

Dipsacus fullonum seed heads. Image by Matt Lavin, used under a CC BY-SA 2.0 license.

Dipsacus fullonum flower head. Image by Chris Evans, University of Illinois, Bugwood.org.

Dipsacus fullonum infestation. Image by Steve Dewey, Utah State University, Bugwood.org.

Stem and fused leaves of *Dipsacus fullonum*. Image by Björn Appel, used under a CC BY-SA 3.0 license.

Dipsacaceae

Dipsacus laciniatus
cutleaf teasel · cut-leaved teasel

ODA Invasive
OIPC Invasive
⚠ Do Not Touch

137

Origin: Native to Europe and western Asia, this species was introduced to North America in the 1700s by European settlers. Like its close relative *Dipsacus fullonum* (common teasel) [pp. 135–136], the spiny seed heads of this plant were historically used to comb wool and it has been grown as an ornamental for gardens and flower arrangements. It has escaped cultivation and established scattered populations, particularly in parts of the northeastern and midwestern United States.

Description: A biennial or short-lived monocarpic perennial that grows and overwinters as a rosette for one or more years, flowers and produces seeds once in its lifetime, and dies after the reproductive event. *Dipsacus laciniatus* is not as widespread as *D. fullonum* in Ohio. This species typically has white flowers and leaves that are deeply pinnately lobed.

Habitat: Forest margins, meadows, prairies, savannas, pastures, river- and streambanks, and along railways and roadsides. Prefers full sun to partial shade with moist and fertile soils.

Height: 50–200 cm (1 ½–7 ft.), occasionally to 3 m (10 ft.).

Foliage: Basal leaves are simple, ovate to lance-shaped, 10–40 cm (4–16 in.) long and 4–10 cm (2–4 in.) wide. Each of the basal leaves has a petiole, can be deeply lobed or unlobed with a coarsely toothed and wavy margin, and has prickly hairs around the margin. Leaves on the flowering stem are opposite, lance-shaped, deeply lobed with lobes that are further divided, becoming smaller and the margins becoming less divided higher up the stem. The paired stem leaves are fused together around the stem, forming a cuplike structure that collects rainwater.

Flowers: Each stem ends in a single ovoid flower head, 5–10 cm (2–4 in.) long and 3–5 cm (1–2 in.) wide, densely crowded with spiky bractlets, and surrounded at the base by a whorl of lance-shaped bracts that spread outward and upward, tapering to a sharp point. These bracts are folded along the midvein and have prickles along the midvein and margins. Flowers are typically white but occasionally pale pink or lavender; they begin opening in a band around the base of the flower head, and flowering progresses sequentially toward the top. Each flower has a spiky bractlet at its base, typically about 2 cm (¾ in.) long but often longer toward the top of the flower head, straight but gradually tapering to a sharp point. The individual flowers are tubular, about 1 cm (½ in.) long, with four rounded lobes, four white stamens that extend above the rest of the flower, and a single pistil. July–September.

Fruit: The fruits are dark brown achenes, 5–8 mm (¼ in.) long and 1–1.5 mm ($^{1}/_{32}$–$^{1}/_{16}$ in.) wide, ridged, covered in flattened hairs, and square in cross section. Each fruit contains a single seed. September–November.

Stems: Stems are upright, hollow, and branching. The stems can have white ridges, and the stem surfaces are covered in white prickles.

Root system: A deep taproot with fibrous secondary roots.

Reproduction: By seed, primarily dispersed in the area around the parent plant but can be moved by flooding, mowing equipment, and the transport and disposal of seed heads in dried floral arrangements. An individual plant can produce 2,000 seeds. Seeds remain viable in the soil seed bank for two years.

Impact and Control: See *Dipsacus fullonum* [pp. 135–136].

Basal leaves of *Dipsacus laciniatus*. Image © Peter M. Dziuk, Minnesota Wildflowers.

Dipsacus laciniatus seed heads. Image by Chris Evans, University of Illinois, Bugwood.org.

Stem and fused leaves of *Dipsacus laciniatus*. Image © Peter M. Dziuk, Minnesota Wildflowers.

Dipsacus laciniatus fruits. Image by Bruce Ackley, The Ohio State University, Bugwood.org.

Dipsacus laciniatus flower heads. Image by Chris Evans, University of Illinois, Bugwood.org.

Dipsacus laciniatus stand. Image © Peter M. Dziuk, Minnesota Wildflowers.

Euphorbiaceae

Euphorbia esula

leafy spurge · green spurge · Hungarian spurge · wolf's milk

Origin: Native to Europe and temperate parts of Asia. This species was accidentally introduced to North America multiple times throughout the 1800s, primarily in contaminated seed stocks and soil used as ship ballast. *Euphorbia esula* has become a serious problem in the western United States, particularly in the northern and central plains regions. It is listed as an invasive species or prohibited noxious weed in 22 states and is considered one of the 100 worst invasive species worldwide by the International Union for Conservation of Nature.

Description: A vigorously colonial perennial with erect stems that arise from a woody root crown. All parts of the plants contain latex and exude a white, milky sap when damaged. There is some suggestion that the invasive North American populations of leafy spurge are *Euphorbia virgata* and that the name *Euphorbia esula* has been misapplied; however there is not yet consensus on this matter. The nonnative *Euphorbia cyparissias* (cypress spurge) is similar in appearance but is smaller and has narrower leaves that are almost needlelike. The nonnative *Euphorbia esula* should not be confused with the native *Euphorbia corollata* (flowering spurge), which has white, petallike bracts surrounding its flowers and leaves that are more upright.

Habitat: Agricultural fields, rangelands, grasslands, prairies, savannas, forest margins, and along roadsides. Prefers full sun and dry soils but can tolerate a variety of conditions.

Height: 20–90 cm (8 in.–3 ft.).

Foliage: Leaves simple, alternate, oblong to linear to lance-shaped, 2–9 cm (¾–4 in.) long and up to 12 mm (½ in.) wide, with rounded or slightly pointed tips. The leaves on the upper part of the stem are typically larger than those on the lower stem, which are often reduced and scalelike. All leaves are bluish green, have a light midvein, and attach directly to the stem. The margins are smooth, and the leaf surfaces are hairless.

Flowers: Each stem is topped with a flat-topped or rounded cluster with 7–20 compound flower structures called cyathia. Occasionally solitary or small clusters of cyathia develop in the axils of upper leaves. Each individual cyathium is surrounded at its base by a pair of yellowish-green, heart-shaped or rounded bracts, 2 cm (¾ in.) in diameter, that resemble petals. Contained within these bracts is a single female flower with a three-chambered globular ovary, 10–25 male flowers, and four yellowish-green nectar glands with a hornlike projection at the end. A whorl of bracts surrounds the base of the terminal cluster. May–September.

Fruit: A green, three-lobed capsule, 5–6 mm (¼ in.) in diameter, with a slightly bumpy surface. Each lobe contains a single seed. When mature, the capsules open explosively, ejecting the seeds into the area surrounding the parent plant. Seeds are mottled brown, ovoid, 2–3 mm ($^1/_{16}$–$^1/_8$ in.) long and 1–1.5 mm ($^1/_{32}$–$^1/_{16}$ in.) in diameter. June–October.

Stems: Stems arise individually from root buds, with branching mainly near the inflorescence. The stems are smooth and become woody when mature.

Root system: The root system is deep, extensive, and persistent. Its taproots can be up to 4 m deep and its lateral roots can produce shoots up to 1 m away from the parent plant.

Reproduction: By seed, dispersed explosively up to 5 m (15 ft.) from the parent plant. There is

139

Euphorbia esula stem and leaves. Image © Katy Chayka, Minnesota Wildflowers.

Euphorbia esula fruits. Image by Matt Lavin, used under a CC BY-SA 2.0 license.

Euphorbia esula flowers. Image by Leslie J. Mehrhoff, University of Connecticut, Bugwood.org.

Euphorbia esula seeds. Image by Julia Scher, Federal Noxious Weeds Disseminules, USDA APHIS PPQ, Bugwood.org.

Leafy bracts at base of flower cluster in *Euphorbia esula.* Image © Peter M. Dziuk, Minnesota Wildflowers.

Euphorbia esula infestation. Image by Leslie J. Mehrhoff, University of Connecticut, Bugwood.org.

also longer distance dispersal via water, wind, and animals. Seeds remain viable in the soil seed bank for up to eight years.

Impact: Forms dense stands that displace native plant species through competition for water and light and the release of allelopathic chemicals. This species dramatically reduces the biological diversity of grasslands and decreases the productivity of pastures and grazing land by 50–75% because cattle and horses will not eat it.

Control: Wear protective clothing, gloves, and eyewear while undertaking control efforts because the sap can cause skin and eye irritation. The extensive root system allows plants to persist and recover quickly after damage. In addition, the root system can exude herbicides after application, making this species difficult to control using conventional methods. Except in very small, isolated patches, cutting and pulling by hand are ineffective and could stimulate spread. Most management programs currently take an integrated approach, using some combination of grazing by sheep and goats, insect biological control agents such as *Aphthona* spp. (flea beetles) and *Spurgia esulae* (gall midge), burning, planting of more competitive native grass or shrub species, mowing and tilling, and using herbicides such as glyphosate, 2,4-D, or picloram.

Fabaceae

Lotus corniculatus

birdsfoot trefoil · common birdsfoot trefoil · garden birdsfoot trefoil · yellow trefoil · birdsfoot lotus · birdsfoot deervetch · bloomfell · cat's clover · crowtoes · ground honeysuckle · bacon and eggs

Origin: Native to Europe, Asia, and northern and eastern Africa. This species was introduced to the United States as early as the mid-1700s. It is not clear whether its introduction was accidental or intentional but it has subsequently been planted widely for forage and erosion control. Despite its tendency to spread beyond areas of cultivation, this species is still sold commercially.

Description: A low-growing, perennial legume with showy yellow flowers. It is similar to many yellow-flowered *Trifolium* and *Medicago* species, but *Lotus corniculatus* can be differentiated by its two to seven large flowers borne in an umbel, compound leaves with five leaflets, and leaflets with smooth margins.

Habitat: Pastures, meadows, river- and streambanks, roadsides, and lawns. Tolerates a wide range of environmental conditions, including drought and high salinity.

Height: Stems 15–80 cm (½–3 ft.) long, usually creeping along the ground.

Foliage: Leaves are alternate, compound with five leaflets, the top three leaflets obovate, each 5–20 mm (¼–1 in.) long and 4–10 mm (⅛–½ in.) wide, with two smaller leaflets that attach directly to the main stem below, pointed at the tip and rounded at the base. The surfaces of the leaflets are mostly hairless, and the margins are smooth.

Flowers: Flowering stalks arise from the leaf axils toward the upper part of the stem, each bearing a flat-topped umbel with two to seven (occasionally more) yellow flowers. Individual flowers are 13–17 mm (½–¾ in.) long and wide, pealike in form, each with a large upper lip, and

four modified petals folded together to form the lower lip. The lower petals enclose the flower's reproductive parts, which comprise ten fused stamens and a pistil with a single style. The petals can be streaked with red, and there are occasionally orange variants. June–October.

Fruit: The fruits are cylindrical pods held in a spreading cluster, each pod 1.5–3 cm (⅝–1 in.) long and 2–3 mm (¹⁄₁₆–⅛ in.) wide, with a persistent style at the tip, green turning brown or purplish black at maturity. Each fruit contains 2–25 ovoid seeds, 1.5–3 mm (¹⁄₁₆–⅛ in.) long and 1–1.5 mm (¹⁄₃₂–¹⁄₁₆ in.) in diameter, light to dark brown and occasionally mottled with dark spots. When dry, the fruit splits open to explosively release its seeds. August–October.

Stems: The lower stems are round in cross section, but the upper stems are nearly square. Stems are upright or creeping along the ground and growing upright at the tips, many branched, smooth or sparsely hairy. Stems become almost woody with age.

Root system: A taproot that can be 1 m (3 ft.) long, with a mat of lateral roots and rhizomes near the soil surface.

Reproduction: By seed, primarily dispersed in the area surrounding the parent plant. Seeds

Native Alternatives

Asclepias verticillata (whorled milkweed)
Astragalus canadensis (Canada milk vetch)
Baptisia tinctoria (horsefly weed)
Desmodium spp. (tick-trefoil)—any native species
Lupinus perennis (sundial lupine)
Tephrosia virginiana (goat's rue)

can remain viable in the soil seed bank for ten years or more. Plants also spread underground by rhizomes and aboveground through stolons that root at the nodes.

Impact: Forms dense and persistent stands that displace native plant species. Fresh leaves contain cyanogenic glycosides, making this species potentially toxic to mammals in high doses.

Control: For small infestations, hand pulling and digging out the rhizomes is recommended. Follow-up treatment will likely be necessary to deplete reserves in any remaining rhizome fragments. Repeated mowing at a frequency greater than once every three weeks can also control this species. Small patches can be covered with black plastic or thick mulch. The herbicides 2,4-D plus mecoprop and clopyralid (which specifically targets legumes) have both shown efficacy in controlling this species.

Lotus corniculatus plants. Image by Rob Routledge, Sault College, Bugwood.org.

Lotus corniculatus fruits. Image © Gerald D. Carr.

Lotus corniculatus leaf and stem. Image by Matt Lavin, used under a CC BY-SA 2.0 license.

Lotus corniculatus seeds. Image by Steve Hurst, hosted by the USDA-NRCS PLANTS Database.

Lotus corniculatus flowers. Image by David Cappaert, Bugwood.org.

Lotus corniculatus infestation. Image by Jan Samanek, Phytosanitary Administration, Bugwood.org.

Fabaceae

Melilotus albus

white sweet clover · sweetclover · honey clover ·
Bokhara clover · white melilot

144

Origin: Native to southern Europe, temperate Asia, and northeastern Africa. *Melilotus albus* was first recorded in North America in the early 1700s, although it may have been present before that time because the genus was reported in 1664 without an indication of species. There is long history of cultivation of *M. albus* for soil improvement, as a forage species, and as a source of nectar for honeybees. It has escaped cultivation and is particularly problematic in the midwestern and central regions of the United States.

Description: An annual or biennial legume with a bushy form and dense spikes of small white flowers. Some authorities have in the past considered *M. albus* and *Melilotus officinalis* (yellow sweet clover) [pp. 147–148] to be the same species, but modern classifications separate them. For this reason, records of the former likely underestimate the distribution of this species. It can be distinguished from the latter and other members of the genus by its white flowers. In addition, *M. albus* is generally taller, has foliage that is slightly grayish, and blooms two to four weeks later than *M. officinalis*.

Habitat: Agricultural fields, pastures, grasslands, meadows, prairies, river- and streambanks, along railways and roadsides, and fencerows. Prefers full sun and slightly moist to dry soils in highly disturbed areas.

Height: 50–250 cm (1 ½–8 ft.).

Foliage: Alternate, compound with three oblong to lance-shaped leaflets, each 1.5–5 cm (⅝–2 in.) long and 4–12 mm (⅛–½ in.) wide, rounded at the tip and base. The center leaflet grows on a stalk above the other two. Leaf surfaces are hairless and the leaflets are toothed along the upper margins. The foliage contains coumarin, which gives it a sweet, haylike scent that is intensified by drying.

Flowers: Small white flowers are densely clustered in spikelike racemes, 5–20 cm (2–8 in.) long, borne at the ends of the upper stems and in branching stems that arise from leaf axils. Each flower is held on a stalk 1–2 mm (½–1 in.) long. Individual pealike flowers are 4–6 mm (⅛–¼ in.) long, each with a large upper lip and four modified petals folded together to form the lower lip. The lower petals enclose the reproductive parts of the flower, which comprise ten fused stamens and a pistil with a single style. The flowers have a mild fragrance. June–October.

Fruit: The fruit is an ovoid pod, 3–5 mm (⅛–¼ in.) long and 2–2.5 mm (1/16 in.) wide, light brown, with a persistent style at the tip. The pod's surface is covered with netlike veins. Each pod contains one or two light brown seeds, ovoid, 2–2.5 mm (1/16 in.) long and 1–1.5 mm (1/32–1/16 in.) in diameter. August–October.

Stems: Each root crown gives rise to 1–10 upright stems, many branched, round in cross section or slightly furrowed, light green, hairless.

Native Alternatives
Baptisia tinctoria (horsefly weed)
Desmodium (tick-trefoil)—any native species
Lespedeza capitata (round-headed bush clover)
Lupinus perennis (sundial lupine)
Tephrosia virginiana (goat's rue)
Thalictrum dasycarpum (meadow rue)
Veronicastrum virginicum (Culver's root)

Melilotus albus leaf. Image by Patrick J. Alexander, hosted by the USDA-NRCS PLANTS Database.

Melilotus albus fruits. Image © Gerald D. Carr.

Stipules at base of *Melilotus albus* leaves. Image by Harry Rose, used under a CC BY 2.0 license.

Melilotus albus seeds. Image by Steve Hurst, hosted by the USDA-NRCS PLANTS Database.

Melilotus albus flowers. Image © Gerald D. Carr.

Melilotus albus plants. Image by Harry Rose, used under a CC BY 2.0 license.

Root system: A taproot with many secondary roots.

Reproduction: By seed, dispersed by water and animals. An individual plant can produce up to 5,000 seeds in a growing season. Seeds can remain viable in the soil seed bank for 20 years or longer.

Impact: This species invades disturbed sites and can establish dense stands that outcompete native grasses and forbs. Because *M. albus* fixes nitrogen, it can alter the nutrient availability in the soil, thereby changing the plant community composition in a site. If hay or silage containing this species develops mold, the coumarin in the foliage can break down into dicoumarin compounds that prevent blood clotting, which can lead to internal hemorrhaging in cattle or other livestock. It is also a weed of wheat crops and can impart its sweet odor to the wheat, causing a condition known as "sweetclover taint." In addition, this species is a host to many viral plant diseases, making it a concern in agricultural settings.

Control: Isolated plants can be pulled by hand, with care taken to remove as much of the taproot as possible. Repeated mowing or cutting close to ground level before fruiting can also help control this species. The herbicides 2,4-D amine, clopyralid, and metsulfuron-methyl can control this species.

Fabaceae

Melilotus officinalis

yellow sweet clover · ribbed melilot · field melilot · cornilla real

Origin: Native to Europe and Asia, although the precise native range is somewhat obscured by taxonomic confusion between this species and *Melilotus albus* (white sweet clover) [pp. 144–146]. It was first recorded in North America in the mid-1800s, although it was likely present before then. The genus was reported in 1664 without an indication of species, and sweet clovers were recorded frequently throughout the 1700s. There is a long history of cultivation of *Melilotus officinalis* as a means of soil improvement, a forage species, and a source of nectar for honeybees. Some sources suggest that it has been planted less frequently than *M. albus* and therefore is less widespread. Nonetheless, it has escaped cultivation and is particularly problematic in the midwestern and central regions of the United States.

Description: A bushy annual or biennial legume with dense spikes of small yellow flowers. In the past, some authorities have considered *M. officinalis* and *M. albus* the same species, but modern classifications separate them. Because of this taxonomic confusion, distribution records may not accurately differentiate the two species. It can be distinguished from *M. albus* because it is generally shorter, has greener foliage, and blooms two to four weeks earlier in the growing season. *Melilotus altissimus* (tall sweet clover) is another nonnative, yellow-flowered species of sweet clover that occurs occasionally in Ohio, but it is taller than *M. officinalis* and has hairy seedpods.

Habitat: Agricultural fields, pastures, grasslands, meadows, prairies, river- and streambanks, along railways and roadsides, and fencerows. Prefers full sun and highly disturbed areas.

Height: 40–150 cm (1–5 ft.).

Foliage: Leaves are alternate, compound with three oblong to lance-shaped leaflets, each 1.5–3 cm (⅝–1 in.) long and 5–15 mm (¼–⅝ in.) wide, rounded at the tip and base. The center leaflet grows on a stalk above the other two. The leaf surfaces are hairless, and the leaflets are closely toothed along the margins. The foliage contains coumarin, which gives it a sweet, haylike scent that is intensified by drying.

Flowers: Small yellow flowers are densely clustered in spikelike racemes, 5–15 cm (2–6 in.) long, borne at the ends of the upper stems and in branching stems that arise from leaf axils. Each flower is held on a short stalk 1.5–2 mm (¹⁄₁₆ in.) long. Individual flowers are 4.5–7 mm (⅛–¼ in.) long, pealike in form, with a large upper lip, and four modified petals folded together to form the lower lip. The lower petals enclose the flower's reproductive parts, which comprise ten fused stamens and a pistil with a single style. The flowers have a mild fragrance. May–September.

Fruit: The fruit is an ovoid pod, 3–5 mm (⅛–¼ in.) long and 2–2.5 mm (¹⁄₁₆ in.) wide, light brown, with a persistent style at the tip. The surface of the pod is covered with netlike veins. Each pod contains one or two light brown seeds that are occasionally marked with purple flecks, ovoid, 2–2.5 mm (¹⁄₁₆ in.) long and 1–1.5 mm (¹⁄₃₂–¹⁄₁₆ in.) in diameter. July–September.

Stems: Each root crown gives rise to multiple upright stems, many branched, round in cross section or slightly furrowed, light green or ribbed red near the base, hairless.

Root system: A taproot with many secondary roots.

Reproduction: By seed, dispersed by water and

147

animals. Individual plants produce up to 3,000 seeds in a growing season. Seeds can remain viable in the soil seed bank for 20 years or more.

Impact, Control, and Native Alternatives: See *Melilotus albus* [pp. 144–146].

Melilotus officinalis leaves. Image by Matt Lavin, used under a CC BY-SA 2.0 license.

Melilotus officinalis fruits. Image by Matt Lavin, used under a CC BY-SA 2.0 license.

Flowering *Melilotus officinalis* plants. Image by Howard F. Schwartz, Colorado State University, Bugwood.org.

Melilotus officinalis seeds. Image by D. Walters and C. Southwick, Table Grape Weed Disseminule ID, USDA APHIS PPQ, Bugwood.org.

Melilotus officinalis flowers. Image by Steve Dewey, Utah State University, Bugwood.org.

Melilotus officinalis infestation. Image by Ohio State Weed Lab, The Ohio State University, Bugwood.org.

Hypericaceae

Hypericum perforatum

common St. Johnswort · perforate St. Johnswort
· St. John's blood · St. John's grass · herb John ·
penny John · Klamath weed · Tipton weed · goat
weed · goatsbeard · gammock · racecourse weed ·
witch's herb · rosin rose · chase-devil · devil's flight

Origin: Native to Europe, western Asia, and northwestern Africa. *Hypericum perforatum* has a long history of cultivation as a medicinal and garden plant. European settlers introduced it to North America many times—both to the eastern and western seaboards, from which it escaped cultivation and spread throughout most of the United States. It is a noxious weed in many western states.

Description: A bushy rhizomatous perennial with showy yellow flowers. There are 15 other species of *Hypericum* in Ohio, all of which are native. It is most easily confused with *Hypericum punctatum* (spotted St. Johnswort), which has conspicuous black dots and streaks on the interiors of its flower petals and sepals with few or no black markings. All other native *Hypericum* species have flower petals without black dots. In contrast, the nonnative *H. perforatum* has black dots along the margins of its flower petals and the sepals are marked with black dots and lines.

Habitat: Forest margins, open woodlands, meadows, pastures, orchards, agricultural fields, and roadsides. Prefers full sun and can tolerate a wide range of soil types.

Height: 30–90 cm (1–3 ft.).

Foliage: Leaves simple, opposite, elliptic, 1–5 cm (½–2 in.) long and 3–15 mm (⅛–⅝ in.) wide, with rounded tips. The leaves lack a petiole and attach directly to the stem. Two or three pairs of veins run parallel to the midvein. The surface of each leaf is scattered with translucent dots of glandular tissue that give it a perforated appearance when held up to the

light. In addition, the leaf's lower surface often has scattered black dots, particularly near the margins. The margins are smooth, and the leaf surfaces are hairless. Lower branches often retain leaves through the winter.

Flowers: Numerous yellow flowers are borne in branching and nearly flat-topped cymes at the ends of stems. Individual flowers are 12–30 mm (½–1 in.) in diameter, with five yellow petals marked with tiny black dots along the margins, 40–90 long yellow stamens with yellow anthers, each marked with a black gland, and a single pistil with a widely divergent three-parted style. June–September.

Fruit: A sticky capsule, ovoid, 6–10 mm (¼–½ in.) long and 3.5–5 mm (⅛–¼ in.) wide, that splits into three sections topped with three persistent styles. The fruit turns a deep reddish brown as it matures and splits open along three seams to release numerous seeds. Seeds are dark brown to black, cylindrical, 1–1.3 mm ($\frac{1}{32}$ in.) long and 0.5 mm ($\frac{1}{64}$ in.) in diameter, with a pitted surface. The seeds have a gelatinous coating that becomes sticky when wet, allowing them to adhere to animals for dispersal.

Stems: Several upright stems arise from the base, branching extensively. The stems are usually woody near the base. The stems' surfaces are green to reddish, hairless, double-edged or

149

Native Alternatives

Hypericum hypericoides (St. Andrew's cross)
Hypericum prolificum (shrubby St. Johnswort)
Hypericum pyramidatum (great St. Johnswort)

Hypericum perforatum leaves. Image © Katy Chayka, Minnesota Wildflowers.

Hypericum perforatum stem. Image by Danny S., used under a PDM designation.

Hypericum perforatum flowers. Image by Robert Vidéki, Doronicum Kft., Bugwood.org.

Hypericum perforatum seeds. Image by Steve Hurst, hosted by the USDA-NRCS PLANTS Database.

Hypericum perforatum fruits. Image by Matt Lavin, used under a CC BY-SA 2.0 license.

Hypericum perforatum infestation. Image by John Tann, used under a CC BY 2.0 license.

ridged with black glands along the ridges, with a distinct ring around the stem at the lower nodes.

Root system: A long taproot to 1 m (3 ft.) deep and extensive creeping rhizomes 1–8 cm (½–3 in.) below the soil surface. The root crown is woody. When they are damaged, the rhizomes produce new root crowns.

Reproduction: By seed, dispersed by wind, water, soil movement, on animals or machinery, and by humans and human activities. An individual plant can produce 15,000 to 30,000 seeds in a growing season. Seeds can remain viable in the soil seed bank for 6–10 years. There is also vegetative spread by rhizomes.

Impact: Forms extensive and dense colonies that displace native species. It is problematic as a pasture weed because it is toxic to livestock, causing photosensitivity and central nervous system damage.

Control: For small populations, pull the plants by hand and dig out the root crowns and rhizomes. Repeated mowing can reduce the density of plants, while fire has been found to stimulate germination and lead to dense seedling growth. Several insect biological control agents have been used in the western United States and Canada to control *H. perforatum,* with wide success. The herbicides glyphosate, picloram, and 2,4-D all control this species.

Lamiaceae

Glechoma hederacea

ground ivy · creeping Charlie · gill-over-the-ground · gill-go-by-the-hedge · lizzy-run-up-the-hedge · robin-run-in-the-hedge · run-away-robin · alehoof · turnhoof · hayhofe · field balm · haymaids · hedgemaids · catsfoot · cat ivy · wild snakeroot

Origin: Native to Europe and parts of Asia. This species was introduced to North America in the 1800s for ornamental and medicinal purposes. It has now spread and become a common weed of lawns and gardens throughout the eastern United States.

Description: A perennial evergreen forb with creeping stems, purple flowers, and heart-shaped leaves with scalloped margins. This plant's spreading nature allows it to form dense mats that crowd out other vegetation. It is similar in appearance to *Lamium amplexicaule* (henbit), another nonnative species in the mint family. However, *Glechoma hederacea* has petioles on its leaves and purple flowers in small clusters in the leaf axils, whereas *L. amplexicaule* has stalkless leaves that clasp the stems and pinkish flowers in whorls in the leaf axils.

Habitat: Primarily in disturbed areas such as lawns, gardens, and roadsides but can also be found in grasslands and woodlands. Prefers damp, shaded areas.

Height: Flowering stems are upright and can reach a height of 30 cm (1 ft.), while creeping stems grow to 70 cm (2 ft.).

Foliage: Leaves are opposite, heart-shaped, 2–3 cm (¾–1 in.) in diameter, with a long petiole and scalloped margins. Leaves are bright green and shiny, with palmate veins. Foliage has a strong mint odor when crushed.

Flowers: Small, lavender, tubular flowers are typically in clusters of two or more in the leaf axils. Individual flowers are bilaterally symmetrical, with a tube 1–2 cm (½–¾ in.) long. There is a notched upper lobe, a larger notched lower lobe, and two smaller side lobes. The lower lip

and inside of the tube are often marked with dark purple spots. Each flower contains two long stamens and two short stamens with white anthers and a single pistil with a divided style. The base of the tube is surrounded by sepals that are fused to form a calyx with five sharp teeth. This calyx persists after the flower has withered. March–July.

Fruit: Four nutlets develop within the persistent calyx. The nutlets are brown, ovoid, 1.5–2 mm (¹⁄₁₆ in.) long and 1 mm (¹⁄₃₂ in.) in diameter, each containing a single seed.

Stems: Square in cross section, slender, scrambling over the ground forming thick mats. The stems, often reddish or purplish, are fringed with long, white hairs at the nodes. Numerous flowering stems grow from each individual stem.

Root system: Shallow, fibrous roots develop at the base of the stem. Adventitious roots form from nodes that come into contact with the soil surface.

Reproduction: By seed, dispersed by gravity. Vegetative spread also occurs through creeping stems and rooting stem fragments.

Impact: In large quantities, this species is toxic to horses, either when consumed fresh or in hay. It is generally not a threat to established native plant communities, but it is a nuisance in lawns and gardens. It can also be aggressive along woodland edges and in other disturbed habitats.

Control: Small populations can be removed by pulling or raking when the soil is damp. Because the species can regenerate from nodes, it is important to remove all of the roots and stems. Application of glyphosate is recommended only in areas with total cover by this species. If the species is intermixed with lawn grasses or native vegetation, spot application of commercial pre-mix of 2,4-D plus mecoprop can be used. To minimize impacts on neighboring species, herbicide treatment should be undertaken when the plant is in flower or after the first hard frost in the autumn. Repeated treatment may be necessary because this species can regenerate from the seed bank.

Glechoma hederacea leaf. Image by Bruce Ackley, The Ohio State University, Bugwood.org.

Nutlets developing inside the persistent calyx in *Glechoma hederacea*. Image by Jouko Lehmuskallio, www.NatureGate.net.

Glechoma hederacea stem. Image by Ohio State Weed Lab, The Ohio State University, Bugwood.

Glechoma hederacea nutlets. Image by Steve Hurst, hosted by the USDA-NRCS PLANTS Database.

Glechoma hederacea flowers. Image by Leslie J. Mehrhoff, University of Connecticut, Bugwood.

Glechoma hederacea plants. Image by Robert Vidéki, Doronicum Kft., Bugwood.org.

Lamiaceae

Leonurus cardiaca

motherwort · common motherwort · lion's tail ·
lion's-ear · cowthwort · throw-wort

⚠ **Do Not Touch**

Origin: *Leonurus cardiaca* is most likely native to southeastern Europe and western and central Asia, although its precise native range is obscured by a long history of cultivation. It was introduced to North America by European settlers for use as a medicinal plant and is now found throughout most of the northeastern and midwestern United States. Despite its tendency to spread beyond areas of cultivation, this species is still sold commercially.

Description: A perennial with distinctively hairy, pale pink flowers. There are two similar nonnative species that occur in Ohio, *Leonurus sibiricus* (Siberian motherwort) and *Chaiturus marrubiastrum* (false motherwort). However, *L. sibiricus* has flowers that are less hairy and deeper pink as well as narrower and more deeply cleft leaves than *L. cardiaca,* whereas the leaves of *C. marrubiastrum* lack the cleft lobes that define the true motherworts.

Habitat: Forest margins, open woodlands, fields, meadows, river- and streambanks, roadsides, and gardens. Prefers partial shade and moist, rich soils.

Height: 40–150 cm (1–5 ft.).

Foliage: Leaves are opposite, variably shaped, with a long petiole and prominent veins. Leaves on the lower part of the stem are 5–10 cm (2–4 in.) long, palmately cleft in three to five parts with deeply lobed and toothed margins. Some leaves resemble maple leaves, while others have a rounded or wedge-shaped base. The leaves become progressively smaller and less divided higher up on the stem. The leaves and petioles are slightly hairy.

Flowers: Pale pink tubular flowers are borne in whorls of 6–15 above the leaf axils in the upper parts of the stems. Individual flowers are bilaterally symmetrical, with a tube 8–12 mm (¼–½ in.) long. The upper lip is undivided and fringed with long white hairs on the outer surface. The lower lip is narrow, folded along its length, with a central lobe and two smaller side lobes. The lower lip and throat of the tube are often marked with purple or reddish dots. Each flower contains two long stamens, two short stamens with dark purple anthers, and a single white pistil with a divided style. The base of the tube is surrounded by slightly hairy sepals that are fused to form a calyx tipped with five extremely sharp teeth. This calyx persists after the flower has withered. June–August.

Fruit: Four nutlets develop within the persistent calyx. These are three-sided, reddish brown to brown, 1.8–2.2 mm (¹⁄₁₆ in.) long and 1.2–1.5 mm (¹⁄₃₂ in.) wide, each containing a single seed. August–October.

Stems: Multiple upright stems arise from the root crown, mostly unbranched but occasionally branching in the upper part of the stem. The stems are square in cross section, and there may be short hairs along the angles and at the nodes.

Native Alternatives

Agastache scrophulariifolia (purple giant hyssop)
Blephilia ciliata (hairy woodmint)
Meehania cordata (creeping mint)
Monarda fistulosa (wild bergamot)
Monarda punctata (spotted beebalm)
Pycnanthemum virginianum (mountain mint)
Scutellaria incana (downy skullcap)
Teucrium canadense (American germander)

154

Root system: Shallow but extensive, with fibrous roots and rhizomes spreading laterally.

Reproduction: By seed and rhizomes.

Impact: An attractive but weedy species that is particularly problematic in woodlands. Although *L. cardiaca* has been used historically as an herbal remedy for heart conditions, anxiety, and female reproductive disorders, some of the oils in the foliage can cause photosensitivity if ingested and are toxic at high dosages.

Control: Wear protective clothing and gloves when handling this plant, because some of the oils in the foliage and stems can cause contact dermatitis. This species can be eliminated by repeated cutting or close mowing.

Fruits developing inside the persistent calyces in *Leonurus cardiaca.* Image by Ohio State Weed Lab, The Ohio State University, Bugwood.org.

155

Left: Leonurus cardiaca leaves. Image by Ohio State Weed Lab, The Ohio State University, Bugwood. org. *Right: Leonurus cardiaca* stem. Image by Stefan Lefnaer, used under a CC BY-SA 4.0 license.

Leonurus cardiaca flowers. Image by Hajotthu, used under a CC BY 3.0 license.

Stand of *Leonurus cardiaca* plants. Image by Chris Evans, University of Illinois, Bugwood.org.

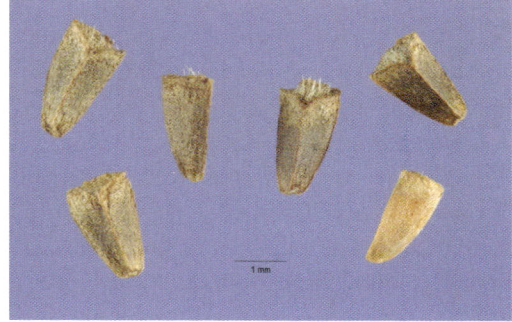

Leonurus cardiaca fruits. Image by Steve Hurst, hosted by the USDA-NRCS PLANTS Database.

Lythraceae

Lythrum salicaria

purple loosestrife · purple lythrum · rainbow weed · spiked loosestrife

Origin: Native to Europe, Asia, and northern Africa. *Lythrum salicaria* was introduced to North America in the early 1800s either in soil used as ship ballast or in contaminated sheep wool. It has also been deliberately planted as an ornamental, a medicinal herb, and a source of nectar for honeybees. This species has escaped cultivation and spread primarily along waterways throughout the northeastern United States and adjacent parts of Canada. The International Union for Conservation of Nature considers *L. salicaria* one of the 100 worst invasive species worldwide.

Description: A perennial with long spikes of showy purple flowers, a bushy form, and a persistent woody rootstock. The native species *Lythrum alatum* (winged loosestrife) occupies a similar habitat but can be distinguished by its shorter stature, square stems with wings along the angles, mostly alternate leaves, and smaller flowers borne singly in the upper leaf axils. Another relative similar in appearance is the nonnative *Lythrum virgatum* (European wand loosestrife). Although it is a popular horticulture plant, *L. virgatum* is also listed as invasive in Ohio. It should not be sold or planted because it can cross-pollinate with *L. salicaria* and produce viable seeds that develop into highly invasive plants.

Habitat: Primarily in wetland habitats, including marshes, river- and streambanks, and along the edges of ponds, lakes, and reservoirs. In addition, this species is found in meadows, pastures, and ditches; along railways and roadsides; and in gardens. Prefers disturbed areas with full sun and wet to moist soils.

Height: 30–200 cm (1–6 ft.).

Foliage: Leaves are simple, opposite or in whorls of three, lance-shaped, 3–10 cm (1–4 in.) long and 1–2 cm (½–¾ in.) wide, with a pointed tip and a heart-shaped or rounded base that attaches directly to the stem. Leaves become progressively smaller farther up the stem. The margins are smooth, and the leaf surfaces may be slightly hairy, particularly on the upper leaves. The foliage turns bright red in the autumn.

Flowers: The inflorescence is a spike, 10–40 cm (4–16 in.) long, densely packed with purple to pink flowers interspersed with alternate, leaflike bracts. Individual flowers are 1–2 cm (½–¾ in.) in diameter, typically with six spreading petals (occasionally with five or seven petals). The petals are marked with a dark vein down the center, and each flower contains 6–12 stamens with purple anthers and a pistil with a green and knobby stigma. The base of the flower is surrounded by slightly hairy sepals that are fused to form a tubular calyx tipped with five teeth. This calyx persists after the flower has withered. July–September.

Native Alternatives

Agastache scrophulariifolia (purple giant hyssop)
Asclepias incarnata (swamp milkweed)
Chamerion angustifolium (fireweed)
Eutrochium purpureum (sweet Joe-Pye weed)
Filipendula rubra (queen-of-the-prairie)
Liatris pycnostachya (prairie blazing star)
Liatris spicata (dense blazing star)
Physostegia virginiana (obedient plant)
Verbena hastata (blue vervain)
Vernonia fasciculata (prairie ironweed)

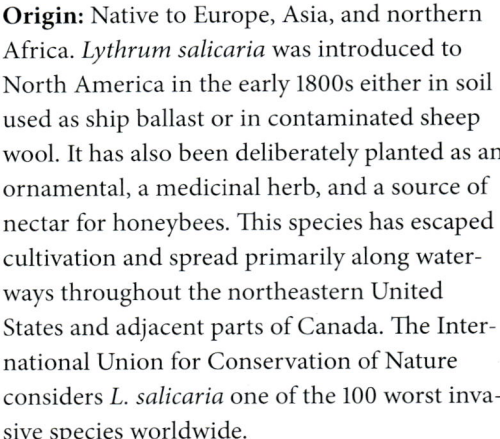

Lythrum salicaria leaves. Image by Leslie J. Mehrhoff, University of Connecticut, Bugwood.org.

Developing fruits on *Lythrum salicaria*. Image by Leslie J. Mehrhoff, University of Connecticut, Bugwood.org.

Lythrum salicaria stem. Image by Rob Routledge, Sault College, Bugwood.org.

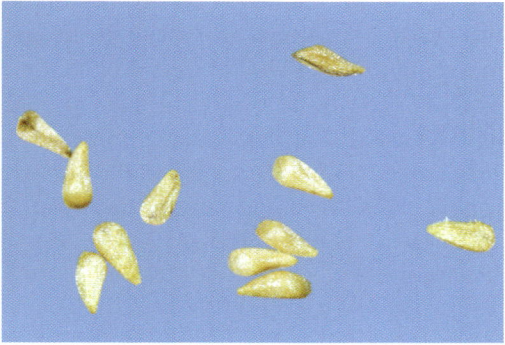

Lythrum salicaria seeds. Image by Ken Chamberlain, The Ohio State University, Bugwood.org.

Lythrum salicaria flowers. Image by Leslie J. Mehrhoff, University of Connecticut, Bugwood.org.

Lythrum salicaria infestation. Image by Leslie J. Mehrhoff, University of Connecticut, Bugwood.org.

157

Fruit: The fruit is an oblong capsule, 3–4 mm (⅛ in.) long, held within the persistent calyx. The capsule contains numerous tiny, tear-drop-shaped seeds, each 1 mm long (¹⁄₃₂ in.) and 0.2–0.4 mm (¹⁄₆₄ in.) wide. August–October.

Stems: Multiple upright stems arise from the root crown, and the stems are highly branched, giving the plants a bushy form. Stems may be round or square in cross section, purplish or reddish in color, and variably hairy. They tend to be woody close to the base.

Root system: This species has both thick, fleshy roots and shallow, fibrous roots that form a dense mass. The root crown becomes woody with age. Plants also develop rhizomes that spread laterally and give rise to new plants.

Reproduction: By seed, dispersed by water, wind, and gravity. An individual plant can produce 100,000–2,700,000 seeds in a growing season. Seeds can remain viable in the soil seed bank for 20 years or longer. Plants also reproduce vegetatively via rhizomes and root or stem fragments.

Impact: Forms dense stands that potentially exclude native plant species. The replacement of native wetland plant communities by *L. salicaria* may reduce habitat suitability for wildlife. It also has the potential to alter decomposition rates, nutrient cycling, and water chemistry in wetland habitats.

Control: Small stands can be pulled by hand or dug out before seed set. Frequent cutting of stems at the ground level is effective in controlling this species, if done consistently over several years. Spot treatment with herbicides such as glyphosate and triclopyr amine is also effective in controlling this species in nonwetland habitats. Herbicides must not be used in wetlands unless by a licensed aquatic herbicide applicator. Biological control using *Galerucella* beetle species has been effective in keeping populations of this invasive species in check.

Onagraceae

Epilobium hirsutum

hairy willowherb · great willowherb · great hairy willowherb · European fireweed · fiddle grass · codlins and cream

Origin: Native to Europe, temperate Asia, and northern Africa. This species, which was first reported in North America in the mid-1800s, may have been introduced accidentally in soil used as ship ballast or intentionally as an ornamental plant. Since its initial introduction, it has been widely planted and has escaped from cultivation to establish scattered populations in parts of the northeastern and northwestern United States.

Description: A semiaquatic perennial plant with pink to dark purple flowers, a bushy form, and hairy stems. *Epilobium hirsutum* occupies a similar habitat to *Lythrum salicaria* (purple loosestrife) [pp. 156–158], with which it sometimes competes, but it has larger flowers with four notched petals. It can also be confused with the native relative *Chamerion angustifolium* (fireweed), which has hairless leaves and four spatula-shaped pink to purple petals that alternate with four narrow purple sepals. Another related species that occurs in northeastern Ohio, the nonnative *Epilobium parviflorum* (smallflower hairy willowherb), is also an extremely aggressive weed that has much smaller light pink flowers.

Habitat: Wetlands, river- and streambanks, low pastures, fields, meadows, ditches, and roadsides. Prefers full sun but can tolerate partial shade once established. Can grow in moist to wet soils and survives having roots submerged by fluctuating water levels.

Height: 50–200 cm (1 ½–6 ft.).

Foliage: Leaves are simple, opposite but with alternate leafy bracts near the flowers, lance-shaped or oblong, 2–12 cm (¾–5 in.) long and 0.5–3.5 cm (¼–1½ in.) wide, each with a pointed tip and a narrow base that clasps the

stem. The margins are sharply toothed, and the leaf surfaces are hairy on both sides.

Flowers: Solitary pink to dark purple flowers are borne on long stalks that extend from the upper leaf axils. Individual flowers are 2–3 cm (¾–1 in.) in diameter, with four petals deeply notched at the tips. Each flower contains eight white stamens and a single pistil ending in a stigma divided into four backward-curving lobes. Four green, lance-shaped sepals at the base of the flower are covered with glandular hairs. June–August.

Fruit: The fruit is a slender tubular capsule, 2.5–9 cm (1–4 in.) long, that splits open along four seams to release its seeds. The seeds are dark brown, oblong, 0.8–1.2 mm ($\frac{1}{32}$ in.) long and 0.2–0.4 mm ($\frac{1}{64}$ in.) wide, with a plume of white hairs at the top. July–September.

Stems: Stems are upright, round in cross section, branched, and covered with straight glandular hairs. Some plants form stolons that creep along the ground.

Root system: A woody root crown, fibrous roots, and thick white rhizomes that spread laterally up to 20 cm (8 in.). Adventitious roots develop near the base of the stem in waterlogged soil.

Reproduction: By seeds, dispersed by wind and water. Vegetative reproduction also occurs via both rhizomes and stolons.

Impact: This species forms dense stands that crowd out other species. It is most problematic in wetland habitats, where it displaces native plant communities.

Control: The native relative *C. angustifolium* is a rare species in Ohio, so before undertaking control efforts take care to ensure that you

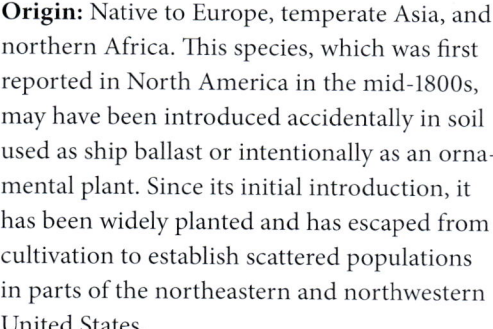

have identified plants correctly. Small stands of *E. hirsutum* can be pulled by hand or dug out before seed set, making sure that roots and rhizomes are removed. Bag and dispose of any plant materials. The herbicide 2,4-D ester can control this species. In wetland areas, chemical control must only be undertaken by a licensed aquatic herbicide applicator. For **Native Alternatives**, see *Hesperis matronalis* [pp. 124–126].

Epilobium hirsutum plant form. Image by Rob Routledge, Sault College, Bugwood.org.

Epilobium hirsutum stem and leaves. Image © Gerald D. Carr.

Epilobium hirsutum flower. Image © Gerald D. Carr.

Epilobium hirsutum seedpods. Image by Rob Routledge, Sault College, Bugwood.org.

Epilobium hirsutum infestation. Image by Jonathan Kington, used under a CC BY-SA 2.0 license.

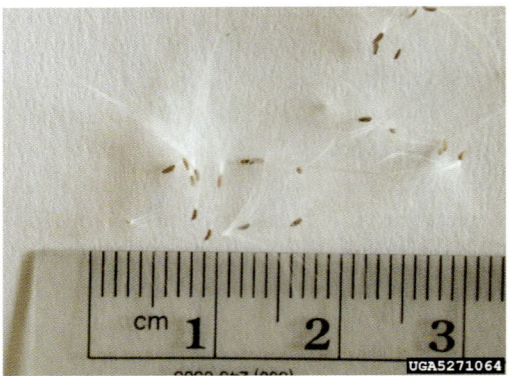

Epilobium hirsutum seeds. Image by Leslie J. Mehrhoff, University of Connecticut, Bugwood.org.

Papaveraceae

Chelidonium majus

greater celandine · rock poppy · nipplewort · swallowwort · tetterwort

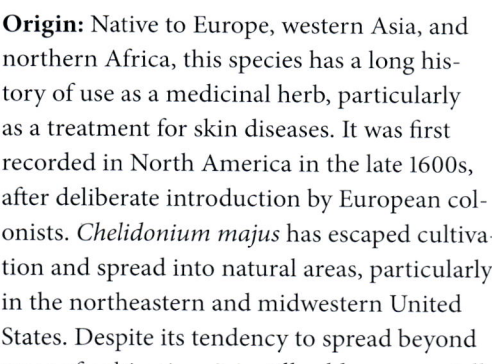

⚠ **Do Not Touch**

161

Origin: Native to Europe, western Asia, and northern Africa, this species has a long history of use as a medicinal herb, particularly as a treatment for skin diseases. It was first recorded in North America in the late 1600s, after deliberate introduction by European colonists. *Chelidonium majus* has escaped cultivation and spread into natural areas, particularly in the northeastern and midwestern United States. Despite its tendency to spread beyond areas of cultivation, it is still sold commercially.

Description: A biennial or short-lived perennial with yellow flowers and grayish-green foliage. When damaged, the leaves, roots, and flowering stems of *C. majus* exude a yellowish-orange sap that contains latex. This species is easily confused with the native *Stylophorum diphyllum* (celandine poppy or wood poppy), which has a similar appearance and a similar common name. However, *S. diphyllum* blooms slightly earlier in the growing season, has larger flowers that are 2–5 cm (¾–2 in.) in diameter, and has fruits that are hairy and ovoid capsules (but flower buds in both species are hairy and ovoid, so be sure to look at fruits rather than buds). Another unrelated species that has a similar common name is the nonnative *Ficaria verna* (lesser celandine) [pp. 180–182], which has yellow flowers with 8–12 petals and heart-shaped leaves.

Habitat: Forest margins, floodplains, wet meadows, along railways and roadsides, fencerows, and gardens. Prefers moist and fertile soils and can tolerate shade.

Height: 30–80 cm (1–3 ft.).

Foliage: There is a rosette of basal leaves with petioles 2–10 cm (¾–4 in.) long and smaller alternate leaves along the flowering stem. Leaves are 10–20 cm (4–8 in.) long and 4–8 cm (2–3 in.) wide, deeply pinnately divided into five to seven oblong segments that are often further lobed, with rounded teeth along the margins. Leaves become progressively smaller and have shorter petioles farther up the stem. The leaves are grayish green, with a smooth upper surface and a hairy lower surface that has a waxy bloom and is covered with fine hairs.

Flowers: Flowering stalks develop opposite leaves in the upper part of the stem, each ending in a flat-topped cluster of three to eight yellow flowers. Individual flowers are 1.3–2 cm (½–¾ in.) in diameter, with four obovate to oblong petals, a pistil with a cylindrical green style, and numerous bright yellow stamens. There are cultivated and naturally occurring double-flowered variants. May–August.

Fruit: The fruit is a slender, podlike capsule, 2–5 cm (¾–2 in.) long and 2–3 mm (1/16–1/8 in.) wide, green and hairless. As the capsule develops, it bulges slightly around the seeds and becomes constricted in the intervals, giving it a bumpy appearance. When the seeds are mature, the two elongate valves split open from the base upward. The seeds are small and black, ellipsoid with one flattened side, 1.2–1.7 mm (1/32–

Native Alternatives

Chrysogonum virginianum (green-and-gold)
Stylophorum diphyllum (celandine poppy or wood poppy)
Trillium cernuum (nodding trillium)
Trillium grandiflorum (great white trillium)
Uvularia grandiflora (large-flowered bellwort)

Chelidonium majus leaves. Image by Robert Vidéki, Doronicum Kft., Bugwood.org.

Cut stem of Chelidonium majus. Image by Beentree, used under a CC BY-SA 3.0 license.

Chelidonium majus roots. Image by Leslie J. Mehrhoff, University of Connecticut, Bugwood.org.

Chelidonium majus flowers. Image by Leslie J. Mehrhoff, University of Connecticut, Bugwood.org.

Left: Chelidonium majus seeds with white elaiosomes. Image by Erutuon, used under a CC BY-SA 2.0 license. Right: Chelidonium majus fruits. Image by Leslie J. Mehrhoff, University of Connecticut, Bugwood.org.

Chelidonium majus plants. Image by Robert Vidéki, Doronicum Kft., Bugwood.org.

¹/₁₆ in.) long and 0.7–1 mm (¹/₃₂ in.) in diameter. The seed surface is covered with netlike veins. Attached to each seed is an elaiosome, which mediates dispersal by ants. June–September.

Stems: Stems ribbed, branching, covered in long white hairs, especially near the base. The stems are brittle, and when broken they exude a yellowish-orange sap that contains latex.

Root system: A shallow crown of orange roots.

Reproduction: Primarily by seed, dispersed by ants and other insects, as well as wind, water, and gravity. Stem and root fragments can also develop into new plants.

Impact: This species spreads aggressively by seed and establishes dense stands that outcompete native species. Although it has been used as a herbal remedy for centuries, it is extremely toxic and, due to the many different isoquinoline alkaloids it produces, can cause liver damage in humans.

Control: Wear protective clothing and gloves while undertaking any control measures, because the sap of this plant can cause skin irritation. Plants can be pulled by hand or dug out. Spot application of glyphosate can be used to control this species.

Plantaginaceae

Linaria vulgaris

yellow toadflax · common toadflax · butter-and-eggs · greater butter-and-eggs · common linaria · flaxweed · Jacob's ladder · ramsted · wild snapdragon

Origin: Native to Europe and northern Asia, this species has a long history of cultivation as an ornamental, medicinal, and dye plant. It was introduced to North America in the late 1600s and has escaped cultivation. *Linaria vulgaris* is now widespread across the United States and is particularly problematic in rangelands in the western regions. Despite its weedy tendencies, it is still available commercially.

Description: A perennial plant with yellow and orange flowers held in a spikelike raceme. Its flowers are similar to those of *Linaria dalmatica* (Dalmatian toadflax), another nonnative species that occurs in Ohio but which is much less common. *Linaria vulgaris* has narrow linear leaves, whereas *L. dalmatica* has extremely broad leaves that clasp the stem. The related native species *Nuttallanthus canadensis* (blue toadflax) has similar foliage, but it has much smaller purple flowers. Based on genetic evidence, *Linaria* and related genera have recently been moved from the Scrophulariaceae to Plantaginaceae, though some references may still place the species in the former family.

Habitat: Forest margins, meadows, prairies, agricultural fields, pastures, along railways and roadsides, fencerows, and gardens. Grows in full sun to partial shade in a range of soil types and conditions.

Height: 30–80 cm (1–3 ft.).

Foliage: The linear or lance-shaped leaves are mainly alternate, but they are spaced closely together on the stem such that they often appear to be opposite or whorled. Leaves are 2–6 cm (¾–3 in.) long and 2–6 mm (1/16–¼ in.) wide, each with a pointed tip and a narrow base that attaches directly to the stem. The margins of the leaves are smooth, and the leaf surfaces are hairless.

Flowers: The yellow and orange flowers are held in spikelike racemes at the ends of the stems. Individual flowers, including the spur, are 2–3.5 cm (¾–1 ½ in.) long. Each flower has an upper lip that is divided into two lobes and folded upward, a lower lip with three lobes that are folded downward, and a long conical nectar spur hanging down from the base. The central part of the lower lip projects upward to form a palate that obstructs the opening to the throat of the flower. The lips and spur are yellow, but the palate is orange. Within the flower there are four stamens and a single pistil. The base of the flower is surrounded by five triangular sepals. June–September.

Fruit: The fruit is an ovoid or spherical capsule, 5–12 mm (¼–½ in.) long and 5–7 mm (¼ in.) wide, with a persistent style at the tip, green turning brown at maturity. The capsule has two cells that open at the top to release numerous seeds. The seeds are dark brown or black, nearly circular and flattened, 1.5–4 mm (1/16–¼ in.) in diameter and 0.5 mm (1/64 in.) thick, with a bumpy surface and surrounded by a papery wing that is notched at one point. July–October.

Stems: Stems are upright, hairless, branching near the top, and slightly woody at the

---

Native Alternatives

Castilleja coccinea (Indian paintbrush)
Chelone glabra (turtlehead)
Collinsia verna (blue-eyed Mary)
Dasistoma macrophylla (mullein foxglove)
Penstemon digitalis (foxglove beardtongue)

Linaria vulgaris stem and leaves. Image by New York State IPM Program at Cornell University, used under a CC BY 2.0 license.

Linaria vulgaris fruits. Image by AnR00002, used under a CC0 1.0 license.

Linaria vulgaris roots. Image by Steve Dewey, Utah State University, Bugwood.org.

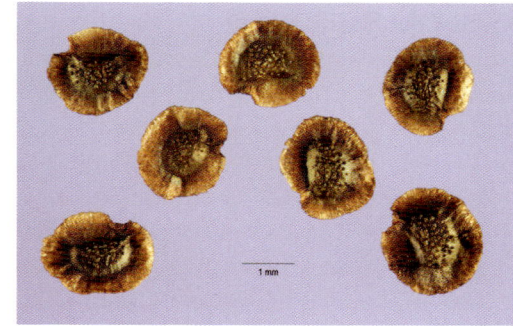

Linaria vulgaris seeds. Image by Steve Hurst, hosted by the USDA-NRCS PLANTS Database.

Linaria vulgaris flowers. Image by William M. Ciesla, Forest Health Management International, Bugwood.org.

Linaria vulgaris infestation. Image by William M. Ciesla, Forest Health Management International, Bugwood.org.

base. Stems arise singly or in clumps from the root crown. Additional stems arise from buds along the lateral roots.

Root system: A taproot and extensive lateral roots.

Reproduction: By seed, dispersed by wind, water, animals, and contaminated crop seed. A single clonal individual (which may include many different stems connected by the same root system) can produce 1,500–30,000 seeds in a growing season. Seeds remain viable in the soil seed bank for up to eight years. Vegetative reproduction also occurs through the creeping lateral roots, which produce new stems from adventitious buds.

Impact: Primarily a weed in pasture and croplands. It is unpalatable to livestock and outcompetes many important forage crops. The root system of this species also serves as winter host for several plant pathogens—including cucumber mosaic virus and broad bean wilt virus—that can transfer to crop plants.

Control: This species is sometimes included in meadow and wildflower seed mixes, so check the contents carefully before purchase or use. Pulling by hand or digging can help control small populations, while hoeing or tilling can be used in larger stands. It is important to disrupt and remove as much of the root system as possible. Some populations of *L. vulgaris* have developed herbicide resistance, so chemical control is not recommended.

Plantaginaceae

Plantago major

broadleaf plantain · greater plantain · great plantain · common plantain · rippleseed plantain · rat-tail plantain

Origin: Native to Europe, northern and central Asia, and northern Africa, *Plantago major* has been spread by human activities to temperate and tropical regions around the world. This species was most likely introduced to North America in the 1600s as a contaminant in crop seed. It is now widespread throughout the United States.

Description: An annual, biennial, or short-lived perennial with a rosette of basal leaves and a dense spike of inconspicuous greenish flowers. A number of other nonnative plantain species occur in Ohio, including *Plantago lanceolata* (buckhorn or narrowleaf plantain), which can be differentiated by its narrower, lance-shaped leaves. The native species *Plantago rugelii* (Rugel's plantain) is also very similar in appearance, but its petioles are purplish at the base and it has slender capsules that split open near the base to release black seeds.

Habitat: Forest margins, open woodlands, shorelines, pastures, orchards, nurseries, agricultural fields, roadsides, lawns, and gardens. Tolerates mowing and trampling. Prefers rich and moist soils but can grow in a wide range of soil types.

Height: Flowering stems 15–40 cm (6–16 in.).

Foliage: All leaves are basal, forming a large rosette. Leaves are broadly elliptic to ovate, 4–20 cm (2–8 in.) long and 1.5–11 cm (⅝–5 in.) wide, narrowing abruptly to a long and broad petiole that is green at the base. The leaves have three to seven prominent veins that run parallel to the margins. Young leaves may be slightly hairy, particularly near the base. Mature leaves are thick and slightly leathery, with hairless surfaces. The leaf margins are smooth or slightly wavy.

Flowers: Each plant produces 1–30 flowering stems from the center of the rosette. The upper part of the stem contains a dense cylindrical spike of inconspicuous greenish flowers, typically 10–20 cm (4–8 in.) long and 6 mm (¼ in.) wide. Each individual flower is ovate, surrounded by overlapping green sepals, 1.5–2.5 mm (¹⁄₁₆ in.) long. Above the sepals there are four triangular petals, 1 mm (¹⁄₃₂ in.) long, papery and translucent light brown. The flowers have a single style covered with white hairs, surrounded by four stamens with 2–3 mm (¹⁄₁₆–⅛ in.) long filaments and purplish anthers that turn yellowish brown with age. At the base of each flower there is a green bract, 1–2 mm (¹⁄₃₂–¹⁄₁₆ in.) long, that curves upward around the flower. July–October.

Fruit: The fruit is an ovoid brown capsule, 4–5 mm (⅛–¼ in.) long and 2–3 mm (¹⁄₁₆–⅛ in.) wide, that splits open along a seam around the center. Each capsule contains 5–22 seeds, light brown, each 1–1.7 mm (¹⁄₃₂–¹⁄₁₆ in.) long and 0.5–1 mm (¹⁄₆₄–¹⁄₃₂ in.) wide, variable in shape, with a roughly textured surface. Seeds are formed shortly after fertilization but may not disperse from the capsule until the following year.

Stems: Flowering stems are upright, leafless, covered with fine white hairs.

Root system: A crown of shallow, fibrous roots.

Reproduction: Primarily by seed. The seed is coated with a substance that becomes sticky when wet, allowing it to adhere to animals for dispersal. An individual plant can produce up to 14,000 seeds in a growing season, and seeds can remain viable in the soil seed bank for 20 years or longer. Vegetative reproduction from root fragments is also possible.

Impact: This widespread weed rapidly colonizes open and disturbed areas. It is primarily a concern to homeowners hoping to grow lawns free of broadleaved species. In agricultural settings, this species may be a reservoir for plant pathogens, including the tobacco mosaic virus.

Control: Grazing or mowing may reduce this species' growth and reproductive output. Individual plants can be pulled by hand. In lawns, the herbicides MCPA and 2,4-D have been found to control this species.

Plantago major leaf. Image by Robert Vidéki, Doronicum Kft., Bugwood.org.

Plantago major fruits. Image by H. Zell, used under a CC BY-SA 3.0 license.

Plantago major rosette. Image by Chris Evans, University of Illinois, Bugwood.org.

Plantago major female (*left*) and male (*right*) flowers. Image © Gerald D. Carr.

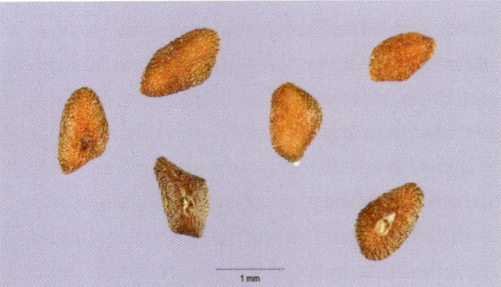

Plantago major seeds. Image by Steve Hurst, hosted by the USDA-NRCS PLANTS Database.

Plantago major infestation. Image by Richard T. Gardner III, Bugwood.org.

Polygonaceae

Fallopia japonica

Japanese knotweed · Asian knotweed · Japanese knotwood · Mexican bamboo · elephant-ear bamboo · Japanese fleeceflower · Himalayan fleece vine · monkeyweed · wild rhubarb · Hancock's curse

Synonyms: *Polygonum cuspidatum, Reynoutria japonica*

169

Origin: Native to Japan, China, Taiwan, and Korea. It was introduced to the United States in 1873 as an ornamental plant and quickly escaped from cultivation. This species is included on lists of invasive species and prohibited noxious weeds throughout the United States, and the International Union for Conservation of Nature considers it one of the 100 worst invasive species worldwide.

Description: A fast-growing, perennial forb with a shrublike form. The semiwoody, hollow stems have enlarged nodes, causing the plant to superficially resemble bamboo. Stems die back each year over winter and resprout from rhizomes in the spring.

Habitat: Found along river- and streambanks, fencerows, utility rights-of-way, waste places, roadsides, and other areas where the soil has been disturbed. It is most common in habitats with full sunlight but can grow in full shade. Tolerant of high temperature, salinity, and drought.

Height: 1–3 m (3–10 ft.).

Foliage: Simple, alternate, 8–15 cm (3–6 in.) long and 5–12 cm (2–5 in.) wide, with a petiole 2–3 cm (¾–1 in.) long. Broadly oval or somewhat triangular, with a flat or squared off base and a pointed tip. Leaf margins are smooth or slightly wavy, and the pinnate veins on the upper leaf surface are conspicuous. The veins on the undersides of leaves are covered in blunt, single-celled hairs that resemble raised bumps. The petiole is swollen at the base where

there is a small gland that functions as an extrafloral nectary.

Flowers: Minute flowers are borne in branching panicles, 8–15 cm (3–6 in.) long, in the upper leaf axils. Individual flowers are 2–8 mm ($\frac{1}{16}$–¼ in.) long and 3 mm (⅛ in.) across, white or greenish white, with five colorful sepals and no petals, eight stamens, and three styles. In most cases, the flowers are functionally unisexual, whereby each male or female flower has the complementary—but sterile—organs of the other sex. Male and female flowers usually grow on different plants. The flowering stalks are slightly hairy, and there are reddish-brown papery sheaths at the point where flowers attach. Flowers produce a strong, sweet fragrance. June–September.

Fruit: Fruits have three papery, light brown wings and are about 8–9 mm (¼–⅜ in.) long. These enclose a glossy dark brown or blackish, three-angled achene that is 2–4 mm ($\frac{1}{16}$–⅛ in.) long and 2 mm ($\frac{1}{16}$ in.) wide, tapering to points at the ends. July–October.

Stems: Young growth can be pink. Older stems are numerous, erect, green streaked with red or purple, often with a waxy bloom. The stems are round in cross section, hollow, and jointed like bamboo.

Native Alternatives
Aralia racemosa (American spikenard)
Aruncus dioicus var. *dioicus* (goatsbeard)
Sambucus racemosa (red elderberry)
Spiraea tomentosa (steeplebush)

Fallopia japonica leaves. Image by John Cardina, The Ohio State University, Bugwood.org.

Fallopia japonica flowers. Image by Leslie J. Mehrhoff, University of Connecticut, Bugwood.org.

Fallopia japonica stems. Image by Chris Evans, University of Illinois, Bugwood.org.

Fallopia japonica fruits. Image by Barbara Tokarska-Guzik, University of Silesia, Bugwood.org.

Fallopia japonica seed. Image by Ken Chamberlain, The Ohio State University, Bugwood.org.

Fallopia japonica flowers. Image © Gerald D. Carr.

Fallopia japonica plants. Image by Leslie J. Mehrhoff, University of Connecticut, Bugwood.org.

Root system: Deep and extensive rhizomes that can spread laterally 20 m (65 ft.) or more. Rhizomes are fleshy and white when young, becoming woody with age, with an orange interior and brown exterior.

Reproduction: By seed and rhizomes. Contrary to a common misconception that seed does not develop in its introduced range, this species does produce seed, and seed viability and germination rates are high. Seeds can be moved by wind, water, contamination in fill dirt, animals, and human activities. Once a plant is established, it often spreads clonally through rhizomes.

Impact: Clones form dense monocultures that outcompete other plant species. It can hybridize with the nonnative, invasive *Fallopia sachalinensis* (giant knotweed) [pp. 172–173] to form the fertile hybrid *Fallopia* x *bohemica* (Bohemian knotweed) [pp. 174–175], which is itself an invasive species.

Control: Dig out plants when young. Once established, this species is very difficult to eradicate, so early detection and removal are crucial. Larger stands can be cut back repeatedly each growing season over several years. Covering stems with black plastic may also be useful. Foliar and cut-stem treatments with glyphosate, triclopyr, and imazapyr have been effective on large populations. However, a method that has been developed more recently in the United Kingdom involves clearing plants and covering the area with wire fence mesh that has 1 cm x 1 cm (½ in. x ½ in.) openings. The mesh needs to be fastened to the ground using landscaping pegs or some other method so it remains in place through the growing season. The plants will repeatedly produce new stems, the stems will be killed by the mesh, and the root reserves will be depleted over time.

Polygonaceae

Fallopia sachalinensis

giant knotweed · giant knotwood · Sakhalin knotweed

Synonyms: *Polygonum schalinense, Reynoutria sachalinensis*

Origin: Native to northern Japan and the Sakhalin and Kurile Islands of Russia. It was introduced to North America as a garden ornamental in 1894 and had escaped cultivation by the 1920s.

Description: A fast-growing, perennial forb with a shrublike form. The semiwoody, hollow stems have enlarged nodes, causing the plant to superficially resemble bamboo. Stems die back each year over winter and resprout from rhizomes in the spring. It is very similar to *Fallopia japonica* (Japanese knotweed) [pp. 169–171] but is larger overall, has much larger, heart-shaped leaves, hairy undersides of leaves, and inflorescences that are much shorter than the surrounding leaves.

Habitat: Found along streams and rivers, utility rights-of-way, waste places, fields, roadsides, and other areas with high levels of human disturbance. It is most prolific in areas with full sun and moist soils.

Height: Typically 2–4 m (6–12 ft.), occasionally to 5 m (15 ft.).

Foliage: Simple, alternate, 15–30 cm (6–12 in.) long and 10–25 cm (4–10 in.) wide, with a petiole 1–4 cm (½–2 in.) long. Ovate with a gradually tapering tip, a heart-shaped base, and rounded basal lobes. Leaf margins are smooth or slightly wavy. The veins on the undersides of the leaves are covered in light-colored hairs that are twisted in structure.

Flowers: Minute white or greenish-white flowers are borne in upright branching panicles, 3–10 cm (1–4 in.) long, in the upper leaf axils. Individual flowers have five colorful sepals and no petals, eight stamens, and three styles. The flowers are functionally unisexual; each male or female flower has the complementary—but sterile—organs of the other sex. Male and female flowers occur on different plants. The sepals of the male flowers are 6–7 mm (¼ in.) long, while those of female flowers are 12–15 mm (½–⅝ in.) long. The flowering stalks are slightly hairy, and there are reddish-brown papery sheaths at the point where flowers attach. August–October.

Fruit: Fruits have three papery whitish or light-brown wings and are about 12–15 mm (½–⅝ in.) long. These enclose a glossy dark-brown, three-angled achene that is 3–5 mm (⅛–¼ in.) long and 1–2 mm ($\frac{1}{32}$–$\frac{1}{16}$ in.) wide, tapering to points at the ends. September–November.

Stems: Young growth can be pink. Older stems are numerous, erect, mostly green with red banding, often with a waxy bloom. The stems are slightly angular in cross section, hollow, and jointed like bamboo.

Root system: Plants develop deep and extensive rhizomes that can grow to be 3 m (10 ft.) deep and extend laterally to 7 m (23 ft.) from the plant. Rhizomes are fleshy and white when young, becoming woody with age, with an orange interior and brown exterior.

Reproduction: By seed and rhizomes.

Impact, Control, and Native Alternatives: See *Fallopia japonica* [pp. 169–171].

Fallopia sachalinensis leaf. Image by Leslie J. Mehrhoff, University of Connecticut, Bugwood.org.

Fallopia sachalinensis leaves. Image by Maja Dumat, used under a CC BY 2.0 license.

Fallopia sachalinensis flowers. Image by Leslie J. Mehrhoff, University of Connecticut, Bugwood.org.

Fallopia sachalinensis fruits. Image by Leslie J. Mehrhoff, University of Connecticut, Bugwood.org.

Red stems and petioles of *Fallopia sachalinensis*. Image by Jan Samanek, Phytosanitary Administration, Bugwood.org.

Fallopia sachalinensis infestation. Image by Tom Heutte, USDA Forest Service, Bugwood.org.

Polygonaceae

Fallopia x bohemica
Bohemian knotweed · hybrid knotweed

Synonyms: *Polygonum x bohemicum, Reynoutria x bohemica*

Origin: A fertile hybrid of *Fallopia japonica* (Japanese knotweed) [pp. 169–171] and *Fallopia sachalinensis* (giant knotweed) [pp. 172–173]. The hybrid was first described in Bohemia—the westernmost region of the Czech Republic—in 1983. It is widespread in North America, but it went largely unnoticed until 2003. Because it is often misidentified as one of the parent species, this hybrid is still underreported.

Description: A fast-growing, shrublike perennial forb. The semiwoody, hollow stems have enlarged nodes, causing the plant to superficially resemble bamboo. Stems die back each year over winter but resprout from rhizomes in the spring. For most characteristics, this hybrid exhibits morphological intermediacy with its parents. It can be distinguished most reliably by the hairs on the veins on the underside of the leaves. These hairs have a broad base and a sharp tip that distinguishes them from *F. japonica* and are much shorter and straighter than those of *F. sachalinensis*.

Habitat: Found along streams and rivers, utility rights-of-way, waste places, fields, roadsides, and other areas with high levels of human disturbance. Prefers moist and nitrogen-rich soils. There is some suggestion that the habitat range of the hybrid is wider than that of either parent species.

Height: Typically 1.5–2.5 m (5–8 ft.), occasionally to 4 m (12 ft.).

Foliage: Simple, alternate, 5–25 cm (2–10 in.) long and 2–18 cm (¾–7 in.) wide, with a petiole 1–3 cm (½–1 in.) long. Ovate, with a tapered tip and a squared or heart-shaped base. Leaf margins are smooth or slightly wavy. The veins on the underside of each leaf are covered with extremely short hairs with a broad base and a pointed tip. Veins and petioles can be reddish purple in immature leaves.

Flowers: Minute white or greenish-white to pink flowers are borne in upright branching panicles, 4–12 cm (2–5 in.) long, in the upper leaf axils. Individual flowers are 4–6 mm (⅛–¼ in.) long and 3 mm (⅛ in.) wide, with five colorful sepals and no petals, eight stamens, and three styles. The flowers are functionally unisexual; each male or female flower has the complementary—but sterile—organs of the other sex. Male and female flowers occur on different plants. The flowering stalks have reddish-brown papery sheaths at the point where flowers attach. July–September.

Fruit: Fruits have three papery, light brown wings and are about 4–6 mm (⅛–¼ in.) long. These enclose a glossy dark brown, three-angled achene that is 2–3 mm (1/16–⅛ in.) long and 1–2 mm (1/32–1/16 in.) wide, tapering to points at the ends. August–October.

Stems: Young growth can be pink. Stems numerous, erect, green with reddish markings, covered with a waxy bloom. The stems are round or slightly angular in cross section, hollow, and jointed like bamboo.

Root system: Plants develop deep and extensive rhizomes. Rhizomes are fleshy and white when young, becoming woody with age, with a white or orange interior and brown exterior.

Reproduction: By seed and rhizomes.

Impact, Control, and Native Alternatives: See *Fallopia japonica* [pp. 169–171].

Fallopia x *bohemica* fruits. Image by Leslie J. Mehrhoff, University of Connecticut, Bugwood.org.

Fallopia x *bohemica* leaves. Image by Robert Vidéki, Doronicum Kft., Bugwood.org.

Fallopia x *bohemica* flowers. Image by Richard T. Gardner III, Bugwood.org.

Fallopia x *bohemica* stems. Image by Robert Vidéki, Doronicum Kft., Bugwood.org.

Polygonaceae

Persicaria longiseta

Oriental lady's thumb · bristly lady's thumb
· Asiatic smartweed · creeping smartweed ·
long-bristled smartweed · bristled knotweed ·
bunchy knotweed · tufted knotweed · Asiatic
waterpepper

Synonyms: *Polygonum cespitosum, Polygonum longisetum*

Origin: Native to eastern Asia, this species was first reported in North America in 1910. It has established populations throughout the eastern and midwestern United States.

Description: An annual plant with creeping stems and spikelike clusters of small, pink, ovoid flowers. It is similar in appearance to several other smartweeds, including the non-native species *Persicaria maculosa* (spotted lady's thumb). *Persicaria longiseta* can be distinguished from its relatives by the long bristles in the flower clusters and at the tops of the membranous ocreae surrounding the stems.

Habitat: Riparian forests, river- and stream-banks, fields, along railways and roadsides, lawns, and gardens. Although often found in moist to damp soils, this species is tolerant of a wide range of soil types, moisture levels, and light availability.

Height: 30–100 cm (1–3 ft.), with longer stems typically creeping along the ground.

Foliage: Leaves simple, alternate, narrowly lance-shaped, 2–8 cm (¾–3 in.) long and 1–3 cm (½–1 in.) wide, with a pointed tip and a narrow base that attaches to the stem with a short petiole. Leaf margins are smooth, and the leaf surfaces are mostly hairless. Leaves are dark green and occasionally have a faint purplish marking shaped like a chevron near the center of the blade. Above the node where the leaf attaches, a papery ocrea surrounds the stem, 5–12 mm (¼–½ in.) long, with long, fine bristles at the top, 4–12 mm (⅛–½ in.) long.

Flowers: At the ends of the stems there are small, pink, ovoid flowers in spikes 1–8 cm (½–3 in.) long and 3–7 mm (⅛–¼ in.) wide. Individual flowers are 2–3 mm ($^1/_{16}$–⅛ in.) long and 2–2.5 mm ($^1/_{16}$ in.) wide, ovoid, consisting of five pink sepals and no petals, often remaining closed through the flowering period. This species is capable of self-fertilization, so sexual reproduction can take place even if the flowers do not open. At the base of each flower there is a small sheath with several 1–6 mm ($^1/_{32}$–¼ in.) long bristles. June–October.

Fruit: Each flower is replaced by a three-angled achene, dark brown to black, shiny, 1.6–2.3 mm ($^1/_{16}$ in.) long and 1.1–1.6 mm ($^1/_{32}$–$^1/_{16}$ in.) wide, tapering to blunt points at the ends.

Stems: Stems reddish brown to green, upright or creeping along the ground, branching, and hairless.

Root system: A shallow and branching taproot. Adventitious roots also develop at nodes along the stem where they come into contact with the soil.

Reproduction: By seed. Seeds remain viable in the soil seed bank for five years or more. This species can also reproduce vegetatively from stem fragments or by rooting at the nodes.

Impact: This species is primarily a weed of disturbed areas, but it forms dense stands that can displace native species. It is also a weed in horticultural and agricultural settings.

Control: Small infestations can be controlled by hand pulling, digging, or frequent mowing. Glyphosate has been used to control this species, although this chemical should not be applied if the plant is growing in riparian areas. This species is highly resistant to 2,4-D.

Persicaria longiseta flowers. Image © Peter M. Dziuk, Minnesota Wildflowers.

Persicaria longiseta leaves. Image by Chris Evans, University of Illinois, Bugwood.org.

Persicaria longiseta infestation. Image by Leslie J. Mehrhoff, University of Connecticut, Bugwood.org.

Persicaria longiseta stem and ocrea. Image © Peter M. Dziuk, Minnesota Wildflowers.

Primulaceae

Lysimachia nummularia

moneywort · creeping yellow loosestrife ·
creeping Jenny · wandering Jenny · running
Jenny · creeping Joan · wandering sailor · herb
twopence · twopenny grass

Origin: Native to Europe and western Asia, *Lysimachia nummularia* was introduced to North America as an ornamental ground cover in the early 1700s. It has escaped from cultivation and spread throughout the northeastern and midwestern United States. Despite its weedy tendencies, this species is still available commercially.

Description: A low-growing perennial with evergreen to semievergreen foliage and a mat-like growth habit. The leaves resemble coins, hence the species name and several of its common names allude to money. On the whole, *L. nummularia* is straightforward to identify, based on its creeping stems, opposite rounded leaves, and relatively large yellow flowers. The nonnative species *Veronica officinalis* (common gypsyweed) and *Glechoma hederacea* (ground ivy) [pp. 152–153] and the native species *Mitchella repens* (partridgeberry) all have opposite leaves and a similar growth habit but have different leaf shapes and flowers. None of the other *Lysimachia* species that are either native or naturalized to Ohio have both creeping stems and rounded leaves.

Habitat: Floodplain forests, forest margins, river- and streambanks, wet meadows, swamps, ditches, roadsides, lawns, and gardens. Can grow in full sun to full shade and prefers moist to wet soils.

Height: Creeping stems 15–50 cm (6–20 in.) long.

Foliage: Leaves simple, opposite, rounded or slightly ovate, 1–3.5 cm (½–1 ½ in.) long and wide, with a rounded or heart-shaped base that attaches to the stem by a short petiole. Leaf margins are smooth, and the leaf surfaces are hairless. The leaf surfaces have faint, scattered, reddish-brown to black glandular dots. The foliage remains green year-round or for most of the year. Some cultivars have yellowish foliage and are marketed as less aggressive, but over time these can revert to the darker green form and spread.

Flowers: Yellow flowers develop in the leaf axils, either solitary or in pairs. Flowers are cuplike, 2–3 cm (¾–1 in.) in diameter, with five rounded petals, five yellow stamens, and a single greenish pistil. The petals are marked with scattered, dark red, glandular dots. At the base of the flower there is a green calyx formed by five triangular sepals, each 5–9 mm (¼–⅜ in.) long. Flowers are only open during daylight. June–August.

Fruit: Fruits rarely develop. The fruit is a spherical or slightly ovoid capsule, 2–3 mm ($\frac{1}{16}$–⅛ in.) in diameter, which develops within the persistent calyx. The capsule contains one to five elliptic seeds, each about 1–1.5 mm ($\frac{1}{32}$–$\frac{1}{16}$ in.) long and 0.6–1 mm ($\frac{1}{64}$–$\frac{1}{32}$ in.) wide. July–September.

Stems: The creeping stems are light green, hairless, and highly branched.

Root system: The roots are slender and fibrous. Adventitious roots form at the nodes where stems come into contact with the soil surface.

Reproduction: Occasionally by seed, most

Native Alternatives
Asarum canadense (wild ginger)
Meehania cordata (creeping mint)
Mitchella repens (partridgeberry)
Phlox stolonifera (creeping phlox)

likely dispersed by water. Most reproduction is vegetative, via stem fragments and creeping stems.

Impact: Not much is known about the direct impact of this species on native plant communities. It does form dense and spreading mats, so it is often assumed to have the potential to displace native species. This is of greatest concern in rare plant communities, especially in riparian areas where *L. nummularia* thrives.

Control: Avoid planting this species in gardens, particularly those near waterways or bordering natural areas. The long stems can be raked up, and then plants can be pulled or dug out to remove the root crown. Close mowing is also effective in controlling spread and reproduction of this species. Spring application of glyphosate or 2,4-D has been shown to provide control in larger stands. In wetland areas, only licensed aquatic applicators should undertake herbicide treatment.

Lysimachia nummularia leaves. Image by Chris Evans, University of Illinois, Bugwood.org.

Pale leaves of the cultivar *Lysimachia nummularia* 'Goldilocks.' Image by David J. Stang, used under a CC BY-SA 4.0 license.

Lysimachia nummularia flowers. Image by Leslie J. Mehrhoff, University of Connecticut, Bugwood.org.

Lysimachia nummularia fruits. Image by Krzysztof Ziarnek, used under a CC BY-SA 4.0 license.

Lysimachia nummularia variety with purple leaves. Image by Salicyna, used under a CC BY-SA 4.0 license.

Lysimachia nummularia infestation. Image by Leslie J. Mehrhoff, University of Connecticut, Bugwood.org.

Ranunculaceae

Ficaria verna

lesser celandine · fig buttercup · figroot buttercup
· fig crowfoot · figwort · pilewort · buttercup ficaria ·
bulbous buttercup · small crowfoot

OIPC Invasive
ODA Invasive
⚠ Do Not Touch

Synonym: *Ranunculus ficaria*

Origin: Native to Europe, western Asia, and northern Africa. This species was introduced to North America as an ornamental and medicinal plant and escaped cultivation. It was first recorded in natural areas in the United States in the mid-1800s. *Ficaria verna* is still available commercially in many places—although, due to its invasive designation, not legally in Ohio—and all varieties should be assumed invasive.

Description: An ephemeral perennial plant that emerges in the winter and produces showy yellow flowers in early spring. The aboveground portions senesce in June and the plants remain dormant for the remainder of the year, with nutrients stored in tuberous roots. It resembles the native species *Caltha palustris* (marsh marigold), which has similar leaves and also produces showy yellow flowers in early spring. The two species can be most reliably distinguished by the three (occasionally four) green sepals at the base of the flowers and typically 8–12 narrow yellow petals of *F. verna*, whereas *C. palustris* only has 5–9 yellow petallike sepals. In addition, *F. verna* produces fig-shaped bulblets underground and bulbils in the leaf axils, both of which are lacking in *C. palustris*. Despite sharing a similar common name, *F. verna* is not closely related to *Chelidonium majus* (greater celandine) [pp. 161–163]; the flowers of the latter species have only four petals.

Habitat: Floodplain forest, river- and streambanks, open woodland, meadows, ditches, roadsides, and gardens. This species grows best in areas with moist soils.

Height: 10–30 cm (4–12 in.).

Foliage: Each plant develops a rosette of basal leaves, 4–9 cm (2–4 in.) long and 4–8 cm (2–3 in.) wide, heart- or kidney-shaped, with petioles that are typically 10–20 cm (4–8 in.) long. Flowering stems have alternate leaves, similar to the basal leaves although smaller and with shorter petioles. Leaves are hairless, glossy, and dark green. The leaf margins can be smooth, wavy, or coarsely toothed.

Flowers: Showy yellow flowers are borne singly on slender stems that emerge from the rosette. Each flower is 2–3 cm (¾–1 in.) in diameter, typically with 8–12 glossy yellow petals and numerous yellow stamens surrounding a central cluster of greenish-yellow pistils. At the base of the flower there are three (occasionally four) green sepals, each 4–8 mm (⅛–¼ in.) long and 3–6 mm (⅛–¼ in.) wide, ovate with blunt tips. March–May.

Fruit: Each flower develops into a cluster of fruits. The fruits are achenes, 3–4 mm (⅛ in.) long and 1.8–2 mm (¹⁄₁₆ in.) wide, flattened, slightly hairy, and beakless. Each fruit contains a single seed. April–June.

Stems: Flowering stems can be upright or creeping along the ground. Stems are hairless. Bulbils are sometimes produced in the leaf axils along the flowering stems. As the plant ages, the axillary bulbils become more apparent.

Native Alternatives
Chrysogonum virginianum (green-and-gold)
Packera aurea (golden groundsel)
Zizia aptera (heartleaf alexanders)
Zizia aurea (golden alexanders)

Ficaria verna leaves. Image by Leslie J. Mehrhoff, University of Connecticut, Bugwood.org.

Three sepals at the base of *Ficaria verna* flower. Image by Salicyna, used under a CC BY-SA 4.0 license.

Developing fruits on *Ficaria verna*. Image by Stefan Lefnaer, used under a CC BY-SA 4.0 license.

Ficaria verna flowers. Image by Leslie J. Mehrhoff, University of Connecticut, Bugwood.org.

Ficaria verna roots and bulblets. Image by Stefan Lefnaer, used under a CC BY-SA 4.0 license.

Aerial bulbils developing in the leaf nodes of *Ficaria verna*. Image by Stefan Lefnaer, used under a CC BY-SA 4.0 license.

Root system: Each plant develops a mass of tuberous roots and fig-shaped bulblets that break away easily.

Reproduction: By seed, although the viability of seeds and seedlings is reported to be low in some varieties. Vegetative reproduction occurs through the spread of root fragments, root bulblets, and aerial bulbils.

Impact: Once established, this species can form a continuous mat of vegetation that dominates the forest floor. Because *F. verna* begins to develop leaves in late winter, it is often well established before other understory plant species emerge. It is primarily a threat to native spring ephemerals, although other forbs and tree seedlings in the understory could also be excluded by dense stands of *F. verna.* In addition, all parts of the plant are toxic if ingested, which is a particular concern for grazing animals.

Control: Given the similarity in appearance between the nonnative *F. verna* and the native *C. palustris,* extreme care should be taken to correctly identify the species before undertaking any control measures. Wear protective clothing and gloves when handling this plant, because contact with damaged or crushed *F. verna* leaves can cause skin irritation. In small infestations, *F. verna* can be pulled by hand or dug out, taking care to remove all roots and bulblets. In areas with large colonies or where soil disturbance is a concern, glyphosate applied to the leaves in the late winter will kill all parts of the plant with minimal disturbance to soil or other plant species. It is important to perform any chemical controls before spring ephemerals emerge and before amphibians are active.

Rosaceae

Duchesnea indica

Indian strawberry · mock strawberry · false
strawberry · yellow-flowered strawberry

Synonym: *Fragaria indica, Potentilla indica*

Origin: Native to temperate and tropical areas
of southern and eastern Asia. This species was
introduced to North America as an ornamental
groundcover, although the timing and location
of its introduction is not clear. It has become a
garden weed and has spread into natural areas
throughout much of the eastern United States.

Description: A perennial with creeping stems,
yellow flowers, and red fruits that resemble small strawberries but lack any real flavor.
Duchesnea indica resembles the native wild
strawberries in the genus *Fragaria,* although
the leaves of *D. indica* are smaller, and its flowers are yellow instead of white. In addition, the
fruits of *D. indica* are held upright, while those
of *Fragaria* species hang downward. It also is
similar in appearance to many *Potentilla* species, although the latter tend to have compound leaves with five or more leaflets.

   Habitat: Forest margins, forest clearings,
meadows, fields, lawns, and gardens. This species prefers partial sun and fertile soils with
moderate levels of moisture.

   Height: Creeping stems can be 30–100 cm
(1–3 ft.) long.

   Foliage: The leaves are alternate, compound
with three ovate leaflets, each 2–4 cm (¾–2 in.)
long and 1–3 cm (½–1 in.) wide, with long petioles that clasp the stem. The margins of the
leaflets have rounded or blunt teeth. The leaflets are strongly pinnately veined, and leaf surfaces are smooth or covered with fine hairs on
the underside. Leaves often persist throughout
the winter.

   Flowers: Yellow flowers are borne singly on
stems 3–6 cm (1–3 in.) long. Each flower is 1–2.5
cm (½–2 in.) in diameter, with five rounded

or slightly notched petals and a ring of 20–30
yellow stamens surrounding a central cluster
of numerous yellow pistils. At the base of the
flower, five narrow triangular sepals alternate
with five leafy bracts that have three to five teeth
along their outer margins. April–August.

   Fruit: Each flower is replaced by a red aggregate fruit, spherical or ovoid, 1–2 cm (½–¾ in.)
in diameter, which resembles a small strawberry but with a bland and spongy interior. The
surface of the aggregate fruit is covered with
ovoid achenes, 1–1.5 mm (¹⁄₃₂–¹⁄₁₆ in.) long and
1 mm (¹⁄₃₂ in.) wide, each containing a single
seed. The triangular sepals turn upward around
the base of the fruit, and the fruits are held
upright at the tops of the stems. June–October.

   Stems: From its base, each plant sends out
multiple stolons that trail along the ground.
Stolons root and produce new plants at the
nodes and tips. The flowering stems and stolons are light green to reddish purple and covered with fine hairs.

   Root system: A main root with numerous
coarse lateral roots branching off it, and adventitious roots that develop along stolons.

   Reproduction: By seed, dispersed by animals. Vegetative reproduction also occurs via
creeping stolons.

Impact: Primarily a weed of lawns, gardens,
and forest margins. Although the fruits are not
toxic, some people have allergic responses after
eating them.

Control: Rake thoroughly to raise the creeping
stems and pull or mow the remaining parts of
the plant. The plant can be controlled with spot
application of 2,4-D during spring.

Duchesnea indica leaves. Image by John Tann, used under a CC BY 2.0 license.

Duchesnea indica plants. Image by William M. Ciesla, Forest Health Management International, Bugwood.org.

Duchesnea indica flower. Image © Gerald D. Carr.

Duchesnea indica fruit. Image © Gerald D. Carr.

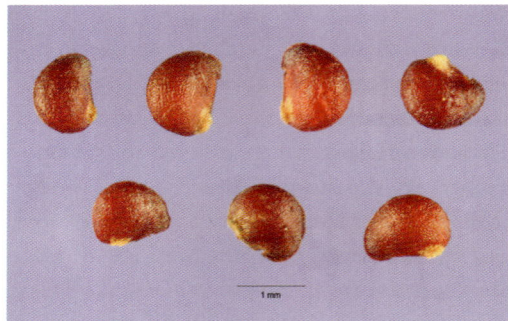

Duchesnea indica achenes. Image by Steve Hurst, hosted by the USDA-NRCS PLANTS Database.

Stand of *Duchesnea indica* plants. Image by Lamiot, used under a CC BY 3.0 license.

Rosaceae

Potentilla recta

sulfur cinquefoil · rough-fruited cinquefoil · upright cinquefoil

Origin: Native to central and southern Europe and central and southwestern Asia. The timing and pathway of its introduction to North America are not clear, but there are records of it growing outside of cultivation by the late 1800s. It is now a common weed across much of the continental United States and Canada.

Description: An upright perennial plant with a woody rootstock, palmately compound leaves, and pale yellow flowers. There are more than 50 species of *Potentilla* in North America—both native and nonnative—that can be easily confused. The distinguishing features of *Potentilla recta* are its upright growth habit, the spreading hairs on its stems, its pale yellow flowers with petals longer than the sepals, and its palmately compound leaves with five to seven leaflets. Its leaves bear a superficial resemblance to *Cannabis sativa* (marijuana), although the latter has narrower leaflets, hairless stems, and small greenish flowers.

Habitat: Forest margins, meadows, fields, pastures, fencerows, along railways and roadsides, gardens, and lawns. Prefers full sun to partial shade and can tolerate dry and infertile soils.

Height: 40–80 cm (1–3 ft.).

Foliage: Each plant first develops a rosette consisting of a few leaves with long petioles. Leaves are palmately compound, with five to seven oblong leaflets, the longest central leaflet 2–10 cm (¾–4 in.) long and 5–35 mm (¼–½ in.) wide and lateral leaflets slightly smaller. The margins of the leaflets are coarsely toothed, the pinnate veins on the lower surfaces are prominent, and the petioles and leaf surfaces are roughly hairy. The leaves on the flowering stem are alternate, similar to the basal leaves but becoming progressively smaller and shorter petioled farther up the stem, sometimes simple or with two or three leaflets near the top of the plant.

Flowers: Pale yellow flowers are borne in open, flat-topped cymes at the top of the stems. Individual flowers are 1.5–2.5 cm (⅝–1 in.) in diameter, with five petals that are deeply notched at the tips. Each flower has approximately 30 yellow stamens surrounding a bright yellow center containing numerous pistils. At the base of the flower are five triangular sepals alternating with five slightly narrower leafy bracts, all of which are covered with fine white hairs. June–August.

Fruit: After flowering, the sepals fold up and form capsulelike receptacles. The fruits contained in the capsules are achenes, each containing a single seed. Each achene is ovate, 1–1.5 mm ($\frac{1}{32}$–$\frac{1}{16}$ in.) long and wide, pale brown, with a netlike pattern covering the surface. July–September.

Stems: In late spring, the plant sends out one or more upright stems that branch frequently near the top. The stems are light green or reddish and covered in long white hairs.

Root system: A taproot with short lateral branches. The woody root crown and stem base persist between growing seasons.

Reproduction: By seed, dispersed by gravity and animals. A single plant can produce approximately 1,600 seeds in a growing season. Seeds remain viable in the soil seed bank for at least three years.

Impact: This species forms dense stands and, due to its perennial root system, is difficult to eradicate. While not toxic, *P. recta* is unpalatable to livestock, allowing it to proliferate in areas with high levels of grazing. Its presence reduces the forage quality in pastures and rangeland.

Control: Hand pulling and hoeing are practical in small populations, but care must be taken to remove as much of the woody root system as possible. Cutting or mowing before flowering can reduce seed production. Spot application of 2,4-D or MCPA is often used in pastures and rangeland in the spring.

Potentilla recta flower. Image © Robert L. Carr.

Potentilla recta leaves. Image by Ohio State Weed Lab, The Ohio State University, Bugwood.org.

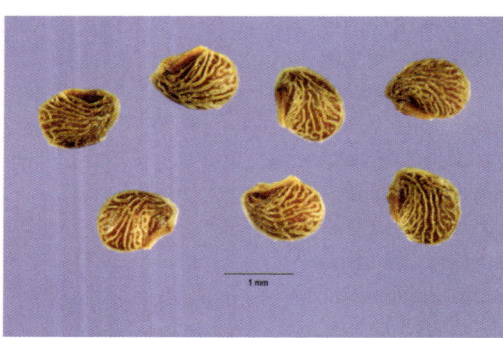

Potentilla recta achenes. Image by Steve Hurst, hosted by the USDA-NRCS PLANTS Database.

Potentilla recta plant. Image by USDA Agricultural Research Service, USDA Agricultural Research Service, Bugwood.org.

Potentilla recta plants in field. Image by Matt Lavin, used under a CC BY-SA 2.0 license.

Potentilla recta receptacles. Image © Peter M. Dziuk, Minnesota Wildflowers.

Rubiaceae

Galium mollugo

smooth bedstraw · hedge bedstraw · great hedge bedstraw · white bedstraw · whorled bedstraw · false baby's breath · cleavers · wild madder

Origin: Native to Europe, temperate parts of northern and western Asia, and northern Africa. It was most likely introduced to North America as an ornamental and was recorded as growing outside of cultivation by the late 1800s. This species now has a widespread distribution throughout the northeastern and midwestern United States.

Description: A long-lived perennial with a sprawling growth habit, whorled leaves, and clusters of tiny white flowers. There are 16 other species in this genus that are also found in Ohio—13 native and 3 nonnative—all of which have the same distinctive arrangement of whorled leaves at intervals along the stem. *Galium mollugo* can be differentiated from other related species by its perennial life history, sprawling habit, lack of hairs on the stem and leaves, whorls of six to eight leaves per node, and white flowers.

Habitat: Forest margins, open woodlands, thickets, fields, pastures, meadows, ditches, along railways and roadsides, and gardens. Grows in full sun to partial shade, typically in areas with a history of disturbance.

Height: 30–120 cm (1–4 ft.).

Foliage: Leaves are obovate to lance-shaped or linear, 1–3 cm (½–1 in.) long and 3–6 mm (⅛–¼ in.) wide, arranged in whorls of six to eight at intervals along the stems. The margins of the leaves are unlobed and can be lined with rough hairs 0.6–1 mm (¹⁄₆₄–¹⁄₃₂ in.) long. Each leaf has a single prominent vein along the center and ends in a rigidly pointed tip.

Flowers: White or greenish-white flowers are borne in loose panicles of numerous flowers, with spreading branches 5–15 cm (2–6 in.) long, arising from the upper leaf axils and at the ends of the stems. Individual flowers are 2–5 mm (¹⁄₁₆–¼ in.) in diameter, with four oblong petals that have blunt or pointed tips. In the center of each flower there are four stamens with yellow to reddish-brown anthers and two styles that are fused at the base. June–August.

Fruit: The dry fruit is smooth, 1–2 mm (¹⁄₃₂–¹⁄₁₆ in.) long and wide, yellowish green to reddish. The fruit splits into two sections when mature, each containing a single seed. The seeds are dark brown, kidney-shaped to spherical, 1–1.5 mm (¹⁄₃₂–¹⁄₁₆ in.) long and 0.5–1 mm (¹⁄₆₄–¹⁄₃₂ in.) wide, with a wrinkled surface. July–September.

Stems: The stems are weak, so they usually grow upright initially and then bend over when they get taller. Multiple stems often grow from the base and form tangled mats of vegetation. The stems are angular with four or five longitudinal ribs, smooth, mostly unbranched except in the upper nodes and flower clusters.

Root system: The root system is perennial, consisting of a highly branched taproot that is 3–5 mm (⅛–¼ in.) in diameter and can grow 50 cm (20 in.) deep. In addition, there are woody rhizomes, 1–4 mm (¹⁄₃₂–⅛ in.) in diameter and up to 5 cm (2 in.) deep, that spread laterally and produce new stems. The roots are typically reddish orange, while the rhizomes tend to be more yellow.

Native Alternatives
Euphorbia corollata (flowering spurge)
Galium boreale (northern bedstraw)
Galium concinnum (shining bedstraw)
Galium triflorum (fragrant bedstraw)
Gillenia trifoliata (bowman's root)

Reproduction: By seed, primarily dispersed by birds, water, and contaminated crop seed. A single plant can produce over 2,000 seeds in a growing season. There is no evidence that seeds remain viable in the soil seed bank beyond the first year. This species also reproduces vegetatively through spreading rhizomes.

Impact: This species, primarily a pasture weed, is unpalatable to livestock and outcompetes many important forage crops.

Control: Plants can be pulled or dug out, taking care to remove both the taproot and rhizomes. Frequent cutting or mowing will reduce seed production. This species has become resistant to several common herbicides—including 2,4-D and MCPA—so those specific products should be avoided. Spot application of triclopyr or picloram to cut stems can help prevent regrowth from the root system, but herbicides should not be used as the sole control agent over the long term.

Galium mollugo stems and leaves. Image © Peter M. Dziuk, Minnesota Wildflowers.

Left: Galium mollugo flowers. *Right: Galium mollugo* fruits. Both images by Ohio State Weed Lab, The Ohio State University, Bugwood.org.

Galium mollugo plants. Image by A. M. Liosi, used under a CC BY-SA 2.5 license.

Galium mollugo infestation in meadow. Image by Leslie J. Mehrhoff, University of Connecticut, Bugwood.org.

Galium mollugo seeds. Image by Ken Chamberlain, The Ohio State University, Bugwood.org.

Scrophulariaceae

Verbascum blattaria

moth mullein · white moth mullein · yellow moth mullein · slippery mullein

Origin: Native to Europe, temperate parts of northern and western Asia, and northern Africa. The timing and pathway of its introduction to North America are not clear, but there are records of *Verbascum blattaria* established in natural areas by 1818. It is now a common weed found throughout much of the United States, and it has been documented in every county in Ohio.

Description: A biennial plant that develops a rosette of leaves in the first growing season and overwinters in the rosette stage. In the second year, plants develop a single stem that bears large white or yellow flowers in a loose raceme.

Habitat: Forest margins, meadows, fields, pastures, along railways and roadsides, fencerows, and gardens. Grows in disturbed areas with full sun. Tolerant of a range of soil types and conditions.

Height: Flowering stems 60–150 cm (2–5 ft.).

Foliage: Basal leaves are oblong or lance-shaped, 10–20 cm (4–8 in.) long and 2–4 cm (¾–2 in.) wide, tapering at the base. The margins are variously toothed, lobed, or deeply divided. The leaves on the flowering stem are alternate, triangular to oblong or lance-shaped, 2–12 cm (¾–5 in.) long and 1–2 cm (½–¾ in.) wide, becoming progressively smaller farther up the stem. The stem leaves have shallowly toothed margins and pointed tips, and they clasp the stem at the base. Both the basal and stem leaves are hairless.

Flowers: There are white- and yellow-flowered forms of this species, both of which are found in Ohio. The flowers are held in a loose raceme, 10–60 cm (4–24 in.) long, at the top of the flowering stem. Flowers typically open sequentially from the bottom of the raceme up, so not all flowers are open at the same time. Individual flowers are 2–3 cm (¾–1 in.) wide, with five white or yellow petals that are fused at the base, five stamens with purple filaments and orange anthers, and a single pistil with a purple style and a green stigma. The upper two petals are slightly smaller than the lower three petals. The stamens are attached to the base of the petals, and the filaments are covered in fine purple hairs. At the base of the flower, the sepals are fused to form a calyx with five narrow, pointed lobes. Both the calyx and the flower stalks are covered in glandular hairs. June–September.

Fruit: The fruit is an ovoid or spherical capsule that develops within the persistent calyx, 5–8 mm (¼–⅜ in.) long and wide, with a persistent style at its tip. The capsule is covered in glandular hairs and is reddish green turning golden brown at maturity. It has two cells and it splits open at the top to release numerous seeds. Seeds are light to dark brown, nearly cylindrical, 0.6–1 mm (¹⁄₆₄–¹⁄₃₂ in.) long and 0.4–0.6 mm (¹⁄₆₄ in.) wide, with ridges on the surface. July–October.

Stems: Stems are upright, arising singly from the base, branching only occasionally near the top. The surfaces are smooth below and covered with glandular hairs around the inflorescence.

Root system: A deep taproot and fibrous secondary roots.

Reproduction: By seed, dispersed by gravity, wind, or animals. A single plant can produce over 1,000 capsules, each of which contains numerous seeds. In a famous and long-running experiment started by William Beal in 1879, the seeds of this species were found to remain viable in the soil seed bank for over 120 years.

Impact: This species is primarily a weed of pastures, fields, and disturbed areas such as roadsides. The long-term viability of seeds in the soil seed bank make this species difficult to eradicate once it has established and set seed.

Control: The key to controlling this species is to prevent seed set. Individual plants can be removed by digging or hoeing, taking care to cut the taproot below the root crown. Larger stands can be mowed close to the ground several times per growing season. While they are still young, plants can be spot treated with 2,4-D or glyphosate.

Overwintering *Verbascum blattaria* rosette. Image by Salicyna, used under a CC BY-SA 4.0 license.

Yellow flowers of *Verbascum blattaria*. Image © Gerald D. Carr.

Left: Verbascum blattaria stem and leaves. Image © Gerald D. Carr. *Right: Verbascum blattaria* fruit. Image © Keir Morse, Minnesota Wildflowers.

Verbascum blattaria seeds. Image by Steve Hurst, hosted by the USDA-NRCS PLANTS Database.

White flowers of *Verbascum blattaria*. Image by Ohio State Weed Lab, The Ohio State University, Bugwood.org.

Glandular hairs on flower stalk and calyx of *Verbascum blattaria*. Image © Peter M. Dziuk, Minnesota Wildflowers.

Scrophulariaceae

Verbascum thapsus

common mullein · great mullein · woolly mullein
· flannel plant · velvet plant · velvet dock ·
mullein dock · big taper

⚠ **Do Not Touch**

Origin: Native to Europe, western and central Asia, and northern Africa. *Verbascum thapsus* has a long history of use as a medicinal herb, and it was likely introduced to North America many times, both intentionally and accidentally. It was so well established by 1818 that it was mistakenly recorded as a native plant. This species, which grows abundantly throughout the continental United States, has been documented in every county in Ohio.

Description: A biennial plant with distinctive velvety leaves, a tall stem, and yellow flowers borne in a dense spike. It forms a rosette of leaves in the first growing season and develops a single flowering stem in the second year.

Habitat: Meadows, fields, pastures, along railways and roadsides, fencerows, and gardens. Grows in disturbed areas with full sun, primarily in sites with dry and stony soils.

Height: Flowering stems 30–250 cm (1–8 ft.).

Foliage: Basal leaves are oblong to obovate, 8–40 cm (3–16 in.) long and 3–14 cm (1–5 ½ in.) wide, with a rounded tip and a tapered base. The leaves on the flowering stem are alternate, similar to the basal leaves, becoming progressively smaller farther up the stem. The lower stem leaves taper to form a winged base that extends down the stem, while the upper stem leaves attach directly. Both the basal and stem leaves are densely covered in woolly white hairs, giving them a grayish-green color and a texture like velvet. The leaf margins can be wavy or edged with small, rounded teeth.

Flowers: Yellow flowers are borne in a dense spike, 20–50 cm (8–20 in.) long and 3 cm (1 in.) in diameter, at the top of the flowering stem. Typically only a few flowers are open at

one time, and each flower only lasts a single day. Individual flowers are 1–2 cm (½–¾ in.) in diameter, with five petals fused at the base, five stamens with yellow filaments and orange anthers, and a single green pistil. The upper two petals are slightly smaller than the lower three, the stamens are attached to the bases of the petals, and the upper three filaments are shorter and covered in fine yellow hairs. At the base of the flower, the sepals are fused to form a calyx with five lobes. June–September.

Fruit: The fruit is an ovoid capsule that develops within the calyx, 6–10 mm (¼–½ in.) long and wide, with a persistent style at its tip. The capsule is green and turns golden brown at maturity. It has two cells and, when mature, splits open at the top to release numerous seeds. Seeds are light to dark brown, nearly cylindrical, 0.6–0.8 mm (¹⁄₆₄ in.) long and 0.4– 0.6 mm (¹⁄₆₄ in.) wide, with ridges on the surface. The clublike spike of fruits often persists over winter and remains standing the next growing season. July–October.

Stems: Stems are stout, upright, arising singly from the base and only occasionally branching near the top. Woolly white hairs densely cover the surfaces of the stems.

Root system: A deep taproot.

Reproduction: By seed, dispersed by gravity, wind, or animals. A single plant can produce 100,000 to 180,000 seeds. Seeds of this species can remain viable in the soil seed bank for several decades.

Impact: Primarily a weed of disturbed sites with low fertility. It is unpalatable to livestock and can form dense patches that outcompete more desirable forage species. The long-term viability of seeds in the soil seed bank make this species difficult to eradicate once it has established and set seed.

Control: The hairs on the stems and leaves can cause contact dermatitis, so wear protective clothing and gloves when undertaking control measures. The key to controlling this species is to prevent seed set. Individual plants can be removed by digging or hoeing, taking care to cut the taproot below the root crown. Larger stands can be mowed close to the ground several times per growing season. Plants can be spot treated with 2,4-D, glyphosate, or triclopyr while in the rosette stage.

Verbascum thapsus rosette. Image by Bonnie Million, Bureau of Land Management, Bugwood.org.

Verbascum thapsus fruits. Image by Bonnie Million, Bureau of Land Management, Bugwood.org.

Left: Verbascum thapsus plants. Image by Chris Evans, University of Illinois, Bugwood.org. *Right: Verbascum thapsus* flower stalks. Image by David Cappaert, Bugwood.org.

Verbascum thapsus seeds. Image by Steve Hurst, hosted by the USDA-NRCS PLANTS Database.

Verbascum thapsus flowers. Image by Forest Starr and Kim Starr, Starr Environmental, Bugwood.org.

Verbascum thapsus stand. Image by Katja Schulz, used under a CC BY 2.0 license.

Solanaceae

Nicandra physalodes

apple-of-Peru · shoofly plant

Origin: Native to South America, *Nicandra physalodes* was most likely introduced to North America as an ornamental species, as well as for medicinal and insecticidal purposes. This species has escaped from cultivation and established scattered populations throughout much of the United States.

Description: An annual plant with a bushy form, light blue to lavender flowers, and dry fruits enclosed in a papery husk. It is similar to members of the genus *Physalis* (ground cherries), of which there are both native and non-native species in Ohio. However, *Physalis* species have smaller white or yellow flowers, and the fruits inside the papery husks are fleshy and edible.

Habitat: Agricultural fields, meadows, ditches, roadsides, and gardens.

Height: 1–2 m (3–6 ft.).

Foliage: Leaves are alternate, ovate, 6–25 cm (3–10 in.) long and 3–18 cm (1–7 in.) wide, with a slender petiole, 1–20 cm (½–8 in.) long, and a rounded to heart-shaped base. The margins of the leaves can be shallowly lobed, bluntly toothed, or wavy. The surfaces of the leaves and petioles are sparsely hairy.

Flowers: Light blue to lavender flowers grow singly from the axils of the upper leaves on stalks 1–4 cm (½–2 in.) long. The flowers are open during the day and close at night. Individual flowers have five petals that are fused into a bell-shaped corolla, 2–4 cm (¾–2 in.) in diameter. The corolla is white in the center, occasionally with purple markings. Each flower contains five stamens with light yellow anthers and a single yellow pistil. The base of the flower is surrounded by five arrow-shaped sepals, 1–3 cm (½–1 in.) long, that form a persistent calyx. July–September.

Fruit: The fruit is a yellow berry, nearly spherical, 1–1.5 cm (½–⅝ in.) in diameter, dry, and divided into three to five cells, each containing numerous seeds. The fruit is surrounded by a husk formed from the calyx, green turning brown and papery at maturity. The seeds are kidney-shaped, 1–1.3 mm ($\frac{1}{32}$ in.) in diameter, flattened, orange to brownish orange, with a pitted surface. August–October.

Stems: The stems are upright, hollow, and much branched. The surfaces are ribbed, typically five-angled in cross section, and hairless.

Root system: A taproot with fibrous secondary roots.

Reproduction: By seed, primarily dispersed by soil and water movement. A single individual can produce 40,000 seeds in a growing season. Seeds remain viable in the soil seed bank for at least five years.

Impact: This rapidly growing species forms a dense canopy, making it very competitive. It is considered a serious weed in a wide range of crop types, including cereals, legumes, and vegetables. In addition, *N. physalodes* is a secondary host for many plant pathogens and pests. It is primarily a concern in agricultural settings but is also frequently found in natural areas adjacent to gardens from which it has escaped.

Control: The seeds of *N. physalodes* can germinate continuously throughout the growing season, making it a difficult species to manage. Young plants can be controlled by light tilling, cutting, mowing, or hoeing. The herbicides 2,4-D and glyphosate can be used for spot treatment.

Nicandra physalodes leaves. Image by Bruce Ackley, The Ohio State University, Bugwood.org.

Nicandra physalodes flower. Image by Boronian, used under a CC BY-SA 3.0 license.

Calyx at base of *Nicandra physalodes* flower. Image by Harry Rose, used under a CC BY 2.0 license.

Nicandra physalodes growing along railway line. Image by Teun Spaans, used under a CC BY-SA 3.0 license.

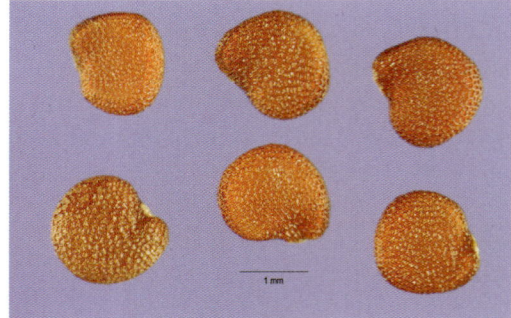

Nicandra physalodes seeds. Image by Steve Hurst, hosted by the USDA-NRCS PLANTS Database.

Nicandra physalodes fruit. Image by Bruce Ackley, The Ohio State University, Bugwood.org.

Solanaceae

Solanum dulcamara

bittersweet nightshade · bitter nightshade
· bittersweet · trailing bittersweet · trailing
nightshade · climbing nightshade · woody
nightshade · blue nightshade · blue bindweed ·
fellenwort · felonwood · poisonberry · poisonflower
· scarlet berry · snakeberry · violet bloom

195

Origin: Native to much of Europe, temperate
Asia, and northern Africa. *Solanum dulcamara*
was most likely introduced to North America
for ornamental or medicinal purposes, and
it had established outside of cultivation by
the mid-1800s. Shipping and agriculture further spread this species; it is now established
through much of the United States except for
the southern regions.

Description: A climbing or trailing perennial with deeply lobed leaves, star-shaped purple flowers, and bright red berries. The stems of
S. dulcamara, which become semiwoody with
age, can persist over winter. There are several
other native and nonnative species of *Solanum*
in Ohio, but *S. dulcamara* is the only one with
a climbing or trailing habit, semiwoody stems,
deeply lobed leaves, and purple flowers.

Habitat: Forests, forest margins, river- and
streambanks, lakeshores, wetland margins,
ditches, roadsides, fencerows, and gardens.
Grows in full sun to full shade. This species prefers moist soils but can tolerate dry conditions.

Height: 1–3 m (3–10 ft.), occasionally taller
when growing with support.

Foliage: Leaves are alternate, ovate or arrow-
shaped, 4–12 cm (2–5 in.) long and 2–9 cm
(¾–3 ½ in.) wide, each tapering to a pointed
tip, often with two or more deep lobes at the
base. The leaves have slender petioles, and the
margins of the leaves are smooth. The leaves
are dark green, sometimes with a purplish
tinge, typically hairless or occasionally with
scattered hairs on the upper surface. The foliage gives off a rank odor when damaged.

Flowers: Branching clusters of 10–25 star-
shaped, purple flowers are produced along the
main stems and in the upper leaf axils. Individual flowers are 8–12 mm (¼–½ in.) in diameter, with five triangular petals that are fused at
the base and spread outward or curve sharply
backward. At the base of each petal are two
greenish-white dots. In the center of the flower
is a pointed column consisting of five fused
stamens with bright yellow anthers and a single style that extends above the anthers. At
the base of the flower there are five smooth
or slightly hairy sepals fused into a calyx with
short triangular lobes. May–September.

Fruit: The fruit is an ovoid berry, 10–15
mm (½–⅝ in.) long and 7–10 mm (¼–½ in.)
in diameter, held on the persistent calyx. The
fruits are green, turning yellow or orange as
they ripen, then bright red at maturity. Each
contains numerous yellow seeds, 2–3 mm
(¹⁄₁₆–⅛ in.) in diameter, nearly circular and
strongly flattened. July–October.

Stems: Stems are much branched and can be
upright, scrambling, or climbing. The stem surfaces are smooth or sparsely hairy and green
to purple in color. Creeping stems can root at
the nodes where they come into contact with
the soil. The lower parts of the stems become
woody with age and persist over winter, while
the leafy herbaceous stems die back.

Root system: A rhizomatous root system that
becomes woody with age.

Reproduction: By seed, primarily dispersed
by birds and mammals. Seeds do not persist
in the soil seed bank. Vegetative reproduction
occurs through spreading rhizomes, the rooting

of stems at the nodes, and through the movement of root and stem fragments.

Impact: Forms dense stands that climb overtop and outcompete native vegetation. All plant parts contain solanine, a toxic glycoalkaloid. The attractive red berries are not as poisonous as those of the related *Atropa belladonna* (deadly nightshade), but they can cause severe gastrointestinal and neurological symptoms if consumed in quantity. In agricultural settings, *S. dulcamara* is problematic as a secondary host for pests and diseases that affect crops in the same plant family (Solanaceae), which includes potatoes, tomatoes, eggplants, and peppers.

Control: Seedlings can be pulled by hand. Once plants are established, stems should be removed and the root system dug out. Regrowth can be controlled by covering the area with a thick layer of mulch or wood chips. If the root system is persistent, spot application of 2,4-D may be effective at killing the rhizome.

Solanum dulcamara leaves. Image by Leslie J. Mehrhoff, University of Connecticut, Bugwood.org.

Solanum dulcamara fruits. Image by Jan Samanek, Phytosanitary Administration, Bugwood.org.

Solanum dulcamara seeds. Image by Steve Hurst, hosted by the USDA-NRCS PLANTS Database.

Flowering *Solanum dulcamara* plant. Image by Leslie J. Mehrhoff, University of Connecticut, Bugwood.org.

Solanum dulcamara flowers. Image by David Cappaert, Bugwood.org.

Semi-woody stems of *Solanum dulcamara* in winter. Image by Leslie J. Mehrhoff, University of Connecticut, Bugwood.org.

Xanthorrhoeaceae

Hemerocallis fulva

tawny daylily · orange daylily · common daylily · fulvous daylily · tiger daylily · ditch lily · railroad daylily · roadside daylily · outhouse lily · wash-house lily · Eve's thread

Origin: Native to eastern Asia, but widely cultivated and established in Europe and other parts of Asia. The roots and flowers have a long history of use as medicine and food. This species was introduced to North America in the 1600s as an ornamental plant. *Hemerocallis fulva* has escaped cultivation and naturalized throughout the eastern United States.

Description: A bulbous perennial with large trumpet-shaped orange flowers and long strap-like leaves. A similar nonnative species—*Hemerocallis lilioasphodelus* (yellow daylily)—is also naturalized in Ohio, but it is much less common and can be differentiated by its yellow flowers. There are also several native *Lilium* species in Ohio that have whorled leaves on their flowering stems and spotted flowers that last longer than a single day, which distinguishes them from *H. fulva*.

Habitat: Forests, forest margins, floodplains, ditches, along railways and roadsides, and meadows. Prefers full sun to partial shade, with moist and fertile soils.

Height: Flowering stems typically 70–150 cm (2–5 ft.) tall.

Foliage: The basal leaves are long and linear, 30–90 cm (1–3 ft.) long and 1–3 cm (½–1 in.) wide, tapering to points at the end. The leaves have parallel veins, are folded along the center, and often bend over and curve toward the ground. Leaves are hairless, and the margins are smooth.

Flowers: A flowering stem develops from the center of the basal leaves. At the end of the stem is a cluster of 5–20 flowers that open one at a time through the season, each lasting a single day. The flowers are tawny orange, 12 cm (5 in.) in diameter, funnel-shaped, with six spreading tepals (three petals and three modified sepals that look like petals). Each individual flower has a single white to pale orange style, 9–10 cm (3 ½–4 in.) long, and six stamens with curving orange filaments, 4.5–6.5 cm (2–3 in.) long, topped with large purplish-black anthers, 5–7 mm (¼ in.) long. Flowers are not fragrant. June–July.

Fruit: The fruit is an ellipsoid capsule, 2–2.5 cm (¾–1 in.) long and 1.2–1.5 cm (½–⅝ in.) in diameter, green turning brown at maturity, that splits open along three seams to release the seeds. Each capsule has three cells that contain rows of dark brown to black seeds, 4–5 mm (⅛–¼ in.) long and 2–3 mm (1/16–⅛ in.) wide, typically with two flat sides and one rounded side like a segment of an orange.

Stems: Flowering stems are upright, round in cross section, and hollow. Unlike the true lilies, the stem of *H. fulva* lacks leaves, but there are a few small lance-shaped bracts that arise individually along the length of the stem.

Root system: The root system is dense, consisting of fleshy, spindle-shaped roots that resemble tubers and thinner fibrous roots. There are also laterally spreading rhizomes.

Reproduction: Primarily by rhizome and root fragments. Many sources suggest that seed is not

Native Alternatives
Lilium canadense (Canada lily)
Lilium michiganense (Michigan lily)
Lilium philadelphicum (wood lily)
Lilium superbum (turk's cap lily)

Hemerocallis fulva leaves. Image by Magnus Manske, used under a CC BY-SA 3.0 license.

Hemerocallis fulva infestation along roadside. Image by David Stephens, Bugwood.org.

Hemerocallis fulva flower. Image by Corey Leopold, used under a CC BY 2.0 license.

Stem and bract of *Hemerocallis fulva.* Image by Dan Tenaglia, Missouri plants.com, Bugwood. org.

Tuberous roots of *Hemerocallis fulva.* Image by Ohio State Weed Lab, The Ohio State University, Bugwood.org.

Hemerocallis fulva seeds. Image by Ken Chamberlain, The Ohio State University, Bugwood.org.

produced, or at least not viable, in this species. However, some seeds are produced, and reproduction by seed may occur infrequently.

Impact: Forms extremely dense clumps that prevent other species from growing. Once it has established, it is difficult to eradicate, due to the extensive root system. This species is most problematic when growing adjacent to natural areas into which it can spread, displacing native species.

Control: Stands of *H. fulva* can be dug up or plowed in the autumn, then raked to remove any tuberous roots. Larger stands can be mowed and covered with mulch or black plastic sheeting to prevent regrowth. Individual plants can be cut and spot treated with glyphosate. Due to the persistent roots, any control measures may require follow-up treatment.

Creepers and Climbers

Creepers and climbers are especially damaging to natural communities because they can form dense blankets of groundcover and overtop taller vegetation such as shrubs and trees. They often physically damage other plants with their weight and make them more susceptible to damage by wind or ice. Many problematic creepers and climbers also have very large root systems or underground storage organs—as found in *Dioscorea polystachya* (Chinese yam) [pp. 231–232]—that make them extremely difficult to eradicate.

Creepers and climbers have relatively weak stems and require support from other plants or structures to climb upward. Most of these plants are not parasitic—the exception in this book being *Cuscuta epithymum* (clover dodder) [pp. 227–228]—and merely use other plants for support rather than extract water or nutrients from them. Creepers and climbers use different strategies to attach to their supports, and this feature can help with species identification. These attachment strategies include

1. Rootlets: adventitious roots that grow into cracks for climbing support, for example, *Hedera helix* (English ivy) [pp. 204–206]. These species significantly damage buildings and other structures.

2. Tendrils: specialized shoots or inflorescences that wrap around other plants or objects for support, for example, *Ampelopsis brevipedunculata* (porcelain berry) [pp. 245–246].

3. Thorns, spines, or prickles: modified plant structures that are woody at maturity and have a sharp point, for example, *Persicaria perfoliata* (mile-a-minute weed) [pp. 241–242]. These can hook onto objects and provide climbing support.

4. Leaf twining: petioles twine around other plants and structures for support, for example, *Clematis terniflora* (sweet autumn clematis) [pp. 243–244].

5. Stem twining: stems wrap around supports. Stems twine directionally, and the direction of twining is fixed within a given species. Sometimes this direction is described as clockwise or counterclockwise, which is confusing, because the same helix can be seen as both clockwise and counterclockwise, depending whether the viewer looks from above or from below. We recommend observing how the stem ascends its vertical support when viewed facing the plant straight on from the front. Some stems ascend from the left to the right, *dextrorse* in modern botanical terminology (Fig. 10a). Contrastingly, some stems ascend from the

right to the left, *sinistrorse* (Fig. 10b). An example of a dextrorse species is *Convulvulus arvensis* (field bindweed) [pp. 225–226] and a sinistrorse species is *Lonicera japonica* (Japanese honeysuckle) [pp. 214–216]. Primarily because this feature can aid in identification, it is important to recognize that species twine in different directions.

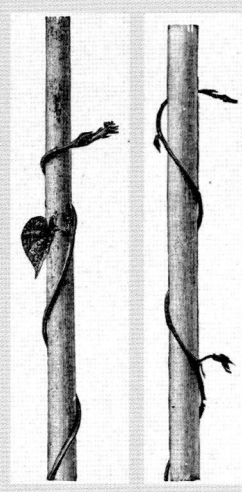

Fig. 10a. *(Left)* A twining stem ascending its support from the left to the right (dextrorse). Image from Bower (1919).

Fig. 10b. *(Right)* A twining stem ascending its support from the right to the left (sinistrorse). Image from Bower (1919).

Apocynaceae

Vinca minor

common periwinkle · lesser periwinkle · small periwinkle · dwarf periwinkle · myrtle · creeping myrtle · running myrtle · vinca

Origin: Native to Europe and western Asia. This species, which has a long history of cultivation, was introduced to North America in the 1700s. It frequently escapes plantings and has become problematic in some natural communities, particularly in the northeastern and midwestern United States. Despite its weedy tendencies, it is still commercially available and commonly used as a groundcover in landscaping.

Description: A low-growing, evergreen, perennial ground cover. It has trailing or prostrate stems that form a dense mat. The stems and petioles exude a milky sap when broken. It is most likely to be confused with the closely related nonnative species *Vinca major* (greater periwinkle). *Vinca minor* has smaller leaves and flowers and hairless leaf margins, whereas *V. major* has larger leaves and flowers and hairy leaf margins.

Habitat: Open woods, forest edges, roadsides, and fields. Most commonly in disturbed areas close to human habitation. Prefers well-drained soils in open to shady sites.

Height: Rarely exceeds a vertical height of 8–15 cm (3–6 in.) but with support occasionally scrambles to 40 cm (16 in.). Stems spread along the ground up to about 1 m (3 ft.).

Foliage: Simple, opposite, 2–5 cm (¾–2 in.) long and 1–2 cm (½–¾ in.) wide, ovate or lance-shaped, each leaf with a blunt tip and a rounded or flat base. Leaves have a leathery texture. The upper surface is dark green and glossy, with a light colored midvein and a light underside, although some horticultural varieties have variegated leaves. Leaf margins are smooth and hairless.

Flowers: Typically blue, lavender, or violet, but some cultivars have white or pink flowers. Each flower is 2–3 cm (¾–1 in.) in diameter, consisting of five spreading, pinwheel-like lobes with blunt tips and a short tubular throat, 8–12 mm (¼–½ in.) long. Flowers are solitary and grow from the leaf axils. April–May, with occasional flowers produced sporadically through summer and into autumn.

Fruit: Fruits rarely develop in cultivated plants. When they do, the fruit is a long and slender green follicle, 1–3 cm (½–1 in.) long and 2–4 mm ($\frac{1}{16}$–⅛ in.) wide, that splits open along one side. The follicles develop in pairs from an individual fertile flower. Each follicle contains 3–5 small brown seeds, 5–6 mm (¼ in.) long by 2 mm ($\frac{1}{16}$ in.) wide. June–July.

Stems: Stems are slender and hairless, green to light brown, fibrous, with a vinelike habit. Stems develop adventitious roots where the nodes come into contact with the ground.

Root system: Woody rhizomes with slender secondary roots that form a dense mat. Also produces underground runners.

Reproduction: Sexual reproduction is not common, but seeds do occasionally develop. Most reproduction is vegetative, through rhizomes and the rooting of stems and stem fragments.

Native Alternatives
Arctostaphylos uva-ursi (bearberry)
Gaultheria procumbens (wintergreen)
Mitchella repens (partridgeberry)
Pachysandra procumbens (Allegheny spurge)
Paxistima canbyi (cliff green)
Waldsteinia fragarioides (barren strawberry)

Impact: Once established, this species can persist for many decades, with single clones spreading vegetatively and forming dense mats that dominate the forest understory. These stands can displace native herbaceous plants and suppress the growth of native tree seedlings, which has long-term consequences for forest regeneration.

Control: This species primarily spreads from cultivated populations, so the most effective method to prevent further establishment is to avoid planting it. Existing garden plantings can be replaced with native alternatives. Established plants can be removed by pulling or digging, taking care to remove all the roots to prevent resprouting. Small stands can be solarized with plastic sheeting. Extensive populations can be handled with repeated mowing, followed with a spot treatment of glyphosate to prevent regrowth.

Vinca minor leaves. Image by Chris Evans, University of Illinois, Bugwood.org.

Leaf variation in cultivated forms of *Vinca minor*. Image by James H. Miller, USDA Forest Service, Bugwood.org.

Infestation by *Vinca minor* in forest understory. Image by Robert Vidéki, Doronicum Kft., Bugwood.org.

Calyx at the base of *Vinca minor* flowers. Image © Gerald D. Carr.

Vinca minor seeds. Image by Steve Hurst, hosted by the USDA-NRCS PLANTS Database.

Vinca minor flowers. Image by Robert Vidéki, Doronicum Kft., Bugwood.org.

Araliaceae

Hedera helix

English ivy · European ivy · common ivy

Origin: Native to Europe, western Asia, and northern Africa. This species, which has a long history of cultivation, was introduced to North America by colonial settlers as early as 1727. *Hedera helix* is widely planted as a groundcover and has escaped cultivation in many eastern, midwestern, and pacific coast states. Although recognized as problematic in many natural ecosystems, it continues to be sold as an ornamental plant.

Description: A woody, evergreen, perennial vine that forms a dense groundcover and can climb trees, buildings, and other structures. It attaches using adventitious roots that exude an adhesive substance, which makes it difficult to remove without causing damage. This species has two distinct growth phases: juvenile (vegetative), during which stems scramble or climb and produce the lobed leaves typically associated with ivy, and adult (reproductive), when the stems produce ovate leaves, flowers, and fruits.

Habitat: Upland forest, forest edges, fields, roadsides, and areas of habitation. Grows in a range of soil pH but prefers slightly acidic, well-drained soils. Tolerates drought and heavy shade.

Height: To 20 cm (8 in.) as a groundcover; vines can climb to 30 m (100 ft.) with support.

Foliage: Simple, alternate, 5–10 cm (2–4 in.) long and 6–13 cm (3–5 in.) wide, with two different forms, depending on the reproductive phase. Juvenile leaves on the climbing shoots are three- or five-lobed, dark green, with whitish or yellowish palmate veins and heart-shaped leaf bases. Adult leaves on the flowering shoots are unlobed, ovate to diamond- or heart-shaped with a sharp point, lighter green with less prom-

inent veins. Leaves of both phases are leathery and covered in a thick, waxy cuticle.

Flowers: Small white or yellow-green flowers, 5–7 mm (¼ in.) in diameter, in starburst-shaped clusters of 8–20 flowers. Flower clusters develop at the ends of the reproductive shoots, either solitary or in groups of three to six clusters on a branching stalk. Each flower has five petals, five stamens, and one style. September–October.

Fruit: A berrylike drupe, dark blue to purple or black, nearly spherical, 5–9 mm (¼–⅜ in.) long and 6–9 mm (¼–⅜ in.) in diameter. Each fruit contains two or three hard stones, 5 mm (¼ in.) long and 3 mm (⅛ in.) wide, each of which encloses a single seed. Fruits develop over winter and mature in April to May of the following year.

Stems: Young shoots and inflorescences are densely covered with hairs. The bark of young shoots is green to reddish brown, whereas climbing and older trailing branches develop light brown, ridged, and furrowed bark. Older trailing stems have been recorded as growing to 10–30 cm (4–12 in.) in diameter.

Root system: Does not develop an extensive underground root system. Creeping juvenile stems develop shallow roots at the nodes where they touch the soil. Climbing juvenile stems produce adventitious roots that attach to structures using root hairs and an adhesive exudate. Stems in the adult phase do not produce roots.

Native Alternatives
Ampelopsis arborea (peppervine)
Bignonia capreolata (crossvine)
Chrysogonum virginianum (green-and-gold)

Left: Juvenile leaves of *Hedera helix.* Image by Chuck Bargeron, University of Georgia, Bugwood. org. *Right: Hedera helix* stem. Image by James H. Miller, USDA Forest Service, Bugwood.org.

Flowers of *Hedera helix.* Image by Jan Samanek, Phytosanitary Administration, Bugwood.org.

Mature leaves of *Hedera helix.* Image by James H. Miller, USDA Forest Service, Bugwood.org.

Hedera helix seeds. Image by Steve Hurst, hosted by the USDA-NRCS PLANTS Database.

Hedera helix fruits. Image by Robert Vidéki, Doronicum Kft., Bugwood. org.

Hedera helix infestation. Image by Chris Evans, University of Illinois, Bugwood.org.

Reproduction: Both vegetative and by seed, which is spread by birds. This species does not form a persistent soil seed bank.

Impact: Forms dense colonies that outcompete understory and shrub species and prevent regeneration by trees. The weight of vines on overgrown trees makes them susceptible to wind damage. The sap of *H. helix* can cause allergic contact dermatitis, and the fruits and other plant parts are mildly toxic. This species is a reservoir for the pathogen *Xylella fastidiosa* (bacterial leaf scorch), a threat to native trees.

Control: Avoid growing *H. helix* near forested areas. Where it has escaped cultivation and needs control, vines can be dug out, cut back, and pulled from trees and the forest floor. To exhaust energy reserves in any roots or plant fragments that remain, repeated cutting may be necessary. Mulching can prevent regrowth at the ground level. Metsulfuron is the most effective herbicide for control of this species. Stump treatments with triclopyr, 2,4-D, and glyphosate offer some control but are not as effective overall.

Asclepiadaceae

Vincetoxicum nigrum

black swallowwort · Louis' swallowwort ·
Louise's swallowwort · black dog-strangling vine
· climbing milkweed

Synonym: *Cynanchum louiseae*

Origin: Native to southwestern Europe. It is likely that this species was introduced to the United States as a garden specimen, but the exact year and place of introduction are not known. It was first recorded as having escaped into natural areas in 1854 in Ipswich, Massachusetts. The species was cultivated and sold as an ornamental for many years, which facilitated its spread and establishment in the northeastern and midwestern United States.

Description: A herbaceous, perennial vine with unbranched stems and a twining habit. It lacks tendrils, and its stems typically twist around themselves or other vegetation for support. Shoots die back every winter and resprout from the persistent root crown the following growing season. Plant parts exude a milky sap when damaged. Two other nonnative species of swallowwort are problematic in the northeastern United States—*Vincetoxicum rossicum* (pale swallowwort) and *Vincetoxicum hirundinaria* (white swallowwort)—but *Vincetoxicum nigrum*'s dark purple to black flowers differentiate it from them. The native *Cynanchum laeve* (smooth swallowwort or honeyvine) has white flowers and leaves with a heart-shaped base.

Habitat: Primarily found in upland forests and open disturbed areas, such as fields, roadsides, forest edges, and abandoned homesteads. Prefers moist and sunny conditions but can grow in dry or heavily shaded areas.

Height: 1–3 m (3–10 ft.).

Foliage: Simple, opposite, 5–12 cm (2–5 in.) long and 2–7 cm (¾–3 in.) wide, oblong to ovate, with a pointed tip and a flat or rounded base. Margins are entire and leaves attach to

the vine by a short petiole. Leaves are glossy and dark green, with a thick waxy coating.

Flowers: Dark purple to black with a pale yellow center, 6 mm (¼ in.) in diameter, with five triangular petals that give the flowers a distinctive star shape. Petals are fleshy and covered with tiny white hairs on the upper surface. Flowers appear in clusters of 6–10 in the axils of leaves. June–September.

Fruit: A slender and tapering pod, 4–8 cm (2–3 in.) long and 1 cm (½ in.) wide, often occurring in pairs. The pods are green, turning light brown at maturity, when they split open lengthwise to release seeds. Seeds are brown, flattened, ovoid, 5–8 mm (¼ in.) long and 3–5 mm ($^1/_{16}$–¼ in.) wide, with a membranous wing along the margin and tufts of white hair at the narrow end that aid in wind dispersal. August–October.

Stems: The slender green stems can be creeping or climbing. Stems twine upward, from left to right. The stems wind around themselves and other plants for support, often creating tangled and impenetrable thickets.

Root system: Extensive root system with fibrous roots and thick rhizomes. Each plant has a perennial root crown with subterranean buds that produce new shoots every growing season and readily resprout after damage.

207

> **Native Alternatives**
> *Asclepias verticillata* (whorled milkweed)
> *Asclepias incarnata* (swamp milkweed)
> *Cynanchum leave* (smooth swallowwort or honeyvine)

Reproduction: By spreading rhizomes, sprouting from the root crown, and abundant wind-dispersed seeds that persist in the soil seed bank for two years.

Impact: This species is problematic in horticultural nurseries, perennial field crops, pasturelands, and natural areas subject to high levels of disturbance. There is evidence that this species is allelopathic, inhibiting the germination and growth of other plants through the production of toxic chemicals. Once established, it forms dense thickets and displaces native species, including the native milkweeds that are host plants for monarch butterflies. Monarchs seldom lay eggs on *V. nigrum,* but they may if there are neighboring populations of the normal host, *Asclepias syriaca* (common milk-weed). When monarch butterflies lay eggs on *V. nigrum,* the larvae do not survive, which could ultimately pose a serious threat to monarch populations in areas invaded by this species.

Control: Control measures are most effective if performed before seedpods mature. Cutting or mowing the shoots as fruit begins to enlarge will prevent maturation of the seed crop. However, resprouting will occur from the root crown, so repeated cutting or mowing will be necessary. To prevent resprouting, manual control should ultimately include the removal of the root crown and rhizomes. Systemic herbicides such as triclopyr and glyphosate are most effective if applied to stems that have been cut after flowering has occurred but before seedpods mature.

Vincetoxicum nigrum leaves. Image by Leslie J. Mehrhoff, University of Connecticut, Bugwood.org.

Vincetoxicum nigrum flowers. Image by Leslie J. Mehrhoff, University of Connecticut, Bugwood.org.

Twining stems of *Vincetoxicum nigrum.* Image by Leslie J. Mehrhoff, University of Connecticut, Bugwood.org.

Developing *Vincetoxicum nigrum* fruits. Image by Leslie J. Mehrhoff, University of Connecticut, Bugwood.org.

Mature *Vincetoxicum nigrum* fruits. Image by Leslie J. Mehrhoff, University of Connecticut, Bugwood.org.

Vincetoxicum nigrum roots. Image by Leslie J. Mehrhoff, University of Connecticut, Bugwood.org.

Vincetoxicum nigrum seeds. Image by Bruce Ackley, The Ohio State University, Bugwood.org.

Infestation of *Vincetoxicum nigrum.* Image by Leslie J. Mehrhoff, University of Connecticut, Bugwood.org.

209

Buxaceae

Pachysandra terminalis

Japanese pachysandra · Japanese spurge ·
carpet box · Chinese fever vine

Origin: Native to Japan, Korea, and northern China. It was introduced to the United States in 1882 as an ornamental. *Pachysandra terminalis* is frequently planted as a groundcover in cultivated landscapes.

Description: A low-growing, evergreen, perennial groundcover. The species name *terminalis* refers to the fact that the flower stalk is at the end of the stem. This is a popular garden plant because it forms a dense carpet of vegetation that fills in bare ground; however, because it spreads aggressively it can become problematic even in a garden setting.

Habitat: Deciduous woodlands, forest, and meadow edges. It grows best in light shade with rich and moist soil but can tolerate deep shade.

Height: 10–30 cm (4–12 in.).

Foliage: Simple, alternate along the stem but resembling whorls at the ends of shoots, 4–10 cm (2–4 in.) long and 1–3 cm (½–2 in.) wide, oval or diamond-shaped, with three prominent veins at the base. The leaf margins are coarsely serrated around the tip. Leaves are somewhat stiff and leathery, glossy, dark green.

Flowers: Separate male and female flowers occur together in a single, terminal spike held above the leaves. A typical spike is about 2–5 cm (¾–2 in.) long, with 15–25 male flowers toward the top and one or two female flowers at the base. Individual flowers have four elongated white petals, each 8 mm (¼ in.) long, and have a sweet fragrance. March–April.

Fruit: Spherical white berries, 1–1.5 cm (½–⅝ in.) in diameter, containing glossy brown, teardrop-shaped seeds, each about 5 mm (¼ in.) long and 3 mm (⅛ in.) wide. Fruits are rarely produced, which suggests that plants may be self-incompatible. Garden stands are often derived from a single clone, making them genetically identical. Fruit and seed production are more likely to occur if several different clones are represented in an area.

Stems: The semiwoody stems remain green. Vertical shoots are upright, but as the stems age they can become brittle and break off or become top heavy and bend over, both of which contribute to the vegetative spread of this species.

Root system: Extensive rhizomes that form dense colonies.

Reproduction: Primarily vegetative through spreading rhizomes, creeping stems, and plant fragments. Rarely by seed.

Impact: Seed production and dispersal are limited. This species can spread from gardens into natural areas, often through the dumping of yard waste. Where *P. terminalis* is established, it forms dense stands and excludes other vegetation.

Control: Avoid planting any new areas with this species, particularly where gardens border on natural areas. Yard waste containing *P. terminalis* cuttings should not be disposed of in natural areas. Where this species is well established, plants can be controlled by digging out the root system or by successive mowing. Mechanical control is typically more effective than chemical control.

Native Alternatives

Asarum canadense (Canadian wild ginger)
Pachysandra procumbens (Allegheny spurge)
Rhus aromatica 'Gro-Low' (fragrant sumac)

210

Pachysandra terminalis leaves. Image by Karan A. Rawlins, University of Georgia, Bugwood.org.

Pachysandra terminalis fruits. Image © Paul S. Drobot.

Pachysandra terminalis stems. Image by Salicyna, used under a CC BY-SA 4.0 license.

Pachysandra terminalis seeds. Image by Steve Hurst, hosted by the USDA-NRCS PLANTS Database.

Pachysandra terminalis flowers. Image by Rob Routledge, Sault College, Bugwood.org.

Pachysandra terminalis infestation in forest understory. Image by Jil Swearingen, USDI National Park Service, Bugwood.org.

Cannabaceae

Humulus japonicus

Japanese hop · wild hop

⚠ **Do Not Touch**

Origin: Native to eastern Asia, including parts of Russia, China, Taiwan, Korea, Vietnam, and Japan. It was introduced to the United States in the mid- to late 1800s for ornamental and medicinal purposes. Records show that it had escaped from cultivation by the early 1900s. There are now scattered populations throughout the eastern and midwestern United States.

Description: A herbaceous, annual, climbing or trailing vine. It scrambles over vegetation, fences, and other structures, or it spreads across open ground to form dense mats. It is similar to the native *Humulus lupulus* (common hops), which is used in beer making, although *H. lupulus* typically has three-lobed leaves with more rounded lobes and petioles that are shorter than the length of the leaf.

Habitat: Commonly found along river- and streambanks or floodplain areas. Also occurs in meadows, fields, forest edges, roadsides, and other areas with high levels of disturbance. Prefers open, sunny habitats with moist soils.

Height: Vines 6–10 m (20–32 ft.) long.

Foliage: Leaves are simple, opposite, 5–15 cm (2–6 in.) long and wide, palmately lobed with five to nine lobes. Leaves are borne on petioles equal to or longer than the leaf, with a triangular bract at the base of each pair of petioles. Leaf margins are coarsely toothed, and the tips of the leaf lobes are sharply pointed. The upper surface of the leaf is sparsely covered with short rough hairs, and the lower surface is covered with longer prickly hairs, particularly along the veins.

Flowers: Species is dioecious, with male and female flowers on separate plants. Both male and female flowers are small, light green to pale red with five petals, in inflorescences borne in the leaf axils. Male flowers grow in a branching, upright inflorescence, 15–25 cm (6–10 in.) long. Female flowers are in drooping, cone-shaped clusters, 7–10 mm (¼–½ in.) long, surrounded by small bracts with densely hairy margins. July–September.

Fruit: A drooping, green, cone-shaped strobile, 1–3 cm (½–1 in.) long. Each female flower develops into a single ovoid, flattened, yellowish-brown achene, 4–5 mm (⅛–¼ in.) long and 3 mm (⅛ in.) wide. August–October.

Stems: Stems are light green to reddish purple, longitudinally ridged, with rows of downward pointing, prickly hairs.

Root system: Shallow, fibrous roots.

Reproduction: By seed, primarily dispersed by wind and water. Because this is an annual species, populations persist through reseeding. The seed bank remains viable for three years.

Impact: This species forms dense stands that exclude other species. This is of particular concern for native riverbank and floodplain vegetation.

Control: Do not plant specimens of *H. japonicus,* and avoid moving seeds between sites. This species is not yet widespread, so the aim should be to prevent spread and eradicate smaller, isolated populations before it becomes a serious problem. Wear protective clothing and gloves when handling this plant, because the prickles on the stems and leaves can cause blisters and dermatitis. Hand pulling and removal of rootstock is best if timed before seeds ripen. Foliar spray with glyphosate can be used before flowering. Stems can be cut and treated with glyphosate from July–September.

Humulus japonicus leaves. Image by Chris Evans, University of Illinois, Bugwood.org.

Humulus japonicus stem and bract at the base of the petioles. Image by Chris Evans, University of Illinois, Bugwood.org.

Female flowers of *Humulus japonicus.* Image by Leslie J. Mehrhoff, University of Connecticut, Bugwood.org.

5 mm

Humulus japonicus seeds. Image by Carole Ritchie, hosted by the USDA-NRCS PLANTS Database.

Male flowers of *Humulus japonicus.* Image by Leslie J. Mehrhoff, University of Connecticut, Bugwood.org.

Humulus japonicus infestation. Image by Leslie J. Mehrhoff, University of Connecticut, Bugwood.org.

Humulus japonicus strobile. Image by Mark A. Garland, hosted by the USDA-NRCS PLANTS Database.

Caprifoliaceae

Lonicera japonica

Japanese honeysuckle · Chinese honeysuckle
· gold-and-silver honeysuckle · white
honeysuckle

Origin: Native to eastern Asia, including parts of China, Japan, and Korea. This species was first recorded in North America in 1806, when it was introduced to Long Island, New York, as an ornamental. It has been widely planted in gardens for its fragrant flowers, which attract hummingbirds, butterflies, and other insects. It has escaped cultivation and become problematic throughout much of the eastern and midwestern United States.

Description: A perennial, woody vine with a trailing or climbing growth habit. Depending on local climate, foliage can be evergreen, semievergreen, or deciduous. Plants climb other vegetation by twining. Its black berries and distinctly separate upper leaves distinguish it from native honeysuckle species. North American honeysuckle vines, in contrast, have red to orange berries and the upper leaves are fused together.

Habitat: Open woodlands, forest edges, wetlands, agricultural fields, along the edges of roadsides and railroads, and other disturbed areas. Prefers full sun and rich, moist soils but can tolerate shade, drought, and nutrient poor soils.

Height: Vines can reach over 10 m (30 ft.).

Foliage: Simple, opposite, ovate to oblong with a tapered point and a rounded or square base, 3–8 cm (1–3 in.) long and 2–3 cm (¾–1 in.) wide. Leaf margin is toothless, but some young leaves may be lobed. The upper surfaces of the leaves are dark green. Both sides of the leaf are hairy when young, becoming hairless with age. Leaves have short petioles, 5–12 mm (¼–½ in.) long.

Flowers: Tubular flowers grow singly or in pairs from the leaf axils. Flowers are white, yellowing with age. The tube is 3–5 cm (1–2 in.) long, with one narrow lobe about 2 cm (¾ in.)

long forming the lower lip and four fused lobes forming a shorter upper lip. Each flower has five yellow-tipped stamens and a white style with a rounded green stigma protruding from the tube. Very fragrant and nectar-rich. April–July.

Fruit: A spherical black berry, 3–4 mm (⅛ in.) in diameter. Like the flowers, the fruits grow singly or in pairs in the leaf axils. Each berry contains two or three small, brown to black, ovate seeds, about 2–3 mm (1/16–⅛ in.) long and 2 mm (1/16 in.) wide. August–October.

Stems: Young stems are green and covered in hairs, becoming purplish brown and hairless with age. Older stems are round in cross-section, have a hollow center, and are covered with brownish bark that peels in long strips. Stems twine upward, from right to left.

Root system: Extensive root system with a taproot and spreading rhizomes.

Reproduction: By seed, primarily dispersed by birds. Vegetative reproduction also occurs through rhizomes and aboveground runners that root at the nodes.

Impact: This species spreads rapidly, and the vines can completely cover native shrubs and trees. The support plants often collapse due to the weight of the vegetation, or they are eventually outcompeted. Dense stands prevent the germination and growth of understory shrubs and trees, which has long-term consequences for forest regeneration.

Native Alternatives
Bignonia capreolata (crossvine)
Lonicera dioica (wild honeysuckle)
Lonicera sempervirens (trumpet honeysuckle)

Lonicera japonica leaves. Image by Bruce Ackley, The Ohio State University, Bugwood.org.

Lobed form of *Lonicera japonica* leaves. Image by Leslie J. Mehrhoff, University of Connecticut, Bugwood.org.

Lonicera japonica stems. Image by Chris Evans, University of Illinois, Bugwood.org.

Lonicera japonica flowers. Image by Leslie J. Mehrhoff, University of Connecticut, Bugwood.org.

Lonicera japonica fruits. Image by Leslie J. Mehrhoff, University of Connecticut, Bugwood.org.

Lonicera japonica seeds. Image by Ken Chamberlain, The Ohio State University, Bugwood.org.

Lonicera japonica infestation. Image by Leslie J. Mehrhoff, University of Connecticut, Bugwood.org.

Control: Do not plant *L. japonica* as an ornamental. This includes cultivars such as 'Halliana' or 'Hall's Prolific,' which are promoted as being less aggressive. Remove any such plants on your property. For small infestations, repeated pulling of entire plants and root systems can be effective. It is important to remove and destroy all plant material after cutting to prevent rooting and reinfestation. Mowing is not recommended, as it stimulates resprouting and leads to denser vegetation. The herbicides glyphosate and triclopyr have been demonstrated to control this species; they should be applied to foliage in the autumn after native plants are dormant but before the first hard freeze.

Celastraceae

Celastrus orbiculatus

oriental bittersweet · Asian bittersweet · Asiatic bittersweet · Chinese bittersweet · Japanese bittersweet · round-leaved bittersweet · oriental staff vine · climbing spindle berry

ODA Invasive
OIPC Invasive

217

Origin: Native to temperate eastern Asia, including China north of the Yangtze River, central and northern Japan, and Korea. It was introduced to North America sometime in the 1860s or 1870s. This species was initially planted as an ornamental and later also promoted for erosion control. It has escaped cultivation and become problematic throughout the northeastern and midwestern United States.

Description: A deciduous, perennial, woody vine that climbs by scrambling and twining. If there is no nearby vegetation or structure on which to climb, the plant will grow as a trailing shrub. The nonnative *Celastrus orbiculatus* closely resembles a native species, *Celastrus scandens* (American bittersweet). The main difference between the two is that *C. scandens* has flowers and fruits in large clusters at the ends of branches, while *C. orbiculatus* has small clusters of flowers and fruits in the axils of the leaves. In addition, *C. scandens* has more elliptical leaves, while *C. orbiculatus* has a round leaf.

Habitat: Undisturbed forest, open woodland, woodland edges, fields, fencerows, and roadsides. Grows best in full sun but can germinate and persist in heavy shade, which allows it to survive in the forest understory.

Height: Typically 6–10 m (20–30 ft.), but climbing vines can reach 20 m (65 ft.).

Foliage: Leaves are simple, alternate, 5–12 cm (2–5 in.) long and 2–8 cm (¾–3 in.) wide, broadly ovate or round with a tapered base and a blunt point at the tip. Leaf shape can be variable, however, so do not use this feature as your only means of identification. Leaves have rounded teeth along the margins. The upper surface of the leaf is glossy, leaves are a rich green during the growing season, and they turn golden yellow in autumn.

Flowers: Flowers within an individual plant usually become unisexual by reduction of male or female parts, giving rise to separate male and female flowers on different plants (making them functionally dioecious). However, a few plants have some flowers that retain both male and female parts in addition to unisexual flowers (polygamo-dioecious). Flowers of both types are small, 3–6 mm (⅛–¼ in.) in diameter, yellowish green, with five petals. Flower clusters develop in the leaf axils along the length of the stem, with three to seven flowers per inflorescence. Male flowers have five stamens almost as long as the petals. Each female flower has a stout style with a three-lobed stigma. May–June.

Fruit: A three-valved capsule, nearly spherical, 6–9 mm (¼–⅜ in.) long and 7–10 mm (¼–½ in.) wide, in clusters in the leaf axils along the length of the stem. Fruits are green during the summer, becoming yellowish orange in the autumn. At maturity, the capsule splits open to reveal seeds enclosed in a fleshy red aril, fused into a single sphere, with each valve containing one or two seeds. Seeds are white to pink, flattened and oval, 3 mm (⅛ in.) long and 2 mm (1/16 in.) wide. July–October, but persisting on the plant into winter.

Stems: Trunk diameter on mature plants can

Native Alternatives

Aristolochia tomentosa (woolly Dutchman's pipe)
Celastrus scandens (American bittersweet)
Lonicera sempervirens (trumpet honeysuckle)

Left: *Celastrus orbiculatus* leaves. *Right:* Stems of *Celastrus orbiculatus* twining around a tree. Both images by Chris Evans, University of Illinois, Bugwood.org.

Celastrus orbiculatus fruits. Image by Leslie J. Mehrhoff, University of Connecticut, Bugwood.org.

Male flowers of *Celastrus orbiculatus*. Image by Leslie J. Mehrhoff, University of Connecticut, Bugwood.org.

Celastrus orbiculatus seeds. Image by Leslie J. Mehrhoff, University of Connecticut, Bugwood.org.

Female flowers of *Celastrus orbiculatus*. Image by Leslie J. Mehrhoff, University of Connecticut, Bugwood.org.

Celastrus orbiculatus infestation. Image by Leslie J. Mehrhoff, University of Connecticut, Bugwood.org.

be as large as 10 cm (4 in.). The bark on young stems and twigs is smooth, light gray to dark brown, with numerous light lenticels. Twigs are round in cross-section and have solid white pith. Mature bark is striated, brown to dark brown. Stems twine upward, from left to right.

Root system: The root system is deep and spreading. The outer surface of the main roots is bright orange.

Reproduction: By seed, dispersed by birds and other wildlife. Vegetative reproduction also occurs through root sprouting. Root fragments are capable of establishing new plants.

Impact: Seedlings form dense stands in the understory, eliminating spring ephemeral wildflowers. Vines of mature plants cover, shade, and outcompete other vegetation. Twining vines can girdle trees or shrubs on which they grow, and the weight of dense vine cover can cause trees to fall over. In addition, the hybridization of this species with the native *C. scandens* has been demonstrated in a research setting. It is not clear how frequently hybridization occurs in nature, but this process could potentially lead to the loss of genetic identity for the native species through introgression.

Control: Seedlings can be pulled by hand, taking care to remove all root fragments. Larger plants can be cut back and the root system dug out. Cut stumps can also be treated with triclopyr or a strong formulation of glyphosate. In areas with extensive cover, a mixture of 2,4-D and triclopyr can be sprayed on the foliage immediately after the first hard frost in the autumn.

Celastraceae

Euonymus fortunei

wintercreeper · climbing euonymus · Chinese
spindle tree · Fortune's spindle · gaiety

Origin: Native to eastern and southeastern Asia, *Euonymus fortunei* was introduced to the United States in 1907 as an ornamental ground-cover. This plant is extremely popular—and arguably overused—in landscaping and has escaped cultivation to establish scattered populations throughout the eastern and central United States.

Description: A perennial, evergreen, woody groundcover or vine. It creates dense mats on the ground and can climb trees high into the canopy by clinging to the bark. There is a huge amount of natural and selected variation within *E. fortunei,* particularly in terms of leaf shape, size, and color. Horticulturist Michael Dirr (2009) lists more than 60 cultivars that have been developed for this species, many of which have golden leaves, white or gold leaf margins and veins, or other variegation in the leaves. Its leaves and habit somewhat resemble the native *Mitchella repens* (partridgeberry) and the non-native *Vinca minor* (common periwinkle) [pp. 202–203]. However, *M. repens* has paired white trumpet-shaped flowers and red berries, and *V. minor* has violet-blue pinwheel-shaped flowers.

Habitat: Common in the cultivated landscape. Occasionally found in open fields but primarily has spread into forest margins and openings. It can grow in full sun to heavy shade and is tolerant of poor soils and a wide pH range.

Height: This species takes different forms, depending on the reproductive phase. Juvenile shoots are vegetative only and can grow as a groundcover or climb as a vine to 12–21 m (40–70 ft.). The mature shoots are reproductive and can develop at the ends of vines or grow as a shrub to 1 m (3 ft.).

Foliage: Leaf shape, size, and color are extremely variable among cultivars and, depending on the reproductive phase, even within an individual. Leaves are simple, opposite, typically 2.5–6 cm (1–3 in.) long and 2–4.5 cm (¾–2 in.) wide, ovate with a rounded base and tapering to a blunt point at the tip. Leaf margins are finely toothed and wavy. Leaves are leathery, smooth, glossy, typically dark green with light-colored veins. Leaves often remain green and persist on the stems over winter, although in extremely cold climates plants shed their leaves.

Flowers: Flowers only occur on mature shoots; the change to a reproductive phase is triggered when the plant receives enough light. Flowers are small, inconspicuous, greenish white, 5 mm (¼ in.) in diameter, with four petals and four long stamens arranged around a greenish pistil. Flowers are in clusters at the end of long, branched stalks that develop from the axils of leaves. June–July.

Fruit: A four-lobed capsule, nearly spherical, 5–10 mm (¼–½ in.) in diameter, white to pink or red, splitting open at maturity to reveal four seeds, each covered in a separate fleshy, orange aril. Like the flowers, fruits are in clusters at the end of long, branched stalks that develop from the axils of leaves. Seeds are white, ellipsoid, 3 mm (⅛ in.) long and 2 mm (1/16 in.) wide. September–November, often persisting on the plant into winter.

Native Alternatives
Arctostaphylos uva-ursi (bearberry)
Gaultheria procumbens (wintergreen)
Mitchella repens (partridgeberry)
Paxistima canbyi (cliff green)

Stems: Stems are green and hairless when young, becoming gray, corky, and streaked with age.

Root system: The root system is extensive, and resprouting can occur from the root crown. Juvenile shoots develop aerial rootlets or trailing roots at nodes to anchor the plants as they creep and climb.

Reproduction: By seed, dispersed by birds and mammals. It also reproduces vegetatively, primarily by rooting at the nodes where stems contact the soil.

Impact: Forms a dense groundcover that displaces native herbaceous plants and tree seedlings. Vines can completely cover shrubs and small trees. Forest gaps created by wind, fire, or damage by insects such as *Agrilus planipennis* (emerald ash borer) may be particularly vulnerable to establishment by this species.

Control: Do not plant new specimens of *E. fortunei*. Remove existing plantings, particularly in properties adjacent to natural areas. Because this species is capable of resprouting from the root crown, any control measure should target removal or killing of the root system. Hand pull smaller plants or use a weed wrench or similar tool. Use clippers to cut vines and then pull them off trees. Cut stems can be spot-treated with glyphosate or triclopyr. Repeated cutting will help control this species.

Left: Euonymus fortunei leaves. *Right: Euonymus fortunei* stem. Both images by James H. Miller, USDA Forest Service, Bugwood.org.

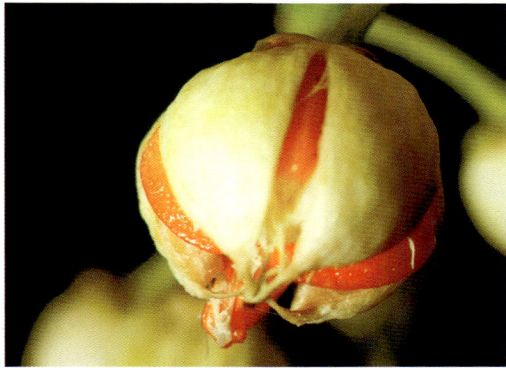

Euonymus fortunei fruit. Image by Alexander Klink, used under a CC BY 3.0 license.

Leaves of *Euonymus fortunei* 'Emerald Gaiety,' a variegated cultivar. Image by A. Barra, used under a CC BY 3.0 license.

Left: Euonymus fortunei flowers. Image by James H. Miller, USDA Forest Service, Bugwood.org. *Right: Euonymus fortunei* growing up the trunk of a tree. Image by Chris Evans, University of Illinois, Bugwood.org.

Convolvulaceae

Calystegia sepium

hedge bindweed · hedge false bindweed · large bindweed · great bindweed · hooded bindweed · bracted bindweed · wild morning glory · hedge convolvulus · hedge-lily · hedge morning glory · lily-bind · bearbind · bellbind · bellbine · belle of the ball · devil's guts · devil's vine · hedgebell · Rutland beauty · bugle vine · heavenly trumpets · old man's nightcap · granny-pop-out-of-bed · bride's gown · wedlock · white witches hat

Prohibited Noxious Weed

Origin: An extremely complex species that is widespread and encompasses at least nine subspecies, many of which are native to North America. One of these—*Calystegia sepium* subsp. *sepium*—is native to Europe and Asia but has been introduced to the United States, most likely as a contaminant in crop seed. Confusion between the native and nonnative subspecies has almost certainly contributed to the underreporting and general lack of awareness of the weedy subspecies. It is extremely problematic in agricultural settings.

Description: A herbaceous, perennial vine with twining stems that climbs over other herbaceous plants, shrubs, and fences or other structures. It resembles the native *Calystegia spithamaea* (false bindweed) but the latter is low-growing and non-vining. Another similar species is the nonnative *Convolvulus arvensis* (field bindweed) [pp. 225–226], which has smaller leaves and flowers. The description that follows focuses on the features of *Calystegia sepium* subsp. *sepium*.

Habitat: Fields, thickets, fencerows, roadsides, and other disturbed areas. Can survive in most soils but does not thrive in deep shade.

Height: Typically to 3 m (10 ft.), occasionally to 4 or 5 m (12–15 ft.).

Foliage: Leaves are simple, alternate, 10–13 cm (4–5 in.) long and 5–8 cm (2–3 in.) wide, with a distinctive arrowhead shape that rap-

idly tapers to a sharp point at the tip. The base of the leaf has a deep, *V*-shaped notch and the basal lobes are angular. Leaves are held on long petioles and the angle between the underside of the leaf and the petiole is about 90˚. The leaf margins are smooth, and there are fine hairs on both the upper and lower leaf surfaces.

Flowers: Flowers are typically white (whereas other subspecies in the complex can have pink to lavender flowers), with five petals fused into a funnel-shaped corolla, 5–7 cm (2–3 in.) long and 3–5.5 cm (1–2 in.) in diameter. There are two small bracts at the base of each flower, scarcely overlapping or overlapping only at base, with reddish margins, tapering to a point at the tip. Each flower has five stamens, 17–25 mm (¾–1 in.) long, and a single swollen stigma. Flowers are solitary and grow from the leaf axils. Individual flowers persist for a single day, usually opening in the morning and closing by late afternoon. May–September.

Fruit: Fruit is an ovoid or nearly spherical capsule, 10–12 mm (½ in.) long and 8–10 mm (¼–½ in.) wide, green drying to brown, splitting into two or four segments to release the seeds. Each capsule contains two to four dull brown to black seeds, 5 mm (¼ in.) long and 3–4 mm (⅛ in.) wide, typically with two flat sides and one side rounded like a segment of an orange. June–October, persisting on dry stems into winter.

Stems: The slender stems are light green or

Left: Calystegia sepium leaves. Image by Robert Vidéki, Doronicum Kft., Bugwood.org. *Right:* Floral bracts on *Calystegia sepium.* Image by André Karwath, used under a CC BY-SA 2.5 license.

Mature *Calystegia sepium* fruit. Image by AnR00002, used under a CC0 1.0 license.

Calystegia sepium flowers. Image © Katy Chayka, Minnesota Wildflowers.

Seeds of *Calystegia sepium* ssp. *sepium.* Image by Steve Hurst, hosted by the USDA-NRCS PLANTS Database.

Developing *Calystegia sepium* fruit. Image by AnR00002, used under a CC0 1.0 license.

Smothering habit of *Calystegia sepium.* Image by AnR00002, used under a CC0 1.0 license.

red, slightly hairy, and round in cross-section. Stems climb by twining upward, from left to right, around other plants or on fences or other structures.

Root system: Develops a large, rhizomatous root system. Aboveground vegetation typically dies back over winter and resprouts from the rhizomes the following spring.

Reproduction: By seed and spreading rhizomes.

Impact: This species is primarily a concern in agricultural systems, where it significantly reduces crop yield. It can also be a problem in gardens and open natural areas, where it can outcompete other vegetation.

Control: Cut vines and dig out roots, preferably before the plant sets seed. Repeated treatment may be necessary. After mechanical removal you can also mulch, cover in black plastic, or overplant with a fast-growing annual crop. Unless you can easily differentiate the vines from nontarget vegetation, herbicide treatment is not recommended.

Convolvulaceae

Convolvulus arvensis

field bindweed · European bindweed · lesser bindweed · field morning glory · small-flowered morning glory · wild morning glory · perennial morning glory · creeping jenny · withy wind · possession vine

Prohibited Noxious Weed

Origin: Native to Europe and Asia, this species was introduced to the United States as early as 1739, most likely as a contaminant in crop seed. It is present throughout the United States and is extremely problematic in agricultural settings. *Convolvulus arvensis* has been classified as a noxious weed in at least 23 states, particularly in the western United States.

Description: A herbaceous, perennial vine with prostrate or climbing stems. It primarily creeps over the ground but also twines around other herbaceous plants, shrubs, and fences or other structures. It is similar to *Calystegia sepium* (hedge bindweed) [pp. 222–224], but has much smaller leaves and flowers.

Habitat: Found in grasslands and along streams or roadsides. It is a major weed of field crops, pastures, and horticulture. Performs best in full sun and is tolerant of drought and poor soils.

Height: To 2 m (6 ft.).

Foliage: Leaves are simple, alternate, 1–5 cm (½–2 in.) long and 5–25 mm (¼–1 in.) wide. Form is variable, typically arrowhead-shaped with a pointed tip but can also be rounded to oval. Leaves held on long petioles, 1–4 cm (½–2 in.) long. The leaf margins are smooth and occasionally slightly hairy.

Flowers: Flowers are white or pink, with five petals fused into a funnel-shaped corolla, 1–2.5 cm (½–1 in.) long and 1–2.5 cm (½–1 in.) in diameter. Each flower has five green to brown sepals surrounding the base, 3–5 mm (⅛–¼ in.) long, and two small bracts on the stem 5–20 mm (¼–1 in.) below each flower. Flowers have a patch of yellow at the throat of the corolla below the reproductive parts, which consist of five stamens and a pistil with a single stigma that divides into two lobes. Flowers grow from the leaf axils with one to three flowers on branching stalks. Individual flowers persist for a single day, usually opening in the morning and closing by late afternoon. May–September.

Fruit: Fruits are ovoid or spherical capsules, 5–10 mm (¼–½ in.) long and 3–5 mm (⅛–¼ in.) wide, green drying to brown, splitting into two or four segments to release the seeds. Each capsule contains one to four dark brown to black seeds with a rough surface, each about 3–4 mm (⅛ in.) long and 2–3 mm (1/16–⅛ in.) wide, typically with two flat sides and one side rounded like a segment of an orange. June–October, persisting on dry stems into winter.

Stems: The slender stems are light green and smooth or slightly hairy. They primarily creep over the ground or climb other plants by twining upward, from left to right.

Root system: Plants develop taproots that can be up to 6 m (20 ft.) deep, as well as shallow horizontal roots that develop into rhizomes. When the rhizomes reach the surface, they establish new crowns, contributing to the species' vegetative spread.

Reproduction: By seed, dispersed by water, livestock, birds, and vehicles or farm machinery. Seeds are estimated to remain viable in the soil seed bank for 20–50 years. Vegetative reproduction also occurs via rhizomes, creeping roots, and root fragments.

Impact: This species forms dense mats of overlapping stems that outcompete other plants for sunlight, moisture, and nutrients. It is one of the most serious weeds of agricultural fields in temperate regions of the world, but it can also be problematic in gardens or disturbed natural areas. The plant contains several alkaloids, making it toxic to livestock, particularly horses.

Control: Cut vines repeatedly, and dig out all the accessible roots, preferably before the plant sets seed. Repeated treatment will be necessary. After mechanical removal. the area can be mulched, covered in black plastic, or overplanted with a fast-growing annual crop. Any herbicide treatment should be applied prior to flowering; 2,4-D (alone or in combination with other herbicides), glyphosate, MCPA, and picloram have been shown to be effective against this species.

Convolvulus arvensis leaves. Image by Robert Vidéki, Doronicum Kft., Bugwood.org.

Twining stems of *Convolvulus arvensis*. Image by AnR00002, used under a CC0 1.0 license.

Convolvulus arvensis flowers. Image by K. George Beck and James Sebastian, Colorado State University, Bugwood.org.

Convolvulus arvensis seeds. Image by Julia Scher, USDA APHIS PPQ, Bugwood.org.

Convolvulus arvensis fruits. Image by Phil Westra, Colorado State University, Bugwood.org.

Prostrate growth of *Convolvulus arvensis* stems. Image by Howard F. Schwartz, Colorado State University, Bugwood.org.

Convolvulaceae

Cuscuta epithymum

clover dodder · legume dodder · common dodder · thyme dodder

Origin: Native to Europe, western Asia, and northern Africa. The history and spread of this species is not well documented, but it was first recorded in Ohio in 1899. Seeds from *Cuscuta* spp. are often contaminants in crop seed, and it is likely that this species was accidentally introduced to the United States this way.

Description: An annual, herbaceous, parasitic vine with threadlike stems that twine around and over host plants in a dense tangle. Dodders have negligible chlorophyll and require a host plant to survive. They attach to the host via haustoria that penetrate the host to extract nutrients and water. This species is parasitic primarily on legumes—including forage crops such as *Trifolium* spp. (clovers) and *Medicago* spp. (alfalfa or bur clovers)—as well as *Thymus* species (thymes). *Cuscuta epithymum* is the most common nonnative *Cuscuta* spp. in Ohio. All *Cuscuta* spp. are on the federal noxious weed list; thus it is illegal to move seed or other material contaminated with their seeds throughout the United States.

Habitat: Found growing in cultivated crops, pastures, fields, meadows, forest margins, and along roadsides. It occurs on a range of soils and climates, provided that suitable host plants are present.

Height: The size of a single individual will vary depending on the host plant. Individual stems can be 1 m (3 ft.) or longer.

Foliage: Often described as leafless because the leaves are reduced to minute scales. Where present, leaves are simple, alternate, 1 mm (1/32 in.) long and 0.4 mm (1/64 in.) wide, oblong, slightly purple to brown.

Flowers: Small pink to purple or white flowers, 3–4 mm (1/8 in.) long and 2.5 mm (1/16 in.) wide, in dense clusters of 7–25 arranged in a compact sphere. Individual flowers have five petals that are fused into a tube at the base and separate into five deep, triangular lobes above. There are five stamens that are shorter than the petals and a single pistil in the center. June–September.

Fruit: A spherical capsule, 2 mm (1/16 in.) in diameter, brown, surrounded by the withered flower, which turns papery. Each capsule contains up to four small seeds, 1 mm (1/32 in.) in diameter, with a flattened ovoid shape, a rough surface, and irregular angles near the base where the seeds develop alongside one another in the ovary. July–October.

Stems: Stems are typically pink to red but can also be yellow or purplish, hairless, very slender, about 1 mm (1/32 in.) in diameter. Stems twine upward, from left to right, and form dense masses around host plants.

Root system: When dodders first germinate, they develop an anchoring root in the soil and then rapidly produce a stem that grows toward—and attaches to—a host plant. The plant then develops a haustorium that penetrates the host's vascular system. The anchoring root dies once the first haustorial connection has been made. An individual dodder can form multiple haustoria with the same or different host plants.

Reproduction: By abundant seed, which can remain viable for up to five years. A single plant has been reported to produce 16,000 seeds. In addition, when growing on perennial hosts, it can form galls within the host where

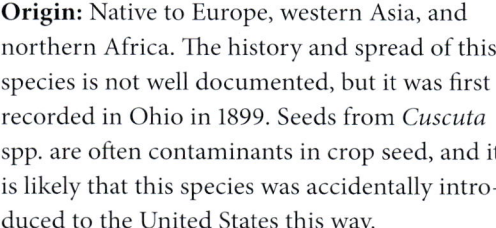

parasitic tissue overwinters. New plants then develop from these galls the following spring.

Impact: All *Cuscuta* spp. drain host resources, which often results in the host plant not being able to reproduce. This contributes to crop loss or a reduction in host vigor.

Control: The most important long-term control for this species is to limit seed production and movement. Because *C. epithymum* is connected to its host plant, the host may need to be removed along with the parasite. Repeated pulling prior to flowering and fruiting can reduce the seed production. Spraying with 2,4-D during the growing season will kill both the host and its parasite.

Cuscuta epithymum infestation. Image by Hans Hillewaert, used under a CC BY-SA 4.0 license.

Cuscuta epithymum stems. Image by Hans Hillewaert, used under a CC BY-SA 4.0 license.

Cuscuta epithymum flowers. Image by Stefan Lefnaer, used under a CC BY-SA 3.0 license.

Cuscuta epithymum seeds. Image by Julia Scher, USDA APHIS PPQ, Bugwood.org.

Convolvulaceae

Ipomoea purpurea

tall morning glory · common morning glory ·
purple morning glory · annual morning glory

Origin: The precise native range is somewhat obscure, but it is reported to have originated in tropical parts of South and Central America. It was introduced to the United States as an ornamental plant sometime in the 1700s. While still cultivated for its showy flowers, this species can grow out of control in gardens, in natural areas, and as an agricultural weed.

Description: A herbaceous, annual vine with twining stems. *Ipomoea purpurea* trails along the ground, wraps around other plants for support, or grows over fences and other structures. It differs from the closely related bindweeds (*Calystegia* and *Convolvulus* species) by its heart-shaped leaves and its seed capsule that splits into three segments.

Habitat: Fields, roadsides, railway lines, riverbanks, waste areas, and arable land. Can grow in a range of soil types but prefers open areas with a history of disturbance.

Height: To 5 m (15 in.).

Foliage: Leaves are simple, alternate, 4–18 cm (2–7 in.) long and 3–16 cm (1–3 in.) wide, heart-shaped. Leaves are held on slender petioles, 2–12 cm (¾–5 in.) long, green to brown and covered with hairs. Leaf margins are smooth or slightly wavy. The veins are prominent on both the upper and lower surfaces, nearly hairless along their upper surfaces but hairy underneath.

Flowers: Flowers are purple, blue, pink, white, or a variegated combination, some with markings that form a five-pointed star. The five petals are fused into a funnel-shaped corolla, 3.5–5.5 cm (1 ½–2 ½ in.) long and 4–6 cm (2–3 in.) in diameter. Each flower has five green to brown, hairy sepals surrounding its base, 10–15

mm (½–⅝ in.) long, with the inner sepals shorter than those on the outside. The reproductive parts consist of five stamens and a single pistil with a stigma that divides into three lobes. Flowers grow from the leaf axils with one to five blossoms on long, branching stalks. Individual flowers persist for a single day, opening in the morning and closing by late afternoon. July–September.

Fruit: Fruit is a spherical capsule, 9–10 mm (⅜–½ in.) in diameter, green drying to brown, splitting into three segments to release the seeds. Each capsule contains four dark brown to black seeds, each about 3–5 mm (⅛–¼ in.) long and 2–4 mm (1/16–⅛ in.) wide, somewhat variable in shape but typically with two flat sides and one side rounded like a segment of an orange. August–October.

Stems: Stems are slender, light green to brown, round in cross-section, covered in brown hairs. This species lacks tendrils and climbs by twining upward, from left to right.

Root system: Root system is fibrous and does not persist from one growing season to the next.

Reproduction: By abundant seed. An individual plant can produce up to 26,000 seeds, and these remain viable in the soil seed bank for many years. Seeds are dispersed by gravity, wind, and water and secondarily by birds or contaminated crop and flower seeds.

229

Native Alternatives

Bignonia capreolata (crossvine)
Clematis virginiana (virgin's bower)
Lonicera dioica (wild honeysuckle)
Lonicera sempervirens (trumpet honeysuckle)

Impact: This species is mainly a weed of agricultural areas and disturbed sites, but it is also an environmental weed that can scramble over and outcompete native species. It is less of a problem than perennial species in the same genus or family, because it does not have persistent roots. However, once established, the abundant seed does allow this species to reproduce and spread within a site. All parts of this plant—including the seeds—can be poisonous if ingested.

Control: Do not plant this species in gardens. Seedlings are easy to pull or dig out. Undertake control measures before the plant sets seed. Cutting is recommended, and it can be followed by the application of a systemic herbicide such as 2,4-D, diquat, or glyphosate.

Ipomoea purpurea leaves. Image by Bruce Ackley, The Ohio State University, Bugwood.org.

Ipomoea purpurea fruits. Image by Chris Evans, University of Illinois, Bugwood.org.

Ipomoea purpurea stem twining from left to right. Image by Bruce Ackley, The Ohio State University, Bugwood.

Ipomoea purpurea seeds. Image by Lindsey Seastone, USDA APHIS PPQ, Bugwood.org.

Ipomoea purpurea flowers. Image by Howard F. Schwartz, Colorado State University, Bugwood.org.

Ipomoea purpurea infestation in an agricultural field. Image by Howard F. Schwartz, Colorado State University, Bugwood.org.

Dioscoreaceae

Dioscorea polystachya

Chinese yam · Chinese potato · cinnamon-vine

Synonym: *Dioscorea batatas*

Origin: Native to eastern Asia, including China, Japan, and Korea. This species has long been used in traditional medicine and cultivated for its edible tubers. It was introduced to the United States in the 1800s for use as an ornamental and food plant. By the 1980s, it had escaped cultivation and established scattered populations throughout the eastern United States.

Description: A herbaceous, perennial vine that climbs by twining. The name *Dioscorea oppositifolia* has been consistently misapplied to this species; it correctly refers to a species native to the Indian subcontinent that has not been recorded in North America. The common name "air potato" is also sometimes used for *Dioscorea polystacha*—in reference to the aerial tubers it produces—but that name is more typically applied to the related nonnative species *Dioscorea bulbifera,* which is invasive in subtropical areas of the United States.

Habitat: Forest edges, woodlands, riverbanks, along roadsides, and other disturbed areas. It can grow in full sun to full shade.

Height: 1–5 m (3–15 ft.).

Foliage: Simple, typically alternate at the base of the stem, becoming opposite or occasionally in whorls of three in the upper nodes, 3–9 cm (1–3 ½ in.) long and 2–7 cm (¾–3 in.) wide. The arrowhead- or heart-shaped leaves are held on a petiole as long as the leaf blade. Each leaf is deeply lobed at the base and pointed at the tip; upper leaves may have slightly three-lobed margins. Leaf margins, petioles, and stems are reddish purple, particularly in new growth. Leaves have seven to nine prominent, parallel veins.

Flowers: Separate male and female flowers occur on different plants; female specimens are rare or absent in North America and male flowers are uncommon. In the native range of this species, both male and female plants produce small, white or greenish-yellow flowers with six petals that are only partially open, in inflorescences that grow from the leaf axils. Each male inflorescence contains one to eight spikes that are 2–8 cm (¾–3 in.) long, with numerous flowers arranged in a zigzag pattern along the stem spaced about 2 mm ($\frac{1}{16}$ in.) apart. Male flowers have six stamens. Where they occur, female inflorescences contain one to three spikes with one to six flowers each, spaced about 1 cm (½ in.) apart. The female flower has a single pistil topped with a three-lobed stigma. Flowers have a spicy fragrance similar to cinnamon. June–September.

Fruit: Within the native range, plants form an ovate or spherical capsule with three prominent ribs, 1.2–2 cm (½–¾ in.) long and wide, green drying to brown, splitting open to release seeds. Seeds are a flattened oval, 5 mm (¼ in.) long and 4 mm (⅛ in.) wide, surrounded by a brown membranous wing. July–November. Fruits have not been observed on this species in North America. Most reproduction is vegetative, through aerial tubers.

Stems: Stems are slender, hairless, green or reddish purple, twining upward, from left to right.

Root system: Each plant produces one or more cylindrical tubers deep underground, growing up to 1 m (3 ft.) long, from which new stems resprout annually.

Reproduction: Given the absence of female plants in North America, it is assumed that reproduction is primarily or entirely vegetative. Persistent underground tubers provide for perennial regeneration of established infestations.

In addition, aerial tubers that resemble small potatoes are produced abundantly in the leaf axils; these tubers, which can float in water, serve as the species' primary mode of dispersal and spread. June–September.

Impact: Forms dense mats that overtop and outcompete native vegetation. It can also weigh down and break the branches of shrubs and trees.

Control: Mowing or cutting can control the spread of this species. Stems should be cut as close to ground level as possible at least once per growing season; treatment is recommended prior to tuber production. Herbicide application is the most effective means for controlling large infestations. Both glyphosate and triclopyr have been shown to provide control. Foliar application should be performed after leaves are fully expanded but before aerial tubers develop. To exhaust the reserves in the underground tubers, repeated treatment is typically necessary over several growing seasons.

Young *Dioscorea polystachya* leaves. Image by Chris Evans, University of Illinois, Bugwood.org.

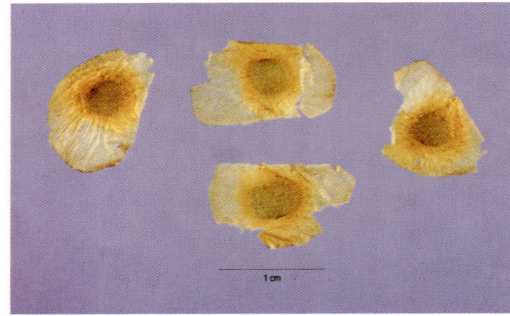

Dioscorea polystachya seeds. Image by Steve Hurst, hosted by the USDA-NRCS PLANTS Database.

Mature *Dioscorea polystachya* leaves. Image by James H. Miller, USDA Forest Service, Bugwood.org.

Dioscorea polystachya leaves and stems. Image by Chris Evans, University of Illinois, Bugwood.org.

Aerial tubers of *Dioscorea polystachya*. Image by Chris Evans, University of Illinois, Bugwood.org.

Dioscorea polystachya infestation. Image by Chris Evans, University of Illinois, Bugwood.org.

Fabaceae

Pueraria montana var. *lobata*

kudzu · kudzu bean · kudzu vine · Japanese
arrowroot · Mason-Dixon vine · porch vine ·
telephone vine · wonder vine

ODA Invasive
OIPC Invasive
**Federal Noxious
Weed**

Origin: Native to eastern Asia—particularly
China, Korea, and Japan—and other countries
in southeastern Asia. This plant has long been
cultivated for food, fiber, and traditional medicine. It became popular in the United States after
being displayed at the Centennial International
Exposition in Philadelphia in 1876. Kudzu was
initially grown as an ornamental shade plant,
but from the 1930s until the 1950s it was widely
promoted for erosion control and livestock forage. It escaped cultivation and has become one
of the most recognizable and iconic weeds in
the southern United States. This species' range
has been steadily expanding northward, and it is
now present in Ohio. In addition to being listed
as invasive, *Pueraria montana* var. *lobata* is considered a noxious weed at the federal level and is
illegal to move throughout the United States.

Description: A semiwoody, deciduous, trailing or climbing perennial vine. It is sometimes
referred to as "the vine that ate the South"
because it grows quickly—as much as 30 cm in
a day and up to 30 m in a growing season—and
can completely engulf structures and overtop
other vegetation.

Habitat: Forests, forest margins, grasslands,
thickets, roadsides, embankments, fencerows,
and abandoned fields. It does best in full sun
and warm climates but can tolerate partial
shade and withstand frost.

Height: Stems 10–30 m (30–100 ft.).

Foliage: Leaves are compound, alternate,
comprising three ovate or diamond-shaped
leaflets, each 8–20 cm (3–8 in.) long and 5–19
cm (2–8 in.) wide. Leaflets are entire or two-
or three-lobed, with fine golden hairs along the

margins and underneath. Petioles are typically
15–20 cm (6–8 in.) long, covered with long
hairs, and swollen at the base.

Flowers: Purple, reddish-purple, or pink
flowers, in upright or occasionally hanging
spikes that are 10–30 cm (4–12 in.) long, growing from the axils of leaves. Each individual
flower is 1–2.5 cm (½–1 in.) long, pealike, with
a large upper lip that has a yellow mark at the
base, and four modified petals folded together
to form the lower lip. The lower petals enclose
the flower's reproductive parts, which comprise
ten fused stamens and a single style. Only a few
flowers are open on a single inflorescence at
one time, with flowers maturing in succession
from the base to the tip. Flowers have a sweet
fragrance, reminiscent of grapes. July–October.

Fruit: Fruits are brown, densely hairy, flattened seedpods, 2–13 cm (¾–5 in.) long and
7–12 mm (¼–½ in.) wide. Each pod contains
1–10 oval to slightly kidney-shaped seeds, 3–5
mm (⅛–¼ in.) long and 3–4 mm (⅛ in.) wide,
reddish brown to greenish gray at maturity.
October–November.

Stems: Stems regrow from stored carbohy-
drates in the roots each spring. Young stems
are green or mottled, covered with light brown
hairs, and climb by twining upward, from left
to right. Older stems become dark brown and
woody. Stems are typically 1–3 cm (½–1 in.) in
diameter, occasionally growing to 10 cm (4 in.).
Creeping stems form new roots and shoots at
the nodes where they contact soil.

Native Alternatives

Wisteria frutescens (Atlantic wisteria)

233

Pueraria montana var. *lobata* leaves. Image by James H. Miller and Ted Bodner, Southern Weed Science Society, Bugwood.org.

Pueraria montana var. *lobata* flowers. Image by Leslie J. Mehrhoff, University of Connecticut, Bugwood.org.

Young stems of *Pueraria montana* var. *lobata*. Image by Leslie J. Mehrhoff, University of Connecticut, Bugwood.org.

Pueraria montana var. *lobata* fruits. Image by Leslie J. Mehrhoff, University of Connecticut, Bugwood.org.

Mature stems of *Pueraria montana* var. *lobata*. Image by Leslie J. Mehrhoff, University of Connecticut, Bugwood.org.

Pueraria montana var. *lobata* seeds. Image by Steve Hurst, hosted by the USDA-NRCS PLANTS Database.

Root system: Plants develop an extensive root system with large, tuberous swellings that can be up to 2 m (6 ft.) long and 20–45 cm (8–18 in.) wide. These provide carbohydrate storage for perennial growth and allow for resprouting after disturbance.

Reproduction: Although seeds are produced, seed set is low in this species, and seedlings are rare. It is possible that some seeds are spread by birds and mammals. Reproduction is primarily vegetative, through the rooting of stem nodes on creeping stems or stem fragments.

Impact: Stands of kudzu reduce biodiversity because they outcompete other vegetation and develop large-scale monocultures. This species can overtop and subsequently kill mature trees.

Kudzu is not currently in all parts of Ohio, but the entire state is within its potential range. It is important to detect new populations early and eradicate them before they become established.

Control: Before undertaking any control efforts, make sure you are dealing with kudzu and not a native trifoliate legume such as *Desmodium rotundifolium* (prostrate tick-trefoil) or *Amphicarpaea bracteata* (hog peanut), or the ubiquitous trifoliate *Toxicodendron radicans* (poison ivy). The most effective management method is to cut below the root crown and remove any stems to prevent rooting from the nodes. This is easiest to do in the spring, before stems have grown.

Excavated root of *Pueraria montana* var. *lobata*. Image by Forest Starr and Kim Starr, Starr Environmental, Bugwood.org.

Pueraria montana var. *lobata* infestation. Image by Kerry Britton, USDA Forest Service, Bugwood.org.

Fabaceae

Securigera varia

crown vetch · purple crown vetch · trailing
crown vetch · axseed

Synonym: *Coronilla varia*

Origin: The original distribution of this species is not entirely clear, but records suggest that it is native to most of Europe, western parts of Asia, and possibly northern parts of Africa. The earliest report of *Securigera varia* in the United States is from 1869, but it is not clear whether it was planted intentionally or introduced accidentally as a contaminant in crop seed. The subsequent spread of this species was deliberate. From the 1950s to the 1980s, it was planted extensively to revegetate or provide erosion control to roadsides, railroad embankments, and mine sites. This species was also planted as an ornamental and used as a cover crop. It has become established and escaped into natural areas throughout much of the United States.

Description: A herbaceous perennial with creeping stems. It is similar in appearance to some other members of the legume family, such as native *Vicia* spp. (vetches) and the non-native *Lotus corniculatus* (birdsfoot trefoil). However, three characteristics in combination distinguish *S. varia* from other species: the compound leaves have an odd number of leaflets, the leaves and flower stalks grow from the main stem, and the pink flowers are arranged in an umbel.

Habitat: Planted extensively along roadsides, railroads, and agricultural sites. Has spread from plantings into prairies, grasslands, meadows, old fields, forest margins, riverbanks and drainage ditches, power line rights-of-way, and waste areas. Prefers full sun but can tolerate partial shade and a range of soil types. Can withstand drought, heavy precipitation, and nutrient-poor soils.

Height: Plants typically reach a height of 30–100 cm (1–3 ft.), but creeping stems can be up to 2 m (6 ft.) long.

Foliage: Leaves are pinnately compound, alternate, 6–15 cm (3–6 in.) long, with 7–25 leaflets per leaf, each leaflet oblong to elliptic, 1–3 cm (½–1 in.) long and 5–10 mm (¼–½ in.) wide. Each leaflet is smooth along the margins and often has a small, toothlike point at the tip.

Flowers: Flowers are variably pink to white, in umbels of 5–25 blossoms, held on stalks 5–15 cm (2–6 in.) long that develop from the axils of the upper leaves. Each flower is about 9–15 mm (⅜–⅝ in.) long, pealike, with a larger upper lip formed by a single petal with two lobes that is typically a darker shade of pink than the rest of the flower, and four modified petals folded together to form the lower lip. The lower petals enclose the reproductive parts of the flower, which comprise ten stamens and a single style. May–September.

Fruit: Each flower develops into a linear seedpod, 2–8 cm (1–3 in.) long and 2–3 mm ($\frac{1}{16}$–⅛ in.) wide, green drying to brown, four-angled with 3–12 joints along its length and pointed at the end. Each joint of the seedpod contains a single reddish brown seed, nearly cylindrical, 3–5 mm (⅛–¼ in.) long by 1–2 mm ($\frac{1}{32}$–$\frac{1}{16}$ in.) wide. August–October.

Native Alternatives

Apocynum androsaemifolium (spreading dogbane)
Asclepias verticillata (whorled milkweed)
Dalea purpurea (purple prairie clover)
Lathyrus venosus (veiny pea)
Vicia americana (American vetch)

Securigera varia leaves. Image by Patrick J. Alexander, hosted by the USDA-NRCS PLANTS Database.

Securigera varia flowers. Image © Gerald D. Carr.

Securigera varia fruits. Image by John Cardina, The Ohio State University, Bugwood.org.

Left: Securigera varia fruits. *Above: Securigera varia* stem. Both images © Gerald D. Carr.

Securigera varia seeds and segment of mature pod. Image by Steve Hurst, hosted by the USDA-NRCS PLANTS Database.

Roadside planting of *Securigera varia.* Image by James H. Miller, USDA Forest Service, Bugwood.org.

Stems: Stems are green, hairless, growing upright or creeping over the ground and nearby vegetation. Does not produce tendrils or twine for support.

Root system: Produces both roots and perennial rhizomes. The taproot is deep and branching. The fleshy rhizomes can grow horizontally to 3 m (10 ft.) or longer, from which new roots and shoots are produced each growing season. The roots form associations with nitrogen-fixing bacteria.

Reproduction: By seed. Soil seed bank viability is estimated to be 2–15 years or more. Vegetative reproduction also occurs via spreading rhizomes and regeneration from stem and rhizome fragments.

Impact: Forms dense stands that exclude native plant species. Due to the extensive rhizome system, it is difficult to remove once established. This plant contains nitroglycosides that are toxic to some mammals—particularly horses and humans—making it a concern in pastures.

Control: For small infestations, hand pulling and digging out of rhizomes is recommended. To deplete reserves in any remaining rhizome fragments, repeated treatment will likely be necessary. Late spring mowing over successive years can also help control this species. Small patches can be covered with black plastic or thick mulch. The herbicides 2,4-D, glyphosate, triclopyr, and clopyralid have all been found to be effective in controlling this species. To prevent regeneration through persistent rhizomes and dormant seeds, multiple applications are required over several years.

Polygonaceae

Fallopia convolvulus

black bindweed · wild buckwheat · climbing buckwheat · climbing knotweed · cornbind · dullseed cornbind · bearbind · devil's tether · pink smartweed

Synonyms: *Polygonum convolvulus, Reynoutria convolvulus*

Origin: Native to Europe, Asia, and northern Africa. The year of introduction to North America is not clear, but it was most likely a contaminant in crop seed. *Fallopia convolvulus* is a serious agricultural weed in most regions of the world. It has been recorded in every state but is less abundant in the southern United States.

Description: A trailing or twining annual vine. Its foliage and growth habit are superficially similar to those of the bindweeds in the genera *Calystegia* and *Convolvulus*. However, *F. convolvulus* has papery ocreae where the leaves attach to the stems and lacks the bindweeds' large, funnel-shaped flowers. It is also similar to two other vining *Fallopia* species native to Ohio. *Fallopia cilinodis* (fringed black bindweed) has a ring of long hairs at the base of the leaf sheath, more pronounced venation in the leaves, and denser clusters of flowers than *F. convolvulus*. *Fallopia scandens* (climbing false buckwheat), has prominent wings on its fruits and shiny black seeds, whereas *F. convolvulus* has nearly wingless fruits and dull black seeds.

Habitat: Occurs most commonly on land cultivated for cereal crops, vegetables, annual flowers, and orchards. It is also found in disturbed areas such as thickets, roadsides, railways, fencerows, and along riverbanks. It grows in a wide range of soil types and is tolerant of drought.

Height: To 250 cm (8 ft.).

Foliage: Leaves are simple, alternate, typically 2–6 cm (¾–3 in.) long and 2–5 cm (¾–2 in.) wide but occasionally to 15 cm (5 in.) long and 10 cm (4 in.) wide, heart-shaped or arrow-shaped, with a pointed tip and widely separated lobes at the base. Each leaf has a petiole 6–15 mm (¼–⅝ in.) long (occasionally to 5 cm [2 in.]), leaf margins are slightly wavy, and there are short hairlike projections along the margins, lower veins, and petioles. There is a papery ocrea, with smooth and hairless margins, above the node where the petiole attaches to the stem.

Flowers: Small, inconspicuous flowers are green and white, often with pink or purplish markings, growing in the axils of leaves in clusters of three to six or in densely flowered terminal stalks 2–15 cm (¾–6 in.) long. Individual flowers are 6 mm (¼ in.) in diameter, with five petallike sepals, and eight stamens encircling a single style. May–October.

Fruit: A three-sided husk, 4–6 mm (⅛–¼ in.) long and 2–3 mm (¹⁄₁₆–⅛ in.) wide, green drying to brown, encloses a single achene. The achene is dull black or brownish black, 2–4 mm (¹⁄₁₆–⅛ in.) long and 1.5–2.5 mm (¹⁄₃₂–¹⁄₁₆ in.) wide, strongly triangular in cross-section, with a blunt base and a pointed tip. June–November.

Stems: Slender green stems, often turning reddish with age, branching freely from the base. Stems trail along the ground or twine upward, from right to left around other plants.

Root system: A slender, branching taproot and fibrous root system that can grow to 80 cm deep. The deep root system allows this plant to survive drought.

Reproduction: This species can potentially produce thousands of seeds per plant, which are then dispersed by farm machinery and water and as a contaminant in cereal crops. The seeds remain viable in the soil for many years.

239

Impact: As a pest of cereal grains and other crops, this species causes significant yield reductions. There is limited information about this species as a weed in natural areas, but it is problematic in disturbed habitats, where it can suppress native vegetation.

Control: In agricultural settings, mechanical methods such as tilling and harrowing have been found to be partially effective at controlling *F. convolvulus*. Many herbicides are used to control this species, but it has developed resistance to 2,4-D, chlorsulfuron, and others. In a garden or natural area, try cutting stems below the root crown. It is critical to remove plants before they set seed. Seeds germinate throughout the growing season, so it is important to monitor continuously for new plants.

Fallopia convolvulus leaves. Image by Lynn Sosnoskie, University of Georgia, Bugwood.org.

Fallopia convolvulus fruits. Image © Robert L. Carr.

Papery ocrea at the nodes of *Fallopia convolvulus*. Image by Stefan Lefnaer, used under a CC BY-SA 4.0 license.

Fallopia convolvulus seeds. Image by Ken Chamberlain, The Ohio State University, Bugwood.org.

Fallopia convolvulus flower. Image by Stefan Lefnaer, used under a CC BY-SA 4.0 license.

Fallopia convolvulus growth habit. Image © Robert L. Carr.

Polygonaceae

Persicaria perfoliata

mile-a-minute weed · mile-a-minute vine · minuteweed · Asiatic tearthumb · giant climbing tearthumb · devil's tearthumb · devil's tail

Synonym: *Polygonum perfoliatum*

Origin: Native to many parts of eastern and southeastern Asia. An accidental introduction through ship ballast was first recorded in Oregon in the 1890s. The species was subsequently introduced to Pennsylvania in the 1930s via contaminated nursery stock. From this second point of introduction, it established and has spread through the northeastern United States.

Description: A fast-growing, herbaceous annual vine that creeps along the ground or climbs over other vegetation with the aid of recurved prickles on the stem and leaves. *Persicaria perfoliata* can be differentiated from two native relatives—*Persicaria arifolia* (halberd-leaved tearthumb) and *Persicaria sagittata* (arrow-leaved tearthumb)—by its nearly triangular leaves and rounded leafy ocreae that encircle the stems at the nodes.

Habitat: Found in forest margins, thickets, roadsides, railroads, fields, wetlands, and streambanks. It prefers moist or wet areas in full sun but can tolerate dry soils and some shade for part of the day.

Height: Can grow to 6 m (20 ft.) or more with support from shrubs and understory trees.

Foliage: Leaves are simple, alternate, 3–8 cm (1–3 in.) long and 3–9 cm (1–3 ½ in.) wide, triangular, light green, with a papery texture. The petioles and the main veins on the undersurface of leaves are covered in recurved prickles.

Flowers: Small, white, inconspicuous flowers in a spike 1–2 cm (½–¾ in.) long with clusters of 10–15 flowers. Flowering stalks are either at the ends of stems or in the axils of leaves, and each inflorescence has a rounded leafy ocrea at the base. Individual flowers are 3–5 mm (⅛–¼ in.) in diameter, with three stigmas, often

remaining closed because they are capable of self-pollination. June–October.

Fruit: Flowers develop into metallic blue, fleshy, berrylike structures, 5 mm (¼ in.) in diameter, each surrounding a single achene. Each achene is shiny black or reddish black, nearly spherical, 2–3 mm (¹⁄₁₆–⅛ in.) in diameter. July–November.

Stems: Stems are light green or reddish, slender, highly branched, and covered in short, recurved prickles that aid in climbing and scrambling. Round leafy ocreae encircle the stems at the nodes.

Root system: Shallow and fibrous.

Reproduction: By seed. The fleshy covering allows seeds to float, and water, birds, mammals, and ants disperse them. Machinery and contaminated nursery stock likely facilitate further spread.

Impact: Forms dense, tangled mats that kill herbaceous vegetation and tree seedlings underneath. This species can grow up to 15 cm (6 in.) a day, which gives it a competitive advantage. Extracts have also been shown to be allelopathic.

Control: To prevent injury by this species' sharp prickles, wear protective clothing and gloves when handling this plant. Stems and roots can be pulled and should be left to dry in the sun for several days before disposal. Repeated mowing or cutting throughout the growing season will prevent flowering and seed production. Foliar applications of glyphosate, clopyralid, and triclopyr have been shown to control this species.

Prohibited Noxious Weed
⚠ Do Not Touch

241

Persicaria perfoliata leaves. Image by Leslie J. Mehrhoff, University of Connecticut, Bugwood.org.

Stems of *Persicaria perfoliata*. Image by Leslie J. Mehrhoff, University of Connecticut, Bugwood.org.

Ocreae at the nodes of *Persicaria perfoliata*. Image by Leslie J. Mehrhoff, University of Connecticut, Bugwood.org.

Persicaria perfoliata flowers. Image by Dalgial, used under a CC BY-SA 3.0 license.

Persicaria perfoliata fruits. Image by Leslie J. Mehrhoff, University of Connecticut, Bugwood.org.

Persicaria perfoliata seeds. Image by Leslie J. Mehrhoff, University of Connecticut, Bugwood.org.

Persicaria perfoliata infestation. Image by Leslie J. Mehrhoff, University of Connecticut, Bugwood.org.

Ranunculaceae

Clematis terniflora

sweet autumn clematis · sweet autumn
virginsbower · Japanese clematis · leatherleaf
clematis · yam-leaved clematis

243

Origin: Native to northeastern Asia, including China, Japan, Korea, Russia, and Taiwan. *Clematis terniflora* was introduced to North America in 1877 as an ornamental plant. It has escaped cultivation and established in many parts of the eastern United States. This species is designated as invasive in many states in the southern and mid-Atlantic regions.

Description: A deciduous, semiwoody, perennial vine that climbs with the aid of twining petioles. It is similar to the native *Clematis virginiana* (virgin's bower or woodbine), but in *C. virginiana* the margins of the leaflets are sharply toothed, while in the nonnative *C. terniflora* they are smooth.

Habitat: Forest margins, thickets, streambanks, fencerows, and roadsides. Prefers full sun or partial shade and can grow in a range of soil types.

Height: Typically 2–3 m (6–10 ft.) but can climb to 9 m (30 ft.).

Foliage: Leaves are opposite, pinnately compound, with three to five leaflets. Leaflets are 5–8 cm (2–3 in.) long and 2–4 cm (¾–2 in.) wide, ovate, with parallel veins and smooth margins. The leaflets' upper surfaces are dark green and shiny, paler underneath. The long petioles and petiolules can bend or twine around neighboring objects for support.

Flowers: Abundant star-shaped white flowers, 2–3 cm (¾–1 in.) in diameter, are produced in flat-topped inflorescences from the axils of the upper leaves. Each flower has four (occasionally five) petallike sepals, numerous long stamens, and five or six pistils with long styles toward the center. Very fragrant. August–September.

Fruit: Flowers develop into a cluster of five or six achenes connected at the heads, each with a persistent style. The achenes are flattened ovoids, 5–9 mm (¼–⅜ in.) long, while the styles are 2–3 cm (¾–1 in.). The styles are covered with long white hairs that spread with age, giving them a feathery appearance. September–October.

Stems: Young stems are smooth and green, while older stems are brown and semiwoody. There are swollen nodes every 15–20 cm (6–8 in.) along the stem. In the spring, new shoots are produced on the previous year's woody stems.

Root system: Plants develop long taproots.

Reproduction: By seed, dispersed by wind. Vegetative regeneration can occur from roots or root fragments.

Impact: This species often occurs at high densities in infested areas, and it has been shown to overtop and outcompete native herbaceous plants, shrubs, and tree seedlings.

Control: Avoid planting new specimens of *C. terniflora,* and remove any existing plants. The sap of this species contains a toxic compound called protoanemonin, which can cause irritation and blistering of the skin and mucous membranes. Wear protective clothing and gloves when handling this plant. Seedlings can be pulled by hand or mowed. Larger plants should be cut back and the root system dug out. Cut stem application of triclopyr can also help prevent regrowth from the taproot.

Native Alternatives
Clematis virginiana (virgin's bower)

Clematis terniflora leaves. Image by Alpsdake, used under a CC BY-SA 4.0 license.

Developing fruits on *Clematis terniflora*. Image by Leslie J. Mehrhoff, University of Connecticut, Bugwood.org.

Clematis terniflora stems. Image by Chris Evans, University of Illinois, Bugwood.org.

Mature fruits on *Clematis terniflora*. Image by Leslie J. Mehrhoff, University of Connecticut, Bugwood.org.

Clematis terniflora flowers. Image by Karan A. Rawlins, University of Georgia, Bugwood.org.

Clematis terniflora infestation. Image by Leslie J. Mehrhoff, University of Connecticut, Bugwood.org.

Vitaceae

Ampelopsis brevipedunculata

porcelain berry · porcelain ampelopsis · Amur peppervine · wild grape · creeper

Origin: Native to northeastern Asia, including parts of China, Japan, Korea, and eastern Russia. This plant was introduced to the United States in 1870 for ornamental use. It has escaped cultivation and become established in a patchy distribution throughout the eastern United States. Despite its weedy tendencies, this species is still commercially available and used in landscaping.

Description: A deciduous, woody, perennial vine that climbs with the aid of tendrils. The leaves and growth habit closely resemble native grape species in the genus *Vitis*. However, *Ampelopsis brevipedunculata* has branching tendrils, its bark does not peel, and the pith in the center of its stems is white, whereas the tendrils of *Vitis* spp. do not branch, its bark peels off in strips, and its pith is brown or tan.

Habitat: Forests, forest margins, fields, pastures, streambanks, pond margins, thickets, railways, and roadsides. Grows in a wide range of light availability and soil types.

Height: To 8 m (26 ft.).

Foliage: Leaves are simple, alternate, 6–12 cm (3–5 in.) long and 3–11 cm (1–5 in.) wide, variable in shape, occasionally heart-shaped but more typically with three (or five) lobes. Leaves are palmately veined and have a coarsely toothed margin. The upper surface is dark green, while the underside is shiny and has coarse hairs along the veins.

Flowers: Small, greenish-white flowers are arranged in umbrella-shaped cymes opposite the leaves. Each flower is 4 mm (⅛ in.) in diameter, with five petals, five stamens, and a single pistil. July–August.

Fruit: A hard, spherical berry, 4–8 mm (⅛–¼ in.) in diameter, ranging in color from blue to purple, pink, green, or yellow, with multiple colors occurring simultaneously on the same plant. The berry's surface is shiny with slightly raised brown speckles. Each fruit contains one to four seeds, each 3–4 mm (⅛ in.) long and 2–4 mm (¹⁄₁₆–⅛ in.) wide, typically with a triangular ovoid shape. September–October.

Stems: Young stems are green and hairy, over time developing into smooth brown bark with white lenticels. Older stems have furrowed brown bark. The stem's pith is white. Branching tendrils develop opposite the leaves on the stem.

Root system: Detailed information on the root system of this species is lacking, but sources suggest that it is extensive and that each plant develops a large, deep taproot.

Reproduction: By seed, dispersed by birds and mammals. The berries float, so they are possibly dispersed by water as well. Seeds are known to have a high germination rate, contributing to the species' weedy nature. Vegetative reproduction also occurs via stem and root fragments.

Impact: This species climbs over and shades out native vegetation. Trees covered in *A. brevipedunculata* vines may be more susceptible to wind and ice damage.

Control: Avoid planting new specimens of *A. brevipedunculata,* and remove any existing

Native Alternatives
Ampelopsis arborea (peppervine)
Ampelopsis cordata (heartleaf peppervine)

plants. Pull young plants by hand, taking care to ensure that as much root material is removed as possible. For mature plants, pull the vines off trees or other vegetation and cut the stem at the base, then treat with triclopyr or glyphosate. Control measures should be performed before the plants have produced fruits.

Ampelopsis brevipedunculata leaves. Image by Pancrat, used under a CC BY-SA 3.0 license.

Young stem of *Ampelopsis brevipedunculata*. Image by Leslie J. Mehrhoff, University of Connecticut, Bugwood.org.

Mature stem of *Ampelopsis brevipedunculata*. Image by Leslie J. Mehrhoff, University of Connecticut, Bugwood.org.

Ampelopsis brevipedunculata flowers. Image by Leslie J. Mehrhoff, University of Connecticut, Bugwood.org.

Ampelopsis brevipedunculata fruits. Image by Leslie J. Mehrhoff, University of Connecticut, Bugwood.org.

Ampelopsis brevipedunculata tendrils. Image by Leslie J. Mehrhoff, University of Connecticut, Bugwood.org.

Ampelopsis brevipedunculata infestation. Image by Leslie J. Mehrhoff, University of Connecticut, Bugwood.org.

Shrubs

Shrubs are perennial, multistemmed woody plants that are usually less than 4–5 m (12–15 ft.) tall. They typically have several stems that arise near the base of the plant, but under certain environmental conditions—such as under high levels of nutrients—some species may be single-stemmed or grow taller than 5 m (15 ft.). Many nonnative shrubs leaf out early in the spring and retain their leaves late into the autumn, which gives them a competitive advantage and often has the consequence of shading out native understory plants. This is especially true of the shrub honeysuckles in the genus *Lonicera* [pp. 255–263], highly invasive plants that often grow in continuous stands. Numerous native shrub species occur in native hardwood forests, but they rarely form a dense shrub layer. Consequently, the introduction of nonnative shrubs often causes a profound change in forest structure, and it decreases the amount of sunlight penetrating the understory.

It is worth noting that most shrubs included in this book are ornamental species that have spread from cultivation. A number of these are still common landscaping plants, including *Berberis thunbergii* (Japanese barberry) [pp. 251–252], *Euonymus alatus* (burning bush) [pp. 264–265], and *Hibiscus syriacus* (rose of Sharon) [pp. 272–273]. When purchasing new plants for gardens, it is important to avoid planting nonnative

species, particularly those that have aggressive tendencies and are likely to spread. There are many beautiful and interesting native shrub species that can be used in landscaping. Furthermore, genera such as *Vaccinium* (including both blueberries and cranberries) are known to support caterpillars of more than 250 lepidopteran species (Tallamy and Shropshire 2009). In addition to being attractive, these native alternatives provide superior food and habitat for pollinators, birds, and other wildlife.

Many of the shrub species covered in this chapter have seeds that are bird-dispersed, which helps to facilitate their spread into natural areas. As with any problem plants, one of the key ways to prevent further spread is to control reproduction and dispersal. Management efforts should therefore target plants before they have ripe fruits that birds or other animals may eat. Control strategies are similar for almost all woody plants, with a recommendation to cut the stem close to the base or to girdle the bark the whole way around the stem. Most shrubs can resprout readily from the rootstock, so it is typically recommended that the root ball be removed by digging out or stump grinding whenever possible. Targeted application of herbicide to a cut stump or girdled stem may also help control subsequent growth from the rootstock.

Adoxaceae

Viburnum opulus var. opulus

cranberry viburnum · European cranberrybush · Guelder rose · water elder · red elder · rose elder · snowball tree · pincushion tree · cramp bark

Origin: A native of Europe, western and central Asia, and northern Africa that has been cultivated since the late 17th century. This species was introduced to North America as an ornamental, and it is still frequently used in landscape plantings. It has escaped from cultivation and spread into natural areas.

Description: This deciduous shrub has multiple stems, often with branches that arch to the ground, creating a rounded habit. *Viburnum opulus* var. *opulus* is a close relative of—and difficult to distinguish from—the native *Viburnum opulus* var. *americanum* (American highbush cranberry). The glands at the top of the petiole are the most reliable feature by which to differentiate the two varieties: the native variety *americanum* has convex upper surfaces, and the glands are mostly taller than they are wide, while the introduced variety *opulus* has concave upper surfaces with a distinct rim, and the glands are wider than they are tall. There are many commercially available var. *opulus* cultivars, but they should not be purchased or planted. Mislabeling of var. *opulus* as the native variety is common, so be extremely careful and only purchase var. *americanum* plants from reputable native plant nurseries.

Habitat: Primarily in disturbed areas close to cultivated landscapes. Found in open woodlands, forest edges, hedgerows, thickets, and areas with damp soils such as swamps, bogs, and streambanks. Grows in both heavy clay and acidic soil and is successful in full sun to partial shade.

Height: 1–5 m (3–15 ft.).

Foliage: Simple, opposite, 5–10 cm (2–4 in.) long and wide, three-lobed with a maplelike shape, palmately veined and coarsely toothed. The petioles are 1–3 cm (½–1 in.) long and have narrow but deep *V*-shaped grooves down the center. At the top of the petiole there is a cluster of two to eight large, oval glands that are concave on top and have distinct rims around the edges.

Flowers: Flowers are white and have five petals each, borne in a flat-topped cyme 5–11 cm (2–4 in.) wide. The flowers have two forms: around the outside of the inflorescence there are slightly irregular, large flowers, 1.5–2.5 cm (⅝–1 in.) wide, and in the center there are numerous smaller flowers, 3–6 mm (⅛–¼ in.) wide. The large flowers are sterile, lacking stamens or pistils, while the small flowers are fertile, with five long stamens. Some cultivars such as 'Snowball' and 'Sterilis' have been bred to have only large flowers in rounded clusters. May–June.

Fruit: A spherical, bright red drupe, 7–10 mm (¼–½ in.) in diameter. Fruits are edible in small

Native Alternatives

Calycanthus floridus (Carolina allspice)
Cornus sericea (red osier dogwood)
Hydrangea arborescens (wild hydrangea)
Ilex verticillata (winterberry holly)
Sambucus racemosa (red elderberry)
Vaccinium corymbosum (highbush blueberry)
Viburnum acerifolium (mapleleaf viburnum)
Viburnum nudum (witherod viburnum)
Viburnum opulus var. *americanum* (American highbush cranberry)

Viburnum opulus var. opulus fruits. Dow Gardens, Bugwood.org.

Viburnum opulus var. opulus leaves. Image by Robert Vidéki, Doronicum Kft., Bugwood.org.

Viburnum opulus seeds. Image by Steve Hurst, hosted by the USDA-NRCS PLANTS Database.

Concave glands on the petioles of Viburnum opulus var. opulus. Image © Peter M. Dziuk, Minnesota Wildflowers.

Variation in Viburnum opulus fruit color. Image by Jan Mehlich, used under a CC BY-SA 3.0 license.

Viburnum opulus var. opulus flowers. Image by Robert Vidéki, Doronicum Kft., Bugwood.org.

quantities but extremely acidic; they are less palatable to wildlife than the native variety *americanum*. Each fruit contains a single flattened seed, nearly circular, 5 mm (¼ in.) across and 1 mm (¹⁄₃₂ in.) thick. Fruits develop August–September but persist on the plant into winter.

Stems: Young, nonwoody shoots are hairless, green to reddish green. Twigs are reddish brown with raised white lenticels. The bark of trunks and older branches is gray, thin, and smooth. Multiple main stems grow from the base, typically to about 5 cm (2 in.) diameter, and are very straight.

Root system: Woody and branching. Root suckering can occur where arching stems touch the ground.

Reproduction: Primarily by seed, dispersed by birds and mammals.

Impact: This species displaces native plants through competition. In many places, the introduced and native varieties hybridize, causing a blending of the genes of the introduced variety into the gene pool of the native variety. This introgression is considered to degrade the gene pool of the native var. *americanum*.

Control: Do not plant var. *opulus* cultivars in your garden. Young plants can be pulled by hand before the root system becomes well established. Older plants should be cut at ground level and treated with glyphosate or triclopyr to kill the roots and prevent resprouting.

Berberidaceae

Berberis thunbergii

Japanese barberry · Thunberg's barberry · red barberry

Origin: Native to Japan and China, this shrub was introduced to the United States in 1875 as an ornamental species. By 1920 it was being promoted as a replacement for the European species *Berberis vulgaris* (common barberry) [pp. 253–254]. *Berberis thunbergii* is still commonly used in landscaping and frequently escapes cultivation.

Description: A low-growing, deciduous shrub with arching branches. Each densely branched stem has a single, narrow spine at each node.

Habitat: Closed canopy forests, open woodlands, woodland edges, fields, streambanks, and wetlands. Extremely adaptable: grows well in full sun to deep shade, can tolerate dry to very moist conditions.

Height: Typically 1 m (3 ft.), sometimes up to 2 m (6 ft.).

Foliage: Simple, alternate, 1–2.5 cm (½–1 in.) long and 3–18 mm (⅛–¾ in.) wide, oval or spatula-shaped, with a smooth margin and rounded tip. Leaf color is green to blue-green, occasionally red or purple in some cultivars. Leaves are borne in clusters of two to six along the branches. This is one of the first shrubs to leaf out in spring.

Flowers: Pale yellow, 5–8 mm (¼ in.) in diameter, solitary in or clusters of two to five. Flowers have six sepals, six inner petals that are smaller than the sepals, and six stamens. Flowers are profuse on the upper parts of the stems. April–May.

Fruit: A glossy, bright red, ovoid berry, 7–11 mm (¼–½ in.) long and 4–7 mm (⅛–¼ in.) wide, singly or in clusters, on slender stalks. Each berry contains a single ovoid seed, 4 mm

(⅛ in.) long by 2 mm ($\frac{1}{16}$ in.) wide. July–October, with fruits persisting until the winter.

Stems: Multiple stems arise from the base and each divides into many branches. The slender, woody stems have a zigzag form and single sharp spines, about 1 cm (½ in.) long, at the base of each cluster of leaves. The deeply grooved outer bark is brown, and the inner bark is yellow.

Root system: A large, shallow root system with both rhizomes and fibrous fine roots spreading out from the crown. The interior of the roots is yellow.

Reproduction: Seeds are dispersed by gravity, birds, and mammals such as rabbits and deer. An individual plant can produce over 1,000 seeds in a single growing season. Vegetative reproduction also occurs, through sprouting from the root crown or rhizomes and the rooting of long stems that touch the ground.

Impact: The combination of vegetative reproduction and seed dispersal by gravity results in dense stands that often cover large areas. These stands shade out understory species such as herbaceous plants and tree seedlings. Stands of this species have been shown to raise the pH of the

Native Alternatives

Aronia melanocarpa (black chokeberry)
Ceanothus americanus (New Jersey tea)
Diervilla lonicera (northern bush honeysuckle)
Ilex verticillata (winterberry holly)
Morella pensylvanica (northern bayberry)
Symphoricarpos albus (snowberry)
Symphoricarpos orbiculatus (coralberry)

soil—making it more alkaline—and alter nitrogen cycling. Furthermore, areas with *B. thunbergii* have higher numbers of *Ixodes scapularis* (deer ticks), which transmit Lyme disease.

Control: Wear gloves to protect against the sharp spines. For individual plants and small populations, hand pulling or digging are recommended. To prevent regrowth, it is essential to remove all root material. If plants have fruits, they should be bagged and carefully disposed of to prevent seed dispersal. In larger populations, mowing or cutting are recommended, ensuring that stems are cut as close to ground level as possible. Cut stems and resprouts can be treated with systemic herbicides like glyphosate and triclopyr.

Berberis thunbergii leaves. Image by James H. Miller, USDA Forest Service, Bugwood.org.

Berberis thunbergii fruits. Image by Chris Evans, University of Illinois, Bugwood.org.

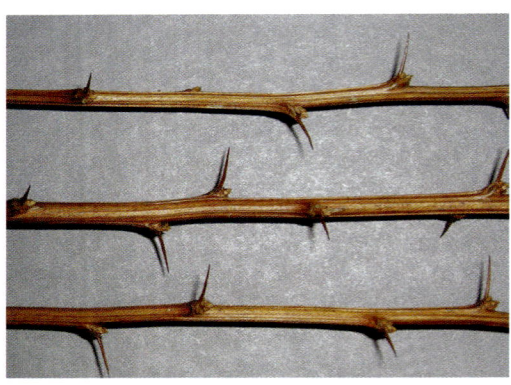

Berberis thunbergii twigs. Image by Leslie J. Mehrhoff, University of Connecticut, Bugwood.org.

Berberis thunbergii seeds. Image by Steve Hurst, hosted by the USDA-NRCS PLANTS Database.

Berberis thunbergii flowers. Image by Leslie J. Mehrhoff, University of Connecticut, Bugwood.org.

Berberis thunbergii infestation. Image by Leslie J. Mehrhoff, University of Connecticut, Bugwood.org.

Berberidaceae

Berberis vulgaris
common barberry · European barberry

ODA Invasive

Origin: Native to Europe, northern Africa, and temperate Asia. This species, which has a long history of cultivation, was introduced to North America by the late 1600s. European colonists widely planted it for food, for yellow dye production, and as hedgerow barriers. There are records suggesting that this species had escaped from cultivation by the mid 1700s. Shortly after the introduction and escape of this species, it was identified as an alternate host for *Puccinia graminis* (black stem rust), which attacks wheat and other cereal crops. Many states—including Ohio—have laws that prevent the sale, transport, and planting of this species or that require the eradication of existing plants.

Description: A deciduous shrub with widely spreading, arching branches. The densely branched stems typically have a three-parted spine at each node.

Habitat: Woodlands, along the edges of rivers, grassland, old fields, pastures, fence lines, and roadsides. It can grow in full sun to shade in a variety of soil types but is particularly successful in limestone soils.

Height: 2–3 m (6–10 ft.).

Foliage: Simple, alternate, ovate to oblong, 2–5 cm (¾–2 in.) long and 1–2 cm (½–¾ in.) wide, with a narrowly tapered base and a rounded tip. The margins of the leaves are finely toothed. Leaf color is dull green, turning a dull red in the autumn. Leaves are in clusters of two to five at each node.

Flowers: Bright yellow flowers, 4–8 mm (⅛–¼ in.) in diameter, in drooping racemes of 10–20 flowers that hang down from the stem. Flowers have six sepals, six inner petals that are smaller than the sepals, and six stamens. Flowers have an unpleasant odor. May–June.

Fruit: Bright red or purple, ovoid berries, 8–12 mm (¼–½ in.) long and 3–5 mm (⅛–¼ in.) wide, in drooping clusters that hang down from the stem. Fruits are edible but very tart. Each berry contains one or two seeds, 5–6 mm (¼ in.) long and 2–4 mm ($^1/_{16}$–⅛ in.) wide. August–October, with fruits persisting until the winter.

Stems: Multiple stems arise from the base, and each divides into many branches that spread widely. Bark is grooved, gray, and smooth, with sharp three-parted (occasionally fewer) spines, 1–2 cm (½–¾ in.) long. The spines are usually round in cross-section but are sometimes flat. Older stems have gray shedding bark, inner bark is yellow.

Root system: A large, shallow root system with both rhizomes and fibrous fine roots spreading out from the root crown.

Reproduction: By seeds, dispersed by birds and mammals. An individual plant can produce 100–5,000 seeds in a growing season, and seeds remain viable in the soil for nine years or more. Vegetative reproduction also occurs, through sprouting from the root crown or rhizomes and the rooting of long stems that touch the ground.

Impact: Can form dense stands that outcompete native vegetation. However, the primary concern is that this species is a reservoir for black stem rust, which affects cereal grains.

Control and Native Alternatives: See *Berberis thunbergii* [pp. 251–252]. For this species, it is critical to prioritize removal of plants growing near agricultural fields.

253

Berberis vulgaris leaves. Image by Leslie J. Mehrhoff, University of Connecticut, Bugwood.org.

Berberis vulgaris stems with spines. Image by Leslie J. Mehrhoff, University of Connecticut, Bugwood.org.

Pendulous racemes of *Berberis vulgaris.* Image by Leslie J. Mehrhoff, University of Connecticut, Bugwood.org.

Berberis vulgaris flowers. Image by Leslie J. Mehrhoff, University of Connecticut, Bugwood.org.

Berberis vulgaris fruits. Image by Leslie J. Mehrhoff, University of Connecticut, Bugwood.org.

Berberis vulgaris seeds. Image by Steve Hurst, hosted by the USDA-NRCS PLANTS Database.

Berberis vulgaris growth habit. Image by Leslie J. Mehrhoff, University of Connecticut, Bugwood.org.

Caprifoliaceae

Lonicera maackii

Amur honeysuckle · bush honeysuckle · late honeysuckle

Origin: Native to temperate eastern Asia, including parts of Russia, China, Japan, and Korea. This species was introduced to North America in the late 1800s as an ornamental. From the 1960s to the 1980s the USDA Soil Conservation Service promoted its planting for soil erosion control and wildlife cover. It has escaped cultivation and become highly invasive throughout much of the United States.

Description: A large, deciduous shrub with multiple stems and arching branches. It is generally larger and more aggressive than the other invasive shrub honeysuckles. Despite its invasive tendencies, there are still several cultivars available commercially.

Habitat: Forest understory and edges, thickets, fencerows, old fields, roadsides, and other disturbed areas. Can grow in a wide range of soil types and in full sun to deep shade. This is the most shade tolerant of the invasive shrub honeysuckles and, because it can survive under tall trees, it has the greatest potential to spread into the forest interior. Tolerant of pollution, drought, and cold.

Height: Typically to 5 m (15 ft.), occasionally taller.

Foliage: Simple, opposite, oval but with a long tapering point that distinguishes it from related species, 4–10 cm (2–4 in.) long and 2–4 cm (¾–2 in.) wide. The leaf margins are toothless. The leaf blades are dark green above and lighter underneath; the veins on the undersides are covered in fine hairs. As with other invasive shrub honeysuckles, the leaves emerge early in the spring a few weeks before native trees and shrubs.

Flowers: Tubular flowers with extremely short flower stalks, growing in pairs from the leaf axils, 1.5–2.5 cm (⅝–1 in.) long. Flowers are white, turning yellow with age, with one narrow lobe forming a small lower lip and four fused lobes forming a larger upper lip. Each flower has five yellow-tipped stamens and a white style with a rounded green stigma protruding from the tube. Very fragrant. May–June.

Fruit: A spherical, semitranslucent, dark red berry, 6–7 mm (¼ in.) in diameter. Like the flowers, the fruits grow in pairs in the leaf axils. Each berry contains several small, pale brown seeds, about 3 mm (⅛ in.) long and 2 mm (1/16 in.) wide. Fruits ripen in October and persist on the plants through winter.

Stems: Individual stems grow to a maximum diameter of 10 cm (4 in.). Young branches and twigs are smooth-textured. Mature bark is grayish brown and appears braided, with flat scaly ridges and narrow grooves, peeling off in vertical strips. As with all invasive shrub honeysuckles, the twigs are hollow in the center, which differentiates them from native honeysuckles.

Native Alternatives

Aesculus pavia (red buckeye)
Aronia melanocarpa (black chokeberry)
Cephalanthus occidentalis (buttonbush)
Cornus amomum (silky dogwood)
Cornus racemosa (gray dogwood)
Diervilla lonicera (northern bush honeysuckle)
Ilex verticillata (winterberry)
Lindera benzoin (spicebush)
Symphoricarpos albus (snowberry)
Symphoricarpos orbiculatus (coralberry)
Viburnum dentatum (arrowwood viburnum)

Lonicera maackii leaves. Image by Chris Evans, University of Illinois, Bugwood.org.

Lonicera maackii seeds. Image by Leslie J. Mehrhoff, University of Connecticut, Bugwood.org.

Lonicera maackii infestation. Image by Richard T. Gardner III, Bugwood.org.

Hollow pith in *Lonicera maackii* twigs. Image by Chris Evans, University of Illinois, Bugwood.org.

Lonicera maackii flowers. Image by Leslie J. Mehrhoff, University of Connecticut, Bugwood.org.

Lonicera maackii bark. Image by Chris Evans, University of Illinois, Bugwood.org.

Lonicera maackii fruits. Image by Chris Evans, University of Illinois, Bugwood.org.

Root system: Shallow, woody root system.

Reproduction: Primarily by seed, dispersed by birds and mammals. There is some vegetative reproduction by adventitious roots.

Impact: There is evidence that *Lonicera maackii* is allelopathic, producing chemicals that inhibit the germination, growth, and survival of other plants. All the invasive shrub honeysuckles create a dense understory layer that alters ecosystem functioning and community structure. The main impact is that it displaces native herbaceous understory plants, tree seedlings, and shrubs. The reduction in growth and diversity of native tree seedlings has long-term consequences for forest regeneration. In addition, studies have found that shrub honeysuckle cover is detrimental to many bird species by providing less nutritious berries and nesting sites that are more vulnerable to predation than in native shrubs. Stands are also associated with larger tick populations, which increases the risk of tick-borne diseases.

Control: Do not plant nonnative shrub honeysuckles as ornamentals, and remove any plants that are on your property. The shallow root system makes it possible for seedlings and younger plants to be pulled or dug out by hand. For larger, more intractable individuals, use one of the lever or wrench tools that are sold for shrub removal. If pulling is a concern because of potential disturbance to the soil layer, cut all stems and treat the stumps with imazapyr or glyphosate. Stem injection of herbicide is also effective in shrub honeysuckles. Once these shrubs have been removed, native alternatives should be replanted.

Caprifoliaceae

Lonicera morrowii

Morrow's honeysuckle · bush honeysuckle

Origin: Native to Japan, Korea, and northeast China. This species was introduced to North America in 1875 as an ornamental shrub. It has escaped from cultivation and is particularly invasive in the northeastern United States.

Description: A densely branched, deciduous shrub with foliage that often reaches to the ground. It can be differentiated from other invasive shrub honeysuckles by the long stalks on which flowers and fruits are formed in combination with hairy leaves and young stems.

Habitat: Upland and riparian forests, edges of swamps, old fields, fencerows, roadsides, and other disturbed sites. The species prefers moderately moist soils with good drainage and can grow in full sun to partial shade.

Height: 2–2.5 m (6–8 ft.).

Foliage: Simple, opposite, oblong to narrowly elliptic, 3–7 cm (1–3 in.) long and 2–3 cm (¾–1 in.) wide, with a rounded or heart-shaped base and a blunt tip. Leaves are bluish green, slightly hairy on the upper surface and densely hairy underneath. Leaf margins are toothless and have a fringe of fine hairs. As with other invasive shrub honeysuckles, the leaves emerge early in the spring a few weeks before native trees and shrubs.

Flowers: Tubular flowers, white turning yellow with age, growing in pairs from the leaf axils. Individual flowers are about 2–2.5 cm (¾–1 in.) long and 2–2.5 cm (¾–1 in.) in diameter; each pair is borne on a hairy flower stalk about 5–15 mm (¼–⅝ in.) long. Flowers have a single narrow lobe forming the lower lip, and an upper lip with four separate, erect lobes that spread apart with age. The outer surfaces of the flower are hairy, especially the tube. Each flower has five yellow-tipped stamens and a white style with a rounded green stigma protruding from the tube. Very fragrant. May–June.

Fruit: A spherical, semitranslucent, orange to red berry, 6 mm (¼ in.) in diameter. Like the flowers, the fruits grow in pairs from the leaf axils, with each pair held on a single stalk about 5–15 mm (¼–⅝ in.) long. Each berry contains a few small, yellow seeds, about 3 mm (⅛ in.) long and 2 mm (¹⁄₁₆ in.) wide. June–August, but can persist through the winter.

Stems: The young twigs are hairy and gray. Mature bark is grayish brown and appears braided, with flat scaly ridges and narrow grooves, peeling off in vertical strips. As with all invasive shrub honeysuckles, the twigs are hollow in the center, which differentiates them from native honeysuckles.

Root system: A shallow, woody root system.

Reproduction: Primarily by seed, dispersed by birds and mammals. There is some vegetative reproduction by adventitious roots.

Impact, Control, and Native Alternatives: See *Lonicera maackii* [pp. 255–257].

258

Lonicera morrowii leaves. Image by Richard T. Gardner III, Bugwood.org.

Lonicera morrowii flowers turning yellow with age. Image by Leslie J. Mehrhoff, University of Connecticut, Bugwood.org.

Lower stem and shallow roots of *Lonicera morrowii*. Image by Leslie J. Mehrhoff, University of Connecticut, Bugwood.org.

Lonicera morrowii fruits. Image by Leslie J. Mehrhoff, University of Connecticut, Bugwood.org.

Lonicera morrowii flowers. Image by Leslie J. Mehrhoff, University of Connecticut, Bugwood.org.

Lonicera morrowii plants. Image by Leslie J. Mehrhoff, University of Connecticut, Bugwood.org.

Caprifoliaceae

Lonicera tatarica

Tatarian honeysuckle · bush honeysuckle

Origin: The native range of this species extends from southern Russia to central Asia. Tatarian honeysuckle was introduced as an ornamental in 1752, making it the first nonnative shrub honeysuckle to be brought to North America. It has escaped cultivation and is now invasive in many parts of the United States.

Description: A woody, deciduous shrub with multiple stems. *Lonicera tatarica* can be differentiated from other nonnative shrub honeysuckles by the combined features of (typically) pink flowers, a lack of hairs on leaves and flowers, and longer stalks for the flowers and fruits (2–2.5 cm [¾–1 in.] long) than in any of the other species. Despite its invasive tendencies, this species is still available commercially.

Habitat: Upland and riparian forests, open woodlands, thickets, floodplains, old fields, roadsides, and other disturbed sites. It can grow in full sun to shade and in moist to dry, gravelly, or sandy soils.

Height: To 3 m (10 ft.).

Foliage: Simple, opposite, ovate to oblong, 3–6 cm (1–3 in.) long and 2–4 cm (¾–2 in.) wide, with a rounded or heart-shaped base and a blunt tip. Leaves are bluish green above and whitish underneath. Leaf margins are smooth and both sides of the leaves are hairless, a distinguishing feature of this species. As with other invasive shrub honeysuckles, the leaves emerge early in the spring a few weeks before native trees and shrubs.

Flowers: Tubular flowers about 2–2.5 cm (¾–1 in.) long and wide, usually pink or red but occasionally white, not fading to yellow with age. Flowers grow in pairs from the leaf axils, each pair borne on a single flower stalk about 2–2.5 cm (¾–1 in.), longer than in other invasive shrub honeysuckles. Each flower has a lower lip formed by a single narrow lobe that is reflexed down, and an upper lip comprising four lobes, the middle two of which are fused near the base and the lateral lobes spreading. The outer surfaces of the flowers are hairless. Each flower has five yellow-tipped stamens and a white style with a rounded green stigma protruding from the tube. Very fragrant. May–June.

Fruit: A spherical, semitranslucent, red to orange or yellow berry, 4–8 mm (⅛–¼ in.) in diameter. Like the flowers, the fruits grow in pairs from the leaf axils, with each pair held on a single stalk about 2–2.5 cm (¾–1 in.) long. Each berry contains three to six brown seeds, about 2.5–3 mm (⅛ in.) long and 2–2.5 mm (¹⁄₁₆ in.) wide. July–August but can persist through the winter.

Stems: The young twigs lack hairs and are light green to reddish brown. Mature bark is light gray and appears braided, with flat scaly ridges and narrow grooves, peeling off in vertical strips. As with all invasive shrub honeysuckles, the twigs are hollow in the center, which differentiates them from native honeysuckles.

Root system: A shallow, woody root system.

Reproduction: Primarily by seed, dispersed by birds and mammals. There is some vegetative reproduction by adventitious roots.

Impact, Control, and Native Alternatives: See *Lonicera maackii* [pp. 255–257]. Although this species is less aggressive than other shrub honeysuckles, it still invades disturbed natural areas, particularly those near urban areas.

Lonicera tatarica leaves. Image by Rob Routledge, Sault College, Bugwood.org.

Lonicera tatarica fruits. Image by Chris Evans, University of Illinois, Bugwood.org.

Bracts at base of flowers in *Lonicera tatarica.* Image by Rob Routledge, Sault College, Bugwood.org.

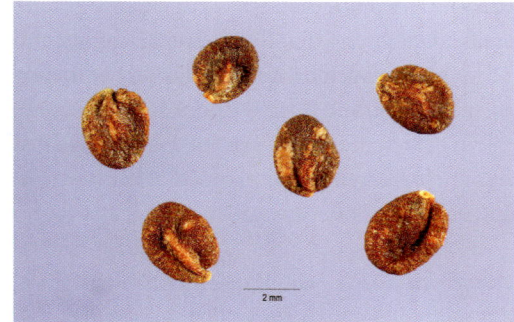

Lonicera tatarica seeds. Image by Steve Hurst, hosted by the USDA-NRCS PLANTS Database.

Lonicera tatarica flowers. Image by Rob Routledge, Sault College, Bugwood.org.

Lonicera tatarica plant. Image by Richard T. Gardner III, Bugwood.org.

Caprifoliaceae

Lonicera x *bella*

Bell's honeysuckle · bella honeysuckle · showy honeysuckle · showy pink honeysuckle · showy fly honeysuckle · showy bush honeysuckle · pretty honeysuckle · whitebell honeysuckle

Origin: This is a fertile hybrid between *Lonicera morrowii* [pp. 258–259] and *Lonicera tatarica* [pp. 260–261]. Both parent species are native to Asia but they are separated geographically and occupy slightly different habitats. The plant was deliberately hybridized in Russia sometime before 1889. There are records of the hybrid being planted in the United States shortly thereafter. It has escaped cultivation and spread into natural areas.

Description: This woody, deciduous shrub has multiple stems with arching or spreading branches. Identification of this plant can be difficult because its features are intermediate between the two parent species and there have been multiple generations of hybridization and backcrossing. The flowers are pink to white in color but they fade to yellow as they wither, and *L. tatarica* flowers do not. The leaves and stems tend to be sparsely hairy. In many places the hybrid has been found to be more common than either parent, and many records have likely misidentified *Lonicera* x *bella* as either *L. morrowii* or *L. tatarica*. It is therefore likely that the distribution map underestimates the actual coverage of the species due to underreporting.

Habitat: Particularly common in disturbed areas, including forest edges, roadsides, utility right-of-ways, and old fields. It is tolerant of a wide range of moisture and light conditions.

Height: Often taller than both the parent species, up to 6 m (20 ft.).

Foliage: Simple, opposite, narrowly elliptic or ovate, 3–7 cm (1–3 in.) long and 1.5–2.5 cm (⅝–1 in.) wide, with a rounded or heart-shaped base and a blunt tip. Leaves are green to blue green above, lighter underneath. Leaf margins are smooth, sometimes with scattered hairs along the edge. The undersides of leaves tend to be slightly hairy, although this feature is variable. As with other invasive shrub honeysuckles, the leaves emerge early in the spring a few weeks before native trees and shrubs.

Flowers: Tubular flowers about 2–2.5 cm (¾–1 in.) long and wide, deep rose to light pink, occasionally white. Flowers fade to yellow with age, distinguishing this plant from the parent *L. tatarica.* Flowers grow in pairs from the leaf axils, each pair borne on a single flower stalk about 5–15 mm (¼–⅝ in.). Flowers have a lower lip formed by a single narrow lobe that is reflexed downward, and an upper lip comprising four lobes that can be fused or singly erect and which spread apart with age. The outer surfaces of the flowers are hairless. Each flower has five yellow-tipped stamens and a white style with a rounded green stigma protruding from the tube. Very fragrant. May–June.

Fruit: A spherical, semitranslucent, bright red berry, 6–13 mm (¼–½ in.) in diameter. Like the flowers, the fruits grow in pairs from the leaf axils, with each pair held on a single stalk about 5–15 mm (¼–⅝ in.) long. Each berry contains two to six seeds, about 3 mm (⅛ in.) long and 2 mm (¹⁄₁₆ in.) wide. August–September.

Stems: The young twigs have sparse hairs and are light green to brown. Mature bark is light gray, with flat scaly ridges and narrow grooves, peeling off in vertical strips. As with all invasive shrub honeysuckles, the twigs are hollow in the center, which differentiates them from native honeysuckles.

Root system: A shallow, woody root system.

Reproduction: Primarily by seed, dispersed by birds and mammals. There is some vegetative reproduction by adventitious roots.

Impact, Control, and Native Alternatives: See *Lonicera maackii* [pp. 255–257].

Lonicera x *bella* leaves. Image by Rob Routledge, Sault College, Bugwood.org.

Lonicera x *bella* flowers fading to yellow with age. Image by Leslie J. Mehrhoff, University of Connecticut, Bugwood.org.

Lonicera x *bella* branches and bark. Image by Leslie J. Mehrhoff, University of Connecticut, Bugwood.org.

Lonicera x *bella* flowers. Image by Joseph Berger, Bugwood.org.

Lonicera x *bella* fruits. Image by Leslie J. Mehrhoff, University of Connecticut, Bugwood.org.

Lonicera x *bella* plant. Image by Leslie J. Mehrhoff, University of Connecticut, Bugwood.org.

Celastraceae

Euonymus alatus

burning bush · winged euonymus · winged
spindletree · winged wahoo

Origin: Native to central and northern China,
Japan, and Korea. *Euonymus alatus* was introduced to North America around 1860 as an
ornamental plant. It has escaped cultivation
and spread into natural areas.

Description: This deciduous shrub has a
broad, rounded, and densely branched crown.
The common name "burning bush" derives
from the scarlet autumn foliage, a feature that
has contributed to its popularity as an ornamental plant. Another characteristic of this
species is the corky ridges on the stems, the
source of the specific epithet *alatus* and the
"winged" descriptor used in many other common names. Although *E. alatus* has been listed
as invasive in several states—including Illinois
and Kentucky—it is still commercially available
and widely planted in many places.

Habitat: Forest understory, open woodland,
forest edges, fields, pastures, prairies, and roadsides. It tolerates different soil types and pH
levels and can grow well in habitats ranging
from full sun to full shade.

Height: Typically 1–4 m (3–12 ft.), occasionally to 6 m (20 ft.).

Foliage: Simple, opposite, elliptic to ovate, 3–10
cm (1–4 in.) long and 1–4 cm (½–2 in.) wide, with
a tapered base and a sharp or blunt tip. Leaf margins are finely toothed. In the autumn, leaves turn
a brilliant purplish red or scarlet.

Flowers: Inconspicuous yellow-green flowers,
6–8 mm (¼ in.) in diameter, in small clusters
of one to three borne in the axils of the leaves.
Each flower has four rounded petals and four
short, yellow stamens arranged around a central,
green, disk-shaped nectary or ovary. May–June.

Fruit: Fruit is a red-purple capsule, 10–15
mm (½–⅝ in.) long and 5–10 mm (¼–½ in.)

wide, that splits open to reveal one or two
seeds (occasionally three or four) enclosed in a
fleshy, orange-red aril. Seeds are white to pink,
ellipsoid, 3 mm (⅛ in.) long and 2 mm (¹⁄₁₆ in.)
wide. September–October.

Stems: Young twigs and branches are green
to brown and have two to four prominent
corky wings along their length. Older stems
have gray-brown bark that splits, giving them a
striped appearance.

Root system: Fibrous, shallow roots form a
dense mat below the plants.

Reproduction: Primarily by seed, dispersed
by birds. Seed drop can also create a dense bed
of seedlings in the shadow of the parent plant.

Impact: Forms dense thickets that displace
native woody and herbaceous plants. Continued widespread use of this plant in landscaping increases the probability that it will escape
from cultivation and become established in
natural areas.

Control: Do not use *E. alatus* cultivars in landscaping. Seedlings can be pulled by hand.
Larger plants can be cut back and the root system dug out. Cut stumps can also be treated
with glyphosate or triclopyr.

Native Alternatives

Aronia melanocarpa (black chokeberry)
Cornus amomum (silky dogwood)
Cornus racemosa (gray dogwood)
Euonymus atropurpureus (eastern wahoo)
Euonymus americanus (strawberry bush)
Ilex verticillata (winterberry)
Rhus aromatica (fragrant sumac)
Vaccinium corymbosum (highbush blueberry)
Viburnum nudum (witherod viburnum)

Euonymus alatus leaves. Image by James H. Miller, USDA Forest Service, Bugwood.org.

Red autumn leaves on *Euonymus alatus*. Image by James H. Miller, USDA Forest Service, Bugwood.org.

Corky wings on stems of *Euonymus alatus.* Image by T. Davis Sydnor, The Ohio State University, Bugwood.org.

Euonymus alatus flowers. Image by Leslie J. Mehrhoff, University of Connecticut, Bugwood.org.

Euonymus alatus fruits. Image by Leslie J. Mehrhoff, University of Connecticut, Bugwood.org.

Euonymus alatus fruits and seeds. Image by Leslie J. Mehrhoff, University of Connecticut, Bugwood.org.

Striped bark on a mature stem of *Euonymus alatus.* Image by Leslie J. Mehrhoff, University of Connecticut, Bugwood.org.

Euonymus alatus plants growing in the forest understory. Image by Leslie J. Mehrhoff, University of Connecticut, Bugwood.org.

Celastraceae

Euonymus europaeus

European spindletree · European Euonymus ·
common spindle

Origin: Native to Europe and western Asia. *Euonymus europaeus* has escaped cultivation and spread into native habitats, particularly in parts of the northeastern United States.

Description: A deciduous shrub to small tree with an upright form when young, broadening to a rounded outline at maturity. It has been used in landscaping primarily for borders or screens, but due to its bright pink fruits and autumn color it does have some ornamental value. This species is considered an emerging invasive in several states.

Habitat: Found in anthropogenic and disturbed habitats, including floodplains, forest edges, thickets, and roadsides. It is tolerant of most soil types, but they must be well drained. Grows well in full sun to partial shade.

Height: Typically 2–6 m (6–20 ft.), some specimens to 8 m (26 ft.).

Foliage: Simple, opposite, ovate to elliptic, 3–8 cm (1–3 in.) long and 1–4 cm (½–2 in.) wide, with a tapered base and tip. Leaf margins are finely toothed. The petiole is 6–12 mm (¼–½ in.) long and grooved along the top. Foliage turns red, orange, or purple in autumn.

Flowers: Inconspicuous greenish-white flowers, 6–10 mm (¼–½ in.) wide, in small cymes of 3–10 flowers, growing from the leaf axils on stalks 1–3.5 cm (½–2 in.) long. Each flower has four widely spreading, narrow white petals and four short, yellow stamens arranged around a central, green, disk-shaped nectary or ovary. April–June.

Fruit: Fruit is a showy pink to red capsule, 1–2 cm (½–¾ in.) wide, usually four-lobed, that splits open to reveal up to four seeds, each enclosed in a fleshy, orange-red aril. Seeds are white to pink, ellipsoid, 4 mm (⅛ in.) long and 3 mm (⅛ in.) wide. September–November.

Stems: Young twigs and branches are smooth and reddish green. Older stems have thin, light-gray to brown bark, occasionally striped with green.

Root system: Fibrous, relatively shallow root system.

Reproduction: Primarily by seed, dispersed by birds. Adventitious roots can develop where branches touch the ground, and there is also some root suckering, both of which can contribute to vegetative spread.

Impact: Although still relatively rare in natural habitats, this species has become locally common in certain areas. Where established, it displaces native species in the forest understory.

Control and Native Alternatives: See *Euonymus alatus* [pp. 264–265].

266

Euonymus europaeus leaves. Image by Robert Vidéki, Doronicum Kft., Bugwood.org.

Aril-covered fruits of *Euonymus europaeus.* Image by Frank Vincentz, used under a CC BY-SA 3.0 license.

Euonymus europaeus flowers. Image by Robert Vidéki, Doronicum Kft., Bugwood.org.

Euonymus europaeus stems. Image by Muriel Bendel, used under a CC BY-SA 4.0 license.

Euonymus europaeus fruits. Image by Robert Vidéki, Doronicum Kft., Bugwood.org.

Euonymus europaeus plants. Image by Robert Vidéki, Doronicum Kft., Bugwood.org.

Elaeagnaceae

Elaeagnus angustifolia

Russian olive · Bohemian olive · Persian olive ·
wild olive · silver berry · oleaster

ODA Invasive
OIPC Invasive

Origin: Native range is from southern Europe to western and central Asia. This species, which has been cultivated in Europe for hundreds of years, was brought to North America for horticultural purposes. The exact year of introduction is not known, but it was likely during the late 1800s. Widely planted for windbreaks and wildlife habitat, it has escaped cultivation and become invasive, particularly in the Great Plains and other parts of the western and midwestern United States.

Description: *Elaeagnus angustifolia,* a deciduous shrub or small tree, has distinctive silvery foliage and a rounded form. Both it and the closely related *Elaeagnus umbellata* [pp. 270–271] have nitrogen-fixing root nodules, which allow them to survive in extremely poor soils. This species can most reliably be differentiated from *E. umbellata* by its narrower leaves, the presence of silvery scales on both the upper and lower leaf surfaces, and by its larger fruiting structures, which are covered with silvery scales.

Habitat: Found in a wide range of open or lightly shaded habitats, including grasslands, old fields, river- and streambanks, lakeshores, roadsides, and urban areas. Can survive on sandy and bare mineral soils. This species is very salt tolerant and moderately shade tolerant.

Height: To 10 m (30 ft.).

Foliage: Simple, alternate, lance-shaped or narrowly elliptic, 3–10 cm (1–4 in.) long and 1–4 cm (½–2 in.) wide. Leaf margins are smooth, and leaves are covered with silver scales on both the upper and lower surfaces. Leaves remain on the plant until late in the autumn and do not change color.

Flowers: Showy, white to yellow, bell-shaped flowers with a narrow tube 12–15 mm (½–⅝ in.) long and four spreading lobes about 12–15 mm (½–⅝ in.) in diameter, silvery on the outer surfaces and yellow within. Each flower has four stamens and a single style. Borne in small clusters of one to three flowers in the leaf axils. Strongly aromatic. June–July.

Fruit: A drupelike fleshy structure with an ovoid shape, 10–17 mm (½–¾ in.) long and 6–8 mm (¼ in.) wide, pale yellow or reddish-brown with a dense covering of silvery scales. These structures are edible and sweet, though with a dry and mealy texture. Each contains a single achene, oblong, 6–13 mm (¼–½ in.) long and 4 mm (⅛ in.) wide. Fruits ripen in August–September but often persist on plants through winter.

Stems: Young twigs are densely covered with silvery scales and can be thorny. Older growth is covered in shaggy dark brown bark that comes off in narrow, fibrous strips. Main stems can be 10–50 cm (4–20 in.) in diameter.

Root system: Woody and extensive, with a deep taproot and well-developed lateral root system. Depending on location and site conditions, roots can develop nitrogen-fixing nodules.

Reproduction: By seed, primarily dispersed by birds. Resprouting also occurs from cut stumps.

Native Alternatives
Carpinus caroliniana (musclewood)
Hamamelis virginiana (witch hazel)
Morella pensylvanica (northern bayberry)
Shepherdia canadensis (russet buffaloberry)
Styrax americanus (American snowbell)

Impact: Can replace native plants and alter vegetation structure. If invasive in riparian areas, it may alter stream hydrology and nutrient cycling. There is evidence that wildlife diversity is lower in stands dominated by *E. angustifolia.*

Control: Seedlings can be pulled by hand, while small saplings can be removed with a weed wrench when the soil is moist. Larger plants should be cut back and the root system dug out, if practical. Cut stumps can also be treated with glyphosate, imazapyr, triclopyr, picloram, metsulfuron-methyl, and 2,4-D.

Elaeagnus angustifolia leaves. Image by AnR00002, used under a CC0 1.0 license.

Left: Fruiting structures of *Elaeagnus angustifolia.* Image by Barry Rice, sarracenia.com, Bugwood.org. *Right:* Fruiting structures of *Elaeagnus angustifolia.* Image by Leslie J. Mehrhoff, University of Connecticut, Bugwood.org.

Spine on *Elaeagnus angustifolia* branch. Image by Leslie J. Mehrhoff, University of Connecticut, Bugwood.org.

Elaeagnus angustifolia achenes. Image by Steve Hurst, hosted by the USDA-NRCS PLANTS Database.

Elaeagnus angustifolia flowers. Image by Joseph Berger, Bugwood.org.

Elaeagnus angustifolia bark. Image by Leslie J. Mehrhoff, University of Connecticut, Bugwood.org.

Elaeagnaceae

Elaeagnus umbellata

autumn olive · autumn berry · autumn elaeagnus · oleaster · spreading oleaster · silverberry

Origin: Native to eastern Asia, including parts of China, Japan, and Korea. This species was introduced to North America in 1830 and widely promoted for wildlife habitat, wind-breaks, and revegetation following strip mining or other ecological disturbance. It has escaped cultivation and become invasive, mostly in the eastern and midwestern United States.

Description: A medium to large deciduous shrub or small tree with a spreading form. Both this species and the closely related *Elaeagnus angustifolia* [pp. 268–269] have nitrogen-fixing root nodules, which allow them to survive in extremely poor soils. *Elaeagnus umbellata* can most reliably be differentiated from *E. angustifolia* by its shorter and broader leaves, the presence of silvery scales only on the lower surfaces of its leaves, and by its spherical, juicy, red fruiting structures.

Habitat: Open woodlands, forest edges, old fields, fencerows, roadsides, and other disturbed areas. Tolerant of drought, salt, and low pH. Does not grow well in wet habitats or deep shade.

Height: To 6 m (20 ft.).

Foliage: Simple, alternate, elliptic, 4–8 cm (2–3 in.) long and 2–4 cm (¾–2 in.) wide, with a blunt tip and tapered base. Leaf margins are smooth and often wavy. Both sides of the leaves are covered with silvery scales when they emerge in spring, but the scales on the upper surfaces wear off, leaving just the underside sil-very in mature leaves. This feature differenti-ates *E. umbellata* from *E. angustifolia*.

Flowers: Showy, white to yellow, bell-shaped flowers with a narrow tube, 8–9 mm (¼–⅜ in.) long, and four spreading lobes about 7

mm (¼ in.) in diameter, white to light yellow. Each flower has four stamens and a single style. Borne in clusters of up to 10 flowers in the leaf axils. Strongly aromatic. May–June.

Fruit: Nearly spherical, drupelike, 4–9 mm (⅛–⅜ in.) in diameter and 5–7 mm (¼ in.) wide, red dotted with silvery scales. Fruiting struc-tures are juicy and edible. Each contains a single achene, oblong, 4–6 mm (⅛–¼ in.) long and 2–3 mm ($\frac{1}{16}$–⅛ in.) wide. August–November.

Stems: Young twigs are green to light brown and densely covered with silvery scales. Can be thorny. The bark of medium to large branches is gray brown, dotted with lighter spots, and relatively smooth. Older stems have split, fur-rowed, gray bark.

Root system: Woody and extensive, with a deep taproot and well-developed lateral root sys-tem. Roots can develop nitrogen-fixing nodules.

Reproduction: By seed, dispersed primarily by birds but also by mammals. Stump sprout-ing also occurs.

Impact: This species grows rapidly and forms a dense shrub layer that displaces native species and alters vegetation structure.

Control and Native Alternatives: See *Elaeagnus angustifolia* [pp. 268–269].

Elaeagnus umbellata leaves. Image by James H. Miller, USDA Forest Service, Bugwood.org.

Fruiting structures of *Elaeagnus umbellata*. Image by Pennsylvania Department of Conservation and Natural Resources–Forestry, Bugwood.org.

Elaeagnus umbellata bark. Image by Leslie J. Mehrhoff, University of Connecticut, Bugwood.org.

Elaeagnus umbellata achenes. Image by Steve Hurst, hosted by the USDA-NRCS PLANTS Database.

Elaeagnus umbellata flowers. Image by Leslie J. Mehrhoff, University of Connecticut, Bugwood.org.

Root nodules on *Elaeagnus umbellata*. Image by Leslie J. Mehrhoff, University of Connecticut, Bugwood.org.

Malvaceae

Hibiscus syriacus

rose of Sharon · althea · althea rose · shrub
althea · Korean rose · Syrian rose · Syrian
ketmia · rose mallow

Origin: Native to eastern Asia, from China to India. Contradicting the specific epithet and some of its common names, this species does not naturally occur in Syria. It was introduced to North America as an ornamental species in the early colonial period.

Description: This upright, deciduous shrub has a pyramidal form. It is commonly used as a foundation plant or to create hedges. Although not yet widely established in natural areas in Ohio, it has been listed as invasive in Tennessee and Georgia. This is a species to avoid planting and monitor closely to prevent spread and further establishment.

Habitat: Forest, forest edges, disturbed ground, and waste areas. It can grow in full sun to partial shade. Tolerant of poor soil, air pollution, heat, and drought.

Height: 2–6 m (6–20 ft.).

Foliage: Simple, alternate, with variable leaf shape—ovate, diamond-shaped, or three-lobed, 5–10 cm (2–4 in.) long and 2–4 cm (¾–2 in.) wide, tapering or rounded at the base. All leaf types are palmately veined and coarsely toothed along the upper margins. Leaf surfaces are smooth except for a few hairs on the veins on the underside. This species is often late to leaf out in the spring, and the plants retain leaves until late in the autumn, when they change to a dull yellow.

Flowers: Large, showy, trumpet-shaped flowers with five petals, 5–10 cm (2–4 in.) in diameter. Depending on cultivar, color is variable, ranging from white, red, purple, mauve, violet, to blue or combinations thereof. Each flower has a prominent central column surrounded by fused stamens that produce yellow pollen, ending in the white female pistils. Indi-

vidual flowers last only one day, but sequential flowering allows this species a long blooming period, from June until September.

Fruit: An ovate, pointed, dry capsule with five valves, 2 cm (¾ in.) long and wide. Capsules open to release numerous brown seeds, each 5 mm (¼ in.) long and wide, with a heart shape or curved teardrop shape and long hairs along the margins. Ripening August–November and persisting on the plant through winter.

Stems: Twigs are light brown with raised dots. The bark of older stems is smooth with brown and gray stripes.

Root system: Forms a deep taproot.

Reproduction: By seed.

Impact: This species self-seeds aggressively and can form a thicket of seedlings that displaces native plant species.

Control: Fertile cultivars are the primary source of seed that spreads into natural areas. Established fertile specimens should be removed from gardens. Several sterile cultivars have been developed; if you want to grow this species, choose one of these. The deep taproot makes it hard to remove after the plant is two or three years old, so seedlings should be pulled out by hand before they establish. Larger plants should be cut back and as much of the taproot removed as possible. Cut stumps can also be treated with glyphosate or triclopyr.

Native Alternatives
Amorpha fruticosa (false indigo bush)
Hibiscus laevis (smooth rose mallow)
Hibiscus moscheutos (swamp rose mallow)
Hydrangea arborescens (wild hydrangea)

Hibiscus syriacus leaves. Image by Chris Evans, University of Illinois, Bugwood.org.

Hibiscus syriacus seeds. Image by Steve Hurst, hosted by the USDA-NRCS PLANTS Database.

Hibiscus syriacus flower. Image by Chrumps, used under a CC BY-SA 3.0 license.

Hibiscus syriacus fruits. Image by Chris Evans, University of Illinois, Bugwood.org.

Hibiscus syriacus planted in a garden. Image by David Stephens, Bugwood.org.

Hibiscus syriacus seedlings. Image by Robert Vidéki, Doronicum Kft., Bugwood.org.

273

Oleaceae

Ligustrum obtusifolium

border privet · Amur privet · regal privet · blunt-leaved privet · obtuse-leaved privet

Origin: Native to Japan, Korea, and northeastern China. This species was introduced to North America in 1860 as a hedge plant. It is still widely used in landscaping and has escaped cultivation to become naturalized in many places.

Description: A fast-growing, deciduous or semi-deciduous shrub with dark green foliage that persists into winter. It is typically multistemmed, with horizontally spreading, arching branches. *Ligustrum obtusifolium* is listed as invasive in several states, including Illinois, Pennsylvania, New Hampshire, Massachusetts, and Connecticut. It is very similar in appearance to *Ligustrum vulgare* but can be distinguished by its more densely hairy twigs and flower stalks, shorter inflorescences, and a narrower flower diameter.

Habitat: Found in disturbed urban and suburban forest remnants, forest gaps and edges, stream valleys, old fields, roadsides, and other areas with disturbed soil. Can grow in full sun to partial shade in a wide range of soils.

Height: Typically 3–4 m (10–12 ft.) at maturity, occasionally to 6 m (20 ft.).

Foliage: Simple, opposite, elliptic to ovate, 2–7 cm (¾–3 in.) long and 5–25 mm (¼–1 in.) wide, with a tapered base and a rounded or blunt point at the tip. Leaf margins are smooth. The upper leaf surface is glossy and hairless, while the lower surface is covered with fine hairs along the midrib. Leaves are held on short petioles, 2–6 mm (¹⁄₁₆–¼ in.) long, often covered in fine hairs. The leaves persist on the plant into winter but drop before spring.

Flowers: Small, tubular, white flowers, borne in a branching panicle 2–5 cm (¾–2 in.) long, in the upper leaf axils and at the tips of branches. Individual flowers have a narrow tube that is 6–8 mm (¼ in.) long and four spreading lobes that open to a diameter of 3–4 mm (⅛ in.), such that the width of the open flower is about half the length of the tube. Flowers are held on hairy stalks less than 3 mm (⅛ in.) long. Each flower has a pistil with a single style and two stamens that extend to the middle of the lobes. Flowers have an unpleasant sweet and musty fragrance. June–July.

Fruit: A small drupe, bluish or purplish black with a dull white bloom on the surface, nearly spherical to broadly ellipsoid, 6–8 mm (¼ in.) long and 4–6 mm (⅛–¼ in.) wide. Like the flowers, fruits are borne in axillary or terminal clusters. The interior of these berries is juicy, and each fruit contains a brown seed, elliptic with a convex dorsal side and flat ventral side, 5–6 mm (¼ in.) long and 3 mm (⅛ in.) wide. Fruits ripen during September and persist on the branches over winter.

Stems: The young twigs are light brown or dull reddish-purple and hairy. Stems have smooth, grayish-brown bark with short, light colored horizontal lenticels. Some plants have short spur branches that resemble thorns.

Root system: Extensive, shallow root system.

Native Alternatives
Aronia melanocarpa (black chokeberry)
Ceanothus americanus (New Jersey tea)
Chionanthus virginicus (fringetree)
Ilex verticillata (winterberry holly)
Lindera benzoin (spicebush)
Morella pensylvanica (northern bayberry)
Prunus virginiana (chokecherry)
Spiraea tomentosa (steeplebush)
Viburnum prunifolium (blackhaw viburnum)

274

Reproduction: By seed, dispersed by birds. Mature plants can produce hundreds of fruits. Vegetative reproduction is also possible through the formation of root and stump sprouts.

Impact: Fruits are toxic to humans and other mammals. It can form dense thickets that displace native species.

Control: Small plants can be removed by mowing, cutting, or digging out. Stems should be cut as close to ground level as possible at least once per growing season. Glyphosate, imazapyr, metsulfuron, or triclopyr can be applied to the cut stumps.

Ligustrum obtusifolium flowers. Image by Leslie J. Mehrhoff, University of Connecticut, Bugwood.org.

Ligustrum obtusifolium leaves. Image by Richard T. Gardner III, Bugwood.org.

Ligustrum obtusifolium bark. Image by Dalgial, used under a CC BY-SA 3.0 license.

Ligustrum obtusifolium fruits. Image by Leslie J. Mehrhoff, University of Connecticut, Bugwood.org.

Ligustrum obtusifolium flowers. Image by Leslie J. Mehrhoff, University of Connecticut, Bugwood.org.

Ligustrum obtusifolium fruits and seeds. Image by Leslie J. Mehrhoff, University of Connecticut, Bugwood.org.

Oleaceae

Ligustrum vulgare
European privet · common privet · wild privet

Origin: A native of Europe, northern Africa, and southwestern Asia, this ornamental shrub has been cultivated for centuries. The precise date of its introduction to North America is not known, but it has been planted widely, particularly in hedgerows. It is still available commercially and used in landscaping but has spread into natural areas across the United States.

Description: A fast-growing, deciduous or semi-deciduous shrub with dark green foliage that persists into winter. The branches are erect and irregularly spreading. This species is listed as invasive in the neighboring states of Indiana, Kentucky, Illinois, and Pennsylvania. It is very similar in appearance to *Ligustrum obtusifolium* but can be distinguished by its less hairy twigs and flower stalks, longer inflorescences, and a wider flower diameter. There are many cultivars with green, white, and yellow fruits as well as variegated foliage. Some features may therefore vary from the parent species as described below.

Habitat: Forest edges and fragments, fencerows, windbreaks, roadsides, and other areas with disturbed soil. Prefers direct sunlight but can grow in shade if the soil has high nutrients. It is tolerant of most soil types and grows well in humid areas.

Height: Typically 3–4 m (10–12 ft.), occasionally to 6 m (20 ft.).

Foliage: Simple, opposite, ovate or lance-shaped, 2–6 cm (¾–3 in.) long and 5–15 mm (¼–⅝ in.) wide, with a rounded or pointed tip. Leaf margins are smooth. The upper and lower leaf surfaces are hairless. The leaves persist on the plant into winter but drop before spring.

Flowers: Small, tubular, white to cream flowers, borne in branching panicles 3–6 cm (1–3 in.) long, in the upper leaf axils and at the tips of branches. Each individual flower has a narrow tube, 4–6 mm (⅛–¼ in.) long, and four spreading lobes that open to a diameter of 4–6 mm (⅛–¼ in.), such that the width of the open flower is about equal to the length of the tube. Each flower has a pistil with a single style and two stamens that are held within the tubes or barely extend beyond the end of the tube. Flowers have an unpleasant sweet and musty fragrance. June.

Fruit: A small drupe, lustrous dark purple or black, spherical to ovoid, 6–8 mm (¼ in.) long and wide. Fruits are borne in axillary or terminal clusters. The interior of these berries is juicy, and each fruit contains one to two brown seeds, elliptic with a convex dorsal side and flat ventral side, 5 mm (¼ in.) long by 3 mm (⅛ in.) wide and 2 mm (¹⁄₁₆ in) deep. Fruits ripen during September and persist on the branches over winter.

Stems: The young twigs are green and covered with very fine hairs, becoming hairless with age. Stems have smooth, whitish-tan to gray bark with small brown lenticels.

Root system: Extensive, shallow root system.

Reproduction: By seed, dispersed by birds. Mature plants can produce hundreds of fruits. Seed viability is about one year. It also reproduces vegetatively by root and stump sprouts.

Impact, Control, and Native Alternatives: See *Ligustrum obtusifolium* [pp. 274–275]. The pollen of this species can cause a severe allergic reaction in some people.

Ligustrum vulgare leaves. Image by Robert Vidéki, Doronicum Kft., Bugwood.org.

Ligustrum vulgare fruits. Image by Robert Vidéki, Doronicum Kft., Bugwood.org.

Ligustrum vulgare twigs and leaf buds. Image by Stefan Lefnaer, used under a CC BY-SA 4.0 license.

Ligustrum vulgare seeds. Image by Steve Hurst, hosted by the USDA-NRCS PLANTS Database.

Ligustrum vulgare flowers. Image by Robert Vidéki, Doronicum Kft., Bugwood.org.

Ligustrum vulgare plants. Image by Robert Vidéki, Doronicum Kft., Bugwood.org.

Rosaceae

Rhodotypos scandens

black jetbead · jetbead · white kerria · jetberry
bush

Origin: Native to central China, Japan, and Korea. This species was introduced to North America in 1866 for ornamental purposes. It has escaped cultivation and only recently come to the attention of land managers as an increasing problem in natural areas.

Description: A loosely branched, deciduous shrub with arching stems. This species is relatively uncommon but has started spreading rapidly in the eastern United States. It is probably underreported because it is not well known. However, it has been designated as invasive in several states, including Illinois, Virginia, and Pennsylvania. Although it is not an overly popular garden plant, seeds and plants are still available commercially. These should be avoided.

Habitat: Meadows, roadsides, forest, forest edges and fragments, and waste areas. It prefers full sun but can withstand shade, allowing it to survive in the understory in intact forest. It tolerates salt, air pollution, soil compaction, and drought.

Height: 1–2 m (3–6 ft.).

Foliage: Simple, opposite, 6–10 cm (3–4 in.) long and 3–5 cm (1–2 in.) wide. Leaves are ovate, each with a rounded base and a long, pointed tip. The leaf margins are doubly toothed. Due to the ribbed veins, the leaf surfaces have a rough appearance. This species is early to leaf out in spring and late to lose leaves in the autumn.

Flowers: White flowers, 5 cm (2 in.) in diameter, consisting of four petals and four green sepals with toothed margins, borne singly at the tips of branches. Each flower has four distinct pistils and numerous stamens. May–June.

Fruit: A shiny, hard, black drupe, ovoid, 1 cm (½ in.) long and 5–8 mm (¼ in.) in diameter, in clusters of four (occasionally fewer) at the tips of branches. Each fruit contains a single seed, 7–9 mm (¼–⅜ in.) long and 4–6 mm (⅛–¼ in.) in diameter. Fruits appear in midsummer, ripen September–December, and persist through the winter.

Stems: Multiple stems grow from the base. Young stems are green and smooth, turning brown with age. Older growth is reddish brown with gray streaks and slightly orange lenticels.

Root system: Rhizomes that readily form root suckers.

Reproduction: By seed, dispersed by birds. Vegetative reproduction also occurs.

Impact: Grows in dense stands that displace native shrubs, shade out herbaceous understory species, and restrict tree seedling establishment. Due to the presence of a cyanide-like chemical called amygdalin, fruits are highly toxic to humans and other mammals.

Control: Do not plant new specimens of *Rhodotypos scandens*. Remove any existing stands, then replant with a native species. Small plants can be removed by hand. Dig out the entire root system to prevent regrowth. For

Native Alternatives

Cornus amomum (silky dogwood)
Cornus racemosa (gray dogwood)
Corylus americana (American hazelnut)
Hydrangea arborescens (wild hydrangea)
Physocarpus opulifolius (ninebark)
Viburnum acerifolium (mapleleaf viburnum)
Viburnum dentatum (arrowwood viburnum)

larger infestations, plants should be cut to the ground in autumn or winter. Spring application of glyphosate or triclopyr to cut stumps can help prevent resprouting.

Rhodotypos scandens fruits. Image by Leslie J. Mehrhoff, University of Connecticut, Bugwood.org.

Rhodotypos scandens leaves. Image by Leslie J. Mehrhoff, University of Connecticut, Bugwood.org.

Rhodotypos scandens seed. Image by Tracey Slotta, hosted by the USDA-NRCS PLANTS Database.

Rhodotypos scandens stems. Image by Katja Schulz, used under a CC BY 2.0 license.

Rhodotypos scandens flower. Image by Leslie J. Mehrhoff, University of Connecticut, Bugwood.org.

Rhodotypos scandens plants spreading into the forest understory. Image by Leslie J. Mehrhoff, University of Connecticut, Bugwood.org.

Rosaceae

Rosa canina

dog rose · wild rose · brier rose · common briar · dog briar · witches' briar

⚠ **Do Not Touch**

Origin: Native to Europe, northwest Africa, and western Asia. This species has long been used in cooking and traditional medicine and as a rootstock for the grafting of ornamental roses. It was introduced to North America in the early 1700s and has escaped from cultivation to become problematic in natural areas.

Description: A perennial, arching shrub with a scrambling habit. The recurved prickles help support this species as it climbs. The common name "dog rose" may come from the historic belief that its roots could cure the bite of rabid dogs, although the name might have preceded—and contributed to—this belief. There is also conjecture that *canina* is a reference to the prickles, which are shaped like dogs' teeth.

Habitat: Woodlands, fields, prairies, thickets, roadsides, river- and streambanks, and in waste areas. Grows in full sun to partial shade in a range of soil conditions, with the exception of very nutrient-poor soil, acidic soil, and peat.

Height: 1–3 m (3–10 ft.), but can scramble to 5 m (15 ft.) with support.

Foliage: Alternate, pinnately compound leaves made up of five to seven leaflets, the entire length of which is 6–11 cm (3–5 in.). Leaflets are ovate, each 1.5–4 cm (⅝–2 in.) long and 1.2–2 cm (½–¾ in.) wide, with toothed margins and a pointed tip.

Flowers: Showy pink to white flowers, 4–6 cm (2–3 in.) in diameter, with five petals and numerous yellow stamens and pistils in a central disk. Flowers are solitary or in clusters of two to four. Sweetly fragrant. June–July.

Fruit: Striking red-orange rosehips, ovoid, 1–2 cm (½–¾ in.) long and 6–16 mm (¼–⅝ in.) wide, borne singly or in clusters of 2–4. Each hip contains 14–23 tan-colored achenes, 5–6 mm (¼ in.) long by 3–4 mm (⅛ in.) wide. Ripen in September–October and persist on the plant into winter, turning black with age.

Stems: The bark of young stems is smooth and green, that of older stems is smooth and brown, gray, or reddish. Stems are covered with stout, flattened prickles that curve downward. The prickles are borne singly or in pairs, 6–7 mm (¼ in.) long and 4–9 mm (⅛–⅜ in.) wide, paired prickles are usually unequal in size.

Root system: Deep, strongly branching, fibrous root system. Can produce suckers.

Reproduction: By seed, dispersed by birds and mammals that eat the fruits. Vegetative reproduction also occurs via sprouting from roots and the development of adventitious roots where stems touch the ground.

Impact: Forms dense thickets that displace native species. It can also decrease the grazing value of pasture and impede the movement of livestock, wildlife, and vehicles.

Control: Wear gloves to protect against the sharp prickles. Smaller individuals can be pulled out using a weed wrench or similar tool. Larger plants can be cut, but they will resprout from the base. Cut stems can be treated with triclopyr or glyphosate. Other herbicides, such as picloram and imazapyr, can be used as a foliar application; application is best in summer, after flower buds appear but before fruit sets.

Native Alternatives
Rosa carolina (pasture rose)
Rosa palustris (swamp rose)
Rosa setigera (climbing prairie rose)
Rubus odoratus (purple-flowering raspberry)
Spiraea alba (meadowsweet)

Rosa canina leaves. Image by Robert Vidéki, Doronicum Kft., Bugwood.org.

Rosehips of *Rosa canina.* Image by Jan Samanek, Phytosanitary Administration, Bugwood.org.

Rosa canina flower. Image by Robert Vidéki, Doronicum Kft., Bugwood.org.

Rosa canina achenes. Image by Steve Hurst, hosted by the USDA-NRCS PLANTS Database.

Hypanthium of a *Rosa canina* flower. Image © Gerald D. Carr.

Rosa canina growth habit. Image by Robert Vidéki, Doronicum Kft., Bugwood.org.

Rosaceae

Rosa multiflora

multiflora rose · baby rose · Japanese rose ·
seven-sisters rose · Eijitsu rose · rambler rose

ODA Invasive
OIPC Invasive
⚠ **Do Not Touch**

Origin: Native to eastern China, Japan, and Korea. This species has been introduced to North America repeatedly, but its first ornamental cultivars were likely planted during the late 1700s. In the late 1800s, it was introduced again as a rootstock for the grafting of ornamental roses. By the early 1900s. it was being promoted for use in erosion control, shelterbelts, living fences, plantings along highways, and as wildlife habitat.

Description: A dense, spreading, perennial shrub with scrambling tendencies. It has stout stems and recurved prickles. This species is invasive throughout most of the northeastern United States, where it is a major problem in both disturbed and natural areas.

Habitat: Grows in a wide range of conditions, allowing it to invade diverse habitats, including open deciduous forest, forest edges, thickets, meadows, river- and streambanks, fencerows, utility rights-of-way, abandoned pastures, roadsides, and other disturbed areas. It tolerates poor soil and partial shade.

Height: Typically 2–3 m (6–10 ft.), can climb to 5 m (15 ft.) with support.

Foliage: Alternate, pinnately compound leaves 5–10 cm (2–4 in.) long, made up of 5–11 leaflets. Each leaflet is ovate, 1.5–4 cm (⅝–2 in.) long and 8–25 mm (¼–1 in.) wide, with sharply toothed margins, a rounded or wedge-shaped base, and a rounded or pointed tip. At the base of the petiole are stipules with bristly teeth and glandular hairs. The upper leaf surface is smooth and glossy, medium to dark green; the underside of the leaf is paler and covered with short hairs.

Flowers: Small white to light pink flowers, 1–2 cm (½–¾ in.) in diameter, with five petals and numerous deep yellow stamens and pistils

in a central disk. Flowers are produced in large, branching panicles. Sweetly fragrant. May–June.

Fruit: Flowers develop into small red to reddish-purple rosehips, spherical or ovoid, 6–8 mm (¼ in.) long and 4–6 mm (⅛–¼ in.) in diameter. The interior of the hip is dry and fleshy, occupied mostly by numerous angular, yellow achenes, each about 2–4 mm ($^1/_{16}$–⅛ in.) long and 2 mm ($^1/_{16}$ in.) wide. Ripen in August–September and persist on the plant throughout winter.

Stems: Multiple stems arise from the root crown. Twigs are red to green, smooth, and round in cross section. Older stems have smooth, light brown bark streaked with green. Stems are covered in stiff, sharp, recurved prickles with flattened bases, usually in pairs but occasionally scattered. The prickles are initially red or green, turning brown with age.

Root system: An extensive, fibrous root system capable of sending up new shoots.

Reproduction: By abundant seeds, dispersed by birds and mammals. These can remain viable in the soil for 10–20 years. Vegetative reproduction also occurs via sprouting from roots and the development of adventitious roots where stems touch the ground.

Impact: This prolific species forms dense thickets that are impenetrable to humans, wildlife, and livestock. Because it replaces palatable forage plants, it is particularly destructive in pastureland. It is also a serious problem in natural areas where it displaces native plants.

Control and Native Alternatives: See *Rosa canina* [pp. 280–281]. The seed bank is persistent in *Rosa multiflora,* so follow-up treatments will be important regardless of the control method used.

Rosa multiflora leaves. Image by Leslie J. Mehrhoff, University of Connecticut, Bugwood.org.

Rosa multiflora stems. Image by James H. Miller, USDA Forest Service, Bugwood.org.

Rosehips of *Rosa multiflora.* Image by Chris Evans, University of Illinois, Bugwood.org.

Rosa multiflora achenes. Image by Steve Hurst, hosted by the USDA-NRCS PLANTS Database.

Rosa multiflora flowers. Image by Leslie J. Mehrhoff, University of Connecticut, Bugwood.org.

Rosa multiflora infestation. Image by Leslie J. Mehrhoff, University of Connecticut, Bugwood.org.

Rosaceae

Rubus phoenicolasius

wineberry · Japanese wineberry · wine raspberry

⚠ Do Not Touch

Origin: Native to China, Japan, and Korea. This species was introduced to North America in the 1890s as breeding stock for blackberry and raspberry cultivars, although the first date of introduction may have been earlier. It had escaped from cultivation and naturalized into natural areas by the 1970s. This species is recorded in most states east of the Mississippi River.

Description: A multistemmed, spiny, perennial shrub with arching stems. The entire plant is covered in purplish-red, glandular hairs, which is a distinguishing characteristic for this species. The perennial root system produces biennial canes that grow vegetatively in the first year and become reproductive in the second year.

Habitat: Early- to late-successional forest, riparian zones, edges of wetlands, old fields, roadsides, vacant lots, and other disturbed areas. It tolerates a wide range of light availability, soil type, and moisture levels.

Height: 1–3 m (3–10 ft.).

Foliage: Alternate, compound leaves with three leaflets, the terminal one ovate to heart-shaped, 4–8 cm (2–3 in.) long and 2–5 cm (¾–2 in.) wide, with irregularly toothed margins, a rounded base, and a short point at the tip. The lateral leaflets are similar but much smaller. Due to a dense cover of white, woolly hairs, the undersides of the leaves are conspicuously white or silvery.

Flowers: Small white to pale pink flowers, 6–10 mm (¼–½ in.) in diameter, with five small petals, five green sepals covered in glandular hairs, and numerous yellow stamens and pistils in the center. The sepals are much longer than the petals, making the flowers appear as if

they have not opened. Flowers are produced in branching panicles, 6–10 cm (3–4 in.) long, in the leaf axils or at the ends of branches. May–June.

Fruit: A bright red aggregate fruit that resembles a raspberry, about 1 cm (½ in.) wide. The fruit is composed of numerous drupelets clustered together, and it separates cleanly from the fleshy receptacle, forming a hollow shell, which differentiates it and other raspberries from species in the blackberry group. Seeds are 3–4 mm (⅛ in.) long and 2 mm ($^{1}/_{16}$ in.) wide. June–July.

Stems: The stems have a few slender prickles and are densely covered in reddish-purple, glandular hairs 3–5 mm (⅛–¼ in.) long. These hairs give the stems a distinctly shaggy appearance. Young stems are green, turning red with maturity.

Root system: A perennial, rhizomatous root system that sends up new canes each year.

Reproduction: By seed, dispersed by birds and mammals. Vegetative reproduction also occurs via sprouting from roots and the development of adventitious roots where the tips of stems touch the ground.

Impact: This species forms dense, impenetrable thickets. It is more aggressive than many native raspberry and blackberry species and it is able to establish even in mature forests.

Control: Wear gloves to protect against the sharp prickles. Small plants can be removed by hand. Dig out the entire root system to prevent regrowth. For larger infestations, foliar or cut-stump application of glyphosate or triclopyr can be effective.

Rubus phoenicolasius leaves. Image by Leslie J. Mehrhoff, University of Connecticut, Bugwood.org.

Rubus phoenicolasius stems in winter. Image by Leslie J. Mehrhoff, University of Connecticut, Bugwood.org.

Rubus phoenicolasius flower. Image by Leslie J. Mehrhoff, University of Connecticut, Bugwood.org.

Rubus phoenicolasius fruits. Image by Leslie J. Mehrhoff, University of Connecticut, Bugwood.org.

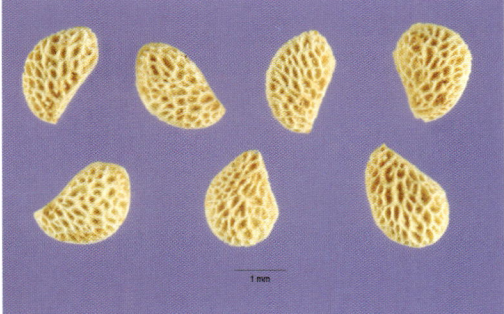

Rubus phoenicolasius seeds. Image by Steve Hurst, hosted by the USDA-NRCS PLANTS Database.

Rubus phoenicolasius growing in the forest understory. Image by Leslie J. Mehrhoff, University of Connecticut, Bugwood.org.

Trees

Trees are typically defined as perennial woody plants with a single stem that is 4–5 m (12–15 ft.) or taller. However, under certain environmental conditions, some tree species are multistemmed and—especially under poor nutrient conditions or inundation by water—may be shorter than 4 m (12 ft.). Correspondingly, some shrubs occasionally grow as single-stemmed and taller than 5 m (15 ft.). Thus, there can be overlap in these categories. Some species, such as *Frangula alnus* (glossy buckthorn) [pp. 298–299] and *Rhamnus cathartica* (European buckthorn) [pp. 300–301], grow either as shrubs or small trees.

As with the shrubs covered in this book, many of the problematic trees reviewed here are ornamental species that have escaped cultivation. Examples include *Pyrus calleryana* (callery pear, often referred to as 'Bradford' pear, after a popular cultivar by that name) [pp. 302–304] and *Betula pendula* (European birch) [pp. 290–291], both of which are attractive but have a tendency to spread beyond the area of planting through the movement of seeds. Some problematic tree species also spread genetically through the transfer of pollen. In the case of *Salix* species [pp. 309–312], the nonnative can hybridize with native willows and create intermediate forms. When nonnative genes transfer into native populations, it reduces the genetic diversity and can contribute to the replacement of the native populations by nonnative ones over time through introgression. Because hybrids are difficult to distinguish from native trees, invasion at the genetic level can be nearly impossible to detect.

Trees are particularly problematic as invasive or weedy species, because when mature they are extremely difficult and potentially dangerous to remove. Where possible, try to control problem tree species by targeting them at the seedling stage. Saplings can often be removed using a weed wrench. Given the hazards associated with felling mature trees, certified professional arborists should always be hired to assist with their removal (search the International Society of Arboriculture's website: www.treesaregood.org/findanarborist). The most effective way to kill a problem tree is to cut it and have the stump ground down to prevent regrowth. Alternatively, the cut stump can be carefully and selectively treated with an appropriate herbicide. The chemicals will be translocated into the roots and will kill the plant at the root level; this is prevents resprouting of the tree from underground reserves. Basal bark treatment is another possible control method: a chemical herbicide is applied directly to the stems of thin-barked woody plants. To facilitate product uptake into the plant cells under the bark, basal oils are generally mixed with the herbicide. This will kill the tree, and the dead tree will need to be removed at a later time. A final option is to girdle the tree. In girdling, the tree is cut in a 5 cm wide strip in the bark around the trunk's circumference to create a break in the vascular tissue. Without the movement of water and food between the roots and leaves, the tree will eventually die. Girdling can take many years to be fully effective—during which trees could continue to reproduce—and dying trees can shed branches, which is extremely dangerous. For this reason, girdling is not recommended for large trees in residential areas. However, it can be useful for targeting smaller trees, particularly in sites without regular human activity. Be aware of the possible hazards and use extreme caution when undertaking any control methods with trees.

Betulaceae

Alnus glutinosa

black alder · European alder · sticky alder ·
common alder

Origin: Native to Europe, southwest Asia, and
northern Africa. This species has been widely
planted for soil stabilization and as an ornamental.

Description: A fast-growing, small- to medium-sized, deciduous tree with a pyramidal or
narrowly oval crown. The buds, young leaves,
and fruiting catkins are coated with a resinous gum that makes them sticky, which is the
source of the species name *glutinosa.*

Habitat: Riparian zones, wetlands, along the
margins of ponds and lakes, forests, and urban
areas. The species prefers moist, damp conditions, but it is adaptable to poor or dry soils.

Height: Typically 10–20 m (30–65 ft.), occasionally to 30 m (100 ft.) or more.

Foliage: Leaves are simple, alternate, 5–12
cm (2–5 in.) long and 5–10 cm (2–4 in.) wide,
broadly oval, tapering to a point at the base.
Leaves have an obvious notch at the tip, differen-
tiating this species from the native shrub alders
in the region. Leaf margins are doubly serrate,
with five to eight principal veins on a side.

Flowers: Monoecious, with separate male
and female flowers on the same plant. The male
flowers are reddish brown, borne in drooping
catkins, 5–12 cm (2–5 in.) long, clustered near
the branch tips in groups of two to five. The
male catkins form in autumn and overwinter
on the tree. Female flowers are borne in green
catkins, 1.5–2.5 cm (⅝–1 in.) long and 1–1.5 cm
(½–⅝ in.) wide, clustered in groups of three to
five. March–May.

Fruit: Female catkins mature into dark
brown or black, woody, cone-shaped strobiles
that contain many fruits. The winged samaras
are circular or oval-shaped, 2.5–3.5 mm (¹⁄₁₆–⅛
in.) long and wide. Fruits mature in autumn.

Stems: Trunk can be single or multistemmed.
The bark of young trees is dark brown and
smooth; older bark is dark gray to black with
shallow fissures.

Root system: The root system is relatively
shallow. Roots develop nodules that contain
nitrogen-fixing bacteria, allowing this species
to survive in nutrient-poor habitats. When the
species grows in wet conditions, the roots often
develop mounds around the base of the tree.

Reproduction: Some vegetative reproduction
occurs through root sprouting, but reproduction is mainly by seed. The fruits float in water,
and the species spreads rapidly along waterways. Wind also contributes to its dispersal.

Impact: Forms dense monocultures that displace native species. Stands of *Alnus glutinosa*
can alter soil nutrient status and affect watercourses by changing sedimentation, oxygen
content, and water flow. In addition, *A. glutinosa* readily hybridizes with native species.

Control: For saplings, the most commonly
used means of control is to cut the stems and
treat the stumps with glyphosate or triclopyr.
Girdling the stems is time consuming but also
effective. Because this species can resprout, follow-up procedures are recommended after
both of these methods.

Native Alternatives
Alnus serrulata (smooth alder)
Betula lenta (sweet birch)
Betula nigra (river birch)
Carpinus caroliniana (musclewood)
Ostrya virginiana (hop-hornbeam)

Leaves of *Alnus glutinosa*. Image by Leslie J. Mehrhoff, University of Connecticut, Bugwood.org.

Alnus glutinosa strobiles. Image by Paul Wray, Iowa State University, Bugwood.org.

Male catkins of *Alnus glutinosa*. Image by Robert Vidéki, Doronicum Kft., Bugwood.org.

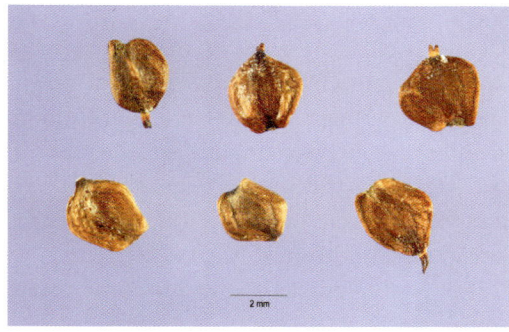

Alnus glutinosa seeds. Image by Steve Hurst, hosted by the USDA-NRCS PLANTS Database.

Female catkins of *Alnus glutinosa*. Image by Piero Amorati, ICCroce–Casalecchio di Reno, Bugwood.org.

Alnus glutinosa trunk and root mounds. Image by Robert Vidéki, Doronicum Kft., Bugwood.org.

Betulaceae

Betula pendula

European birch · European white birch ·
European weeping birch · silver birch · common
birch · warty birch

Origin: Native to Europe, parts of Asia, and northwest Africa. It was introduced to North America as an ornamental species and has escaped cultivation in some areas.

Description: A fast-growing, medium-sized deciduous tree with a pyramidal or oval crown and white peeling bark. Its thin branches droop slightly, thus the species name *pendula*.

Habitat: Roadsides, forest fragments, and areas of habitation. Grows in a wide range of light availability. Prefers well drained, acidic soils.

Height: Typically 15–25 m (50–80 ft.), occasionally to 30 m (100 ft.).

Foliage: Leaves are simple, alternate, 3–7 cm (1–3 in.) long and 2.5–5 cm (1–2 in.) wide, ovate or roughly diamond shaped. Leaf margins are doubly serrate, with forward-curving teeth, and 5–18 lateral veins per side. Both surfaces of the leaves are dotted with glands that produce a sticky resin.

Flowers: Monoecious, with separate male and female flowers on the same plant. The male flowers are yellowish brown, borne in drooping catkins, 3–6 cm (1–3 in.) long, clustered near the branch tips, usually in groups of three but also singly or in pairs. Female flowers are green, borne in cylindrical catkins, 2–4 cm (¾–2 in.) long and 7 mm (¼ in.) wide. April–May.

Fruit: Female flowers develop into abundant winged samaras on the catkins. Each fruit contains a seed about 1–2 mm ($\frac{1}{32}$–$\frac{1}{16}$ in.) wide surrounded by papery wings. Fruits mature in late summer and are dispersed by wind.

Stems: Trunk is slender, typically to about 40 cm (16 in.) in diameter, with creamy to silvery white bark streaked with dark gray horizontal lenticels. The bark does not exfoliate as much as the native *Betula papyrifera* (white birch). Mature trees become furrowed and blackish gray near the bottoms of the trunks. Twigs and new growth are golden brown and covered with small resinous dots.

Root system: Varies according to soil conditions. A taproot is formed in dry places, while shallow roots develop in wet sites.

Reproduction: Through abundant seed, dispersed by wind.

Impact: Escaped populations of this species are not common, and little is known about their impact on native species. However, *Betula pendula* has been shown to persist or become locally naturalized near areas of cultivation in the United States, making it a species worth knowing about and keeping a close eye on. It is likely to compete with its native congeners, such as the *Betula populifolia* (gray birch). In addition, *B. pendula* is highly susceptible to the beetle *Agrilus anxius* (bronze birch borer), which can spread and harm native birch species.

Control: Avoid planting this species. If trees have become established in a natural area, they can be girdled, and then a herbicide such as triclopyr can be applied to the exposed wood. Alternately, herbicide can be injected, applied to cut stems, or used in a basal application. For **Native Alternatives**, see *Alnus glutinosa* [pp. 288–289].

Betula pendula leaves. Image by Robert Vidéki, Doronicum Kft., Bugwood.org.

Betula pendula fruits developing on female catkin. Image by Tom DeGomez, University of Arizona, Bugwood.org.

Female catkins of *Betula pendula*. Image by T. Davis Sydnor, The Ohio State University, Bugwood.org.

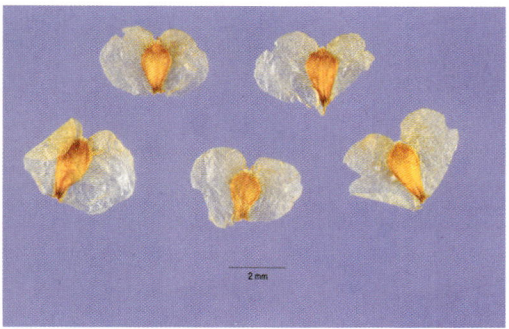

Winged fruits of *Betula pendula*. Image by Steve Hurst, hosted by the USDA-NRCS PLANTS Database.

Male catkins of *Betula pendula*. Image by Norbert Frank, University of West Hungary, Bugwood.org.

White bark on the trunks of *Betula pendula*. Image by Robert Vidéki, Doronicum Kft., Bugwood.org.

Fagaceae

Quercus acutissima

sawtooth oak

Origin: Native to Asia, including Japan, China, Korea, and the Himalayas. This species was first introduced to North America in 1862 as an ornamental and has since been used widely as a landscape plant. Because it produces abundant acorns from an early age (5–10 years), *Quercus acutissima* has also been promoted as a food source for wildlife. It has spread from plantings and has established populations in natural areas.

Description: This fast-growing, medium-sized deciduous tree has a pyramidal or round crown that casts deep shade. It has traits somewhat intermediate between red and white oaks.

Habitat: Commonly planted in gardens, parking lots, and highway medians. Establishes in forests, fields, and along roadsides. The species prefers moist, well-drained soils in full sun, but it adapts to a wide range of soils. Tolerant of heat, air pollution, compacted soil, and dry conditions.

Height: 12–18 m (40–60 ft.).

Foliage: Leaves are simple, alternate, broadly lance-shaped, 8–20 cm (3–8 in.) long and 3–6 cm (2–3 in.) wide, with bristle tips at the end of each vein. Mature leaves are dark green and glossy. When leaves turn brown in the autumn, they do not drop off but are retained on the tree over the winter.

Flowers: Monoecious, with separate male and female flowers on the same plant. Male flowers are yellowish green, borne in drooping catkins, 10–15 cm (4–6 in.) long. Female flowers are small, green, and inconspicuous, on short stalks along the branches. March–May.

Fruit: Large acorns 2–3 cm (¾–1 ½ in.) long and 2 cm (¾ in.) broad, half to two-thirds sur-rounded by a cap covered in curving scales. Acorns take two years to develop, but they ripen earlier in the season than most other oaks, in late summer and early autumn.

Stems: Trunk up to 1.5 m (60 in.) in diameter, flaring out at the base. The bark of mature trees is gray and deeply ridged, with a corky texture.

Root system: Both surface roots and long taproots.

Reproduction: By abundant seed, dispersed by mammals and birds.

Impact: This species has been found to escape plantings and establish in nearby forests. There is no clear evidence that it significantly alters the function of natural systems. However, due to its abundant and early acorn production, it has the potential to displace native oaks. Although the species has been promoted as a source of food for wildlife, several studies suggest that the acorns of this species are less nutritious than native acorns. It is not clear whether this species can hybridize with native oaks.

Control: Seedlings and saplings can be pulled by hand or dug out. Mature trees can be cut down and a nonselective herbicide such as gly-phosate or triclopyr painted on the cut stumps. Any native oak or hickory is a good alternative and will provide food and habitat to diverse wildlife.

Native Alternatives
Carya ovata (shagbark hickory)
Quercus bicolor (swamp white oak)
Quercus imbricaria (shingle oak)
Quercus muehlenbergii (chinquapin oak)

Quercus acutissima leaves. Image by David Stephens, Bugwood.org.

Quercus acutissima fruits. Image by Steve Hurst, hosted by the USDA-NRCS PLANTS Database.

Male flowers of *Quercus acutissima*. Image by Joseph LaForest, University of Georgia, Bugwood.org.

Quercus acutissima bark and leaves. Image by David J. Moorhead, University of Georgia, Bugwood.org.

Quercus acutissima fruit developing from female flower. Image by David Stephens, Bugwood.org.

Quercus acutissima tree. Image by David Stephens, Bugwood.org.

Moraceae

Morus alba

white mulberry · common mulberry · Chinese
white mulberry · silkworm mulberry

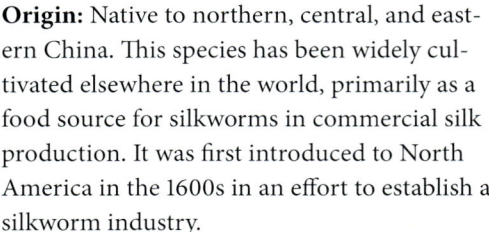

 ⚠ **Do Not Touch**

Origin: Native to northern, central, and eastern China. This species has been widely cultivated elsewhere in the world, primarily as a food source for silkworms in commercial silk production. It was first introduced to North America in the 1600s in an effort to establish a silkworm industry.

Description: This fast-growing, small- to medium-sized deciduous tree has a short trunk, a rounded crown, and a dense canopy with spreading branches. Like other species in this family, *Morus alba* produces a white sap containing latex, which is mildly toxic and can cause skin irritation.

Habitat: Commonly escapes cultivation and spreads into old fields, roadsides, forest edges, fencerows, urban environments, and other disturbed areas. It prefers a warm, moist, well-drained loamy soil in a sunny position but can tolerate intermediate levels of shade and drought.

Height: 15 m (50 ft.).

Foliage: Leaves are simple, alternate, 5–15 cm (2–6 in.) long and 4–9 cm (2–4 in.) wide, of variable shape, ranging from ovate to mitten-shaped or irregularly lobed, with blunt teeth along the margins. Leaves are shiny green on the top surface and paler with slightly hairy veins on the underside. Three prominent veins arise from the rounded or heart-shaped base.

Flowers: This species has separate male and female flowers that can occur on the same tree (monoecious) or, more typically, are on separate trees (dioecious). Male flowers are small, whitish or yellowish green, in 2.5–5 cm (1–2 in.) long catkins that grow from the axils of leaves. This species is notable for its heavy pollen production and rapid release of pollen. Female flowers are inconspicuous and crowded in short spikes. March–April.

Fruit: An ovoid collective fruit (syncarp) that superficially resembles a blackberry, 1–2.5 cm (½–1 in.) long, white, pink, purple or blackish-purple. Each fruit segment contains a single brown seed, 1–2 mm ($\frac{1}{32}$–$\frac{1}{16}$ in.) long and 1 mm ($\frac{1}{32}$ in.) wide. July–August.

Stems: Trunk diameter of mature trees is typically 20–40 cm (8–16 in.). The young bark, inner bark, and roots are orange-brown or yellowish-brown. Older bark is light brown to gray, smooth or with narrow, irregular fissures.

Root system: A wide, spreading root system, with both a taproot and branching lateral roots.

Reproduction: By seed, readily dispersed by birds and other animals that consume the fruits.

Impact: Grows quickly and can outcompete native species. It is particularly problematic in urban and disturbed environments. In addition, it is susceptible to sooty canker disease, which it transmits to the native *Morus rubra* (red mulberry). It also hybridizes with—and replaces—*M. rubra*. The pollen of *M. alba* is a severe allergen.

Control: Wear protective clothing and gloves when undertaking control measures because the sap is very sticky and contact with the skin can cause severe irritation. Seedlings can be pulled by hand or dug out. Large trees can be girdled or cut and the stump painted with a systemic herbicide such as glyphosate or triclopyr.

Variation in shape of *Morus alba* leaves. Image by Jaknouse, used under a CC BY 3.0 license.

Morus alba fruit. Image by Chris Evans, University of Illinois, Bugwood.org.

Morus alba seeds. Image by Steve Hurst, hosted by the USDA-NRCS PLANTS Database.

Male flowers of *Morus alba*. Image by Robert Vidéki, Doronicum Kft., Bugwood.org.

Bark on the trunk of a mature *Morus alba* tree. Image by T. Davis Sydnor, The Ohio State University, Bugwood.org.

Female flowers of *Morus alba*. Image by Suyash Dwivedi, used under a CC BY-SA 4.0 license.

Morus alba tree. Image by T. Davis Sydnor, The Ohio State University, Bugwood.org.

Paulowniaceae

Paulownia tomentosa

princess tree ·empress tree · royal paulownia ·
foxglove-tree · kiri tree

Origin: Native to central and western China, *Paulownia tomentosa* was introduced to North America in the 1840s as an ornamental species. It has been commercially farmed for lumber and used to rehabilitate mine sites and other highly disturbed areas. This species is not currently listed as invasive, because it is not yet found throughout all of Ohio. However, it is extremely aggressive and has been found to exhibit invasive tendencies in the southern part of the state, where it is most abundant.

Description: A fast growing, small- to medium-sized deciduous tree with a rounded growth form. It is sometimes mistaken for native *Catalpa speciosa* and *Catalpa bignonioides,* which also have large, heart-shaped leaves.

Habitat: Primarily found in disturbed habitats such as forest edges, steep rocky slopes, and along roadways and streambanks. It grows best on moist, well-drained soils, but it can establish itself in sites with high soil acidity, drought, and low soil fertility. It is not tolerant of shade.

Height: 15 m (50 ft.), occasionally taller.

Foliage: Simple, arranged in opposite pairs or whorls of three, leaves of mature plants typically 15–40 cm (6–16 in.) long and 10–30 cm (4–12 in.) wide, but leaves of saplings and stump sprouts can be up to 90 cm (35 in.) long and wide. Leaves heart-shaped or three-to-five angled, with a long petiole, hairy on the underside.

Flowers: Showy, fragrant flowers in terminal, upright clusters 15–30 cm (6–12 in.) long. Flowers are pale violet, tubular, 5–7 cm (2–3 in.) long, ending in five unequal lobes. Flowering begins before the leaves emerge in the spring, April–May.

Fruit: The fruit is a woody capsule, ovoid, 3–5 cm (1–2 in.) long and 2–4 cm (¾–2 in.) wide, with a sharp point at the tip. Capsules are sticky and green when first produced, becoming brown at maturity. Each capsule has four compartments that contain several thousand seeds. Seeds are tiny, 2–3 mm (¹⁄₁₆–⅛ in.) long and 0.5 mm (¹⁄₆₄ in.) wide, flattened and winged. Capsules open to release seeds in September–October but persist on the branches through the winter.

Stems: Bark is rough, grayish brown, with shallow vertical fissures. Twigs have prominent white lenticels, are flattened at the nodes, and are hairy at the tips. The pith in the twigs can be chambered or hollow.

Root system: Depending on soil conditions, can be relatively shallow to deep and well developed. The root system typically spreads horizontally and lacks a strong taproot. Roots can develop adventitious buds and send up new shoots.

Reproduction: By seed and sprouting from roots or stumps. A mature tree is capable of producing 20 million seeds in a single growing season; these are dispersed by wind and water. Adventitious buds on the stems and roots allow this species to survive fire, cutting, and other major disturbances.

Impact: This species grows rapidly in disturbed natural areas and displaces native species. It is particularly a concern in forests, steep

Native Alternatives

Catalpa speciosa (northern catalpa)
Cornus alternifolia (pagoda dogwood)
Liriodendron tulipifera (tulip poplar)
Magnolia acuminata (cucumber tree magnolia)
Styrax americanus (American snowbell)

rocky slopes, and scoured riparian areas that provide habitat for rare plants.

Control: Avoid planting new specimens of *P. tomentosa.* Seedlings can be pulled by hand, but since root fragments will readily resprout, the entire root must be removed. Large trees can be girdled or cut and the stump painted with a systemic herbicide like glyphosate or triclopyr. To kill stump sprouts, follow-up treatment will likely be required. Alternatively, professional arborists can grind stumps.

Paulownia tomentosa leaves. Image by James H. Miller, USDA Forest Service, Bugwood.org.

Paulownia tomentosa seeds. Image by Steve Hurst, hosted by the USDA-NRCS PLANTS Database.

Paulownia tomentosa bark. Image by Chris Evans, University of Illinois, Bugwood.org.

Paulownia tomentosa flowers. Image by Leslie J. Mehrhoff, University of Connecticut, Bugwood.org.

Paulownia tomentosa fruits. Image by Leslie J. Mehrhoff, University of Connecticut, Bugwood.org.

Paulownia tomentosa seedling. Image by Leslie J. Mehrhoff, University of Connecticut, Bugwood.org.

Rhamnaceae

Frangula alnus

glossy buckthorn · alder buckthorn · breaking buckthorn · glossy false buckthorn · columnar buckthorn · fen buckthorn · tall hedge buckthorn

Synonym: *Rhamnus frangula*

Origin: Native to Europe, western Asia, and northern Africa. This species was introduced to North America as an ornamental, primarily for use in hedgerows. It is likely that *Frangula alnus* was grown in the United States before 1800, but escaped populations were first recorded in the early 1900s.

Description: A tall shrub to small tree with long and arching branches. This species is typically multistemmed, but it can grow as a single stem reaching a diameter of around 20 cm. Unlike other species in the buckthorn group, *F. alnus* lacks thorns.

Habitat: Particularly aggressive in wetland habitats, including swamps, bogs, marshes and fens. Also found in forest understories and edges, old fields, fencerows, and roadsides. It prefers moist, acidic soils and can survive in full sun to deep shade.

Height: 7–9 m (23–30 ft.).

Foliage: Leaves are simple, alternate, 3–8 cm (1–3 in.) long and 2.5–4 cm (1–2 in.) wide, oval-shaped but ending in a pointed tip. Leaf margins are smooth but wavy. Leaves have prominent venation, with 8–10 pairs of parallel veins that curve up toward the tip.

Flowers: Groups of two to eight flowers are produced in the leaf axils. Flowers are greenish white and star-shaped, 3–5 mm (⅛–¼ in.) in diameter, with five triangular petals. Flowers appear after the leaves have emerged. Most flowering occurs from May through July, but sporadic flowering continues throughout the growing season.

Fruit: Produces spherical, fleshy fruits, 6–10 mm (¼–½ in.) in diameter, that ripen from green to red to dark purple or black. Each fruit contains two or three seeds, each about 5 mm (¼ in.) long. Fruiting mostly July–September.

Stems: Bark is brown or gray, with elongated white lenticels. The sapwood is yellow, and the heartwood is pink to orange. The rust-colored, hairy terminal buds form in pairs.

Root system: An extensive, shallow root system.

Reproduction: Primarily by seed, dispersed by frugivorous birds and mammals.

Impact: This quick-growing species has prolific seed production. It can form dense thickets in wetlands or the forest understory, displacing native plants. It is also problematic because it serves as an alternate host for both alfalfa mosaic virus and the fungus *Puccinia coronata,* which causes oat rust disease.

Control: It is recommended that seedlings be pulled or dug as soon they are detected. Seeds remain viable in the seed bank for many years,

Native Alternatives
Aronia melanocarpa (black chokeberry)
Cephalanthus occidentalis (buttonbush)
Cornus amomum (silky dogwood)
Cornus racemosa (gray dogwood)
Diervilla lonicera (northern bush honeysuckle)
Hamamelis virginiana (common witch hazel)
Sambucus canadensis (common elder)
Sambucus pubens (red-berried elder)
Staphylea trifolia (American bladdernut)
Styrax americanus (American snowbell)

so watch for more seedlings to appear. Larger plants can be cut back or girdled, but to deplete the root reserves repeated treatment is often necessary. Herbicides such as glyphosate or triclopyr can be selectively applied in combination with cutting and girdling.

Frangula alnus leaves. Image by Leslie J. Mehrhoff, University of Connecticut, Bugwood.org.

Frangula alnus seeds. Image by Steve Hurst, USDA NRCS PLANTS Database, Bugwood.org.

Frangula alnus flowers. Image by Rob Routledge, Sault College, Bugwood.org.

Frangula alnus bark. Image by Leslie J. Mehrhoff, University of Connecticut, Bugwood.org.

Frangula alnus fruits. Image by Rob Routledge, Sault College, Bugwood.org.

Frangula alnus seedlings. Image by Leslie J. Mehrhoff, University of Connecticut, Bugwood.org.

Rhamnaceae

Rhamnus cathartica

European buckthorn · common buckthorn ·
purging buckthorn · Hart's thorn · European
waythorn · rhineberry

Origin: Native to Europe, northwest Africa, and western Asia. This species was introduced to North America as an ornamental in the early 1800s, if not before. It had begun to invade natural habitats by the early 1900s.

Description: A profusely branching deciduous shrub or small tree, with woody spines at the ends of twigs and in the forks of branches. It is most commonly multistemmed but occasionally grows as a single stem with a trunk diameter up to 25 cm. Due to the presence of several different quinones, the fruit, leaves, and bark of this species are mildly toxic. The specific epithet *cathartica* is derived from the species' purgative effect.

Habitat: Open woods, pastures, fencerows, roadsides, floodplains, and riparian forests. It primarily grows in sites with well drained, neutral to basic soils.

Height: 2–8 m (6–26 ft.).

Foliage: Simple leaves with finely serrated leaf margins, opposite or subopposite, oval-shaped, 3–9 cm (1–4 in.) long and 2–4 cm (¾–2 in.) wide. Leaf veins branch in an alternate fashion from the midrib and curve inward as they approach the leaf tip.

Flowers: Dioecious, with separate male and female flowers on different plants. Both male and female flowers are greenish yellow, have four petals, and are about 4 mm (⅛ in.) in diameter. Flowers form in clusters of two to six in the leaf axils. May–June, with sporadic flowering during the remainder of the growing season.

Fruit: Spherical fruits, 5–10 mm (¼–½ in.) in diameter, dark purple or black. Each fruit contains three or four seeds. Seeds are 4–5 mm (⅛–¼ in.) long and triangular in cross-section. August–September.

Stems: The bark is a smooth gray-brown, becoming rough and scaly at maturity. Twigs and young stems have prominent lenticels. The sapwood is yellow, and heartwood is pink to orange. The terminal buds are in pairs that resemble the hoof of a buck—giving rise to the common name "buckthorn"—with a narrow spine protruding between the two buds.

Root system: Rooting varies depending on site characteristics. The root system tends to be extensive but not deep.

Reproduction: Primarily by seed, dispersed by frugivorous birds and mammals. Regrowth from root fragments is possible.

Impact: This fast-growing species can form extensive monotypic stands, particularly in the understory of upland forests. Dense shade from these stands eliminates native plant species. There is also evidence that *Rhamnus cathartica* is positively associated with invasive European earthworms; control of *R. cathartica* can decrease invasive earthworm density to the benefit of native plant communities. As is its close relative *Frangula alnus* (glossy buckthorn) [pp. 298–299], this species is an alternate host for both alfalfa mosaic virus and the fungus *Puccinia coronata,* which causes oat rust disease.

Control: Hand pulling seedlings is recommended. Larger plants can be cut back or girdled, but they often resprout, so to deplete the root reserves repeated treatments will likely be necessary. Selective application of herbicides such as glyphosate or triclopyr can be used in combination with cutting and girdling. Foliar spraying of glyphosate or triclopyr is effective but has a high risk of drifting onto nontarget vegetation so should be avoided unless there is no other practical option. Seeds can stay viable in the soil for several years so repeated treatments will be necessary regardless of the control method used. For **Native Alternatives**, see *Frangula alnus* [pp. 298–299].

Left: Bark of young sapling of *Rhamnus cathartica*. Image by Chris Evans, University of Illinois, Bugwood.org. *Right:* Bark of mature tree of *Rhamnus cathartica.* Image by Robert Vidéki, Doronicum Kft., Bugwood.org.

Left: Leaves of *Rhamnus cathartica.* Image by Leslie J. Mehrhoff, University of Connecticut, Bugwood.org. *Right:* Male flowers of *Rhamnus cathartica.* Image by Robert Vidéki, Doronicum Kft., Bugwood.org.

Leaf buds of *Rhamnus cathartica.* Image by Leslie J. Mehrhoff, University of Connecticut, Bugwood.org.

Female flowers of *Rhamnus cathartica.* Image by Leslie J. Mehrhoff, University of Connecticut, Bugwood.org.

Rhamnus cathartica fruits. Image by Jan Samanek, Phytosanitary Administration, Bugwood.org.

Rhamnus cathartica seeds. Image by Steve Hurst, USDA NRCS PLANTS Database, Bugwood.org.

Rosaceae

Pyrus calleryana
Callery pear · Bradford pear

302

Origin: Native to China, Vietnam, Japan, Taiwan, and Korea. In 1916 the US Department of Agriculture introduced *Pyrus calleryana* to help develop resistance to fire blight in common pears, *Pyrus communis.* This species was used for many years as a rootstock in the commercial pear industry. By the 1950s, its ornamental value was recognized and many cultivars were developed, including the very popular 'Bradford.' It is now one of the most widely planted ornamental trees in the urban landscape in the United States. Individual cultivars are largely self-incompatible, but different cultivars can cross-pollinate to produce viable seeds. Because *P. calleryana* is planted at such high densities and there is now so much genetic diversity in the numerous available cultivars, sexual reproduction has become common and this species has spread into natural areas.

Description: This deciduous tree has been used as an ornamental because of its fast growth, upright form, and teardrop or rounded crown. It has showy white flowers in the early spring, and its foliage provides reliable autumn color. Naturalized individuals of *P. calleryana* are sometimes quite thorny, even though the parent cultivars are not.

Habitat: Cultivars are widely planted in urban areas. Naturalized populations are found in disturbed habitats such as roadsides, fallow fields, degraded open woodlands, woodland edges, fencerows, and thickets. Grows best in full sun with moist, well-drained soils, but it can tolerate partial shade. It is highly resistant to disease and pests and reasonably tolerant of drought, heat, and pollution.

Height: 10–18 m (30–60 ft.).

Foliage: Simple, alternate, 4–8 cm (2–4 in.) long and 3.5–6 cm (1 ½–3 in.) wide, heart-shaped to oval with pointed tips. Leaf margins are wavy, with finely serrated margins. Leaves are leathery, glossy and dark green on top, paler green underneath. Foliage turns yellow, orange, red, pink, or purple in autumn, depending on the cultivar.

Flowers: Showy flowers in clusters of 6–12 develop in the leaf axils. Flowers are 2–2.5 cm (¾–1 in.) in diameter, with five white petals, 15–20 stamens, and two or three styles. Flowers are insect pollinated and give off an unpleasant scent. Flowering occurs in early spring before the leaves expand fully, April–May.

Fruit: A spherical fruit, about 1 cm (½ in.) in diameter, green to brown, with pale dots on the surface. The fruit is hard, almost woody, but autumn frost softens it. Each fruit typically contains two to six dark brown seeds, each about 4–6 mm (⅛–¼ in.) long and 2–4 mm (¹⁄₁₆–⅛ in.) wide. Fruits develop in late spring or early summer and remain on the tree until autumn.

Stems: Trunks can grow to about 60 cm (24 in.) in diameter but are typically narrower. The young bark is gray or reddish brown, smooth,

Native Alternatives
Amelanchier arborea (downy serviceberry)
Amelanchier laevis (Allegheny serviceberry)
Cercis canadensis (eastern redbud)
Chionanthus virginicus (fringetree)
Nyssa sylvatica (black tupelo)
Prunus americana (American plum)
Prunus pensylvanica (pin cherry)
Prunus virginiana (chokecherry)
Styrax americanus (American snowbell)

with horizontal lenticels. Mature bark is gray-ish brown with vertical ridges and furrows. Thorny spur shoots are often present.

Root system: Trees typically have shallow but dense root systems; thus, intense root competition makes it difficult for other plants to grow underneath.

Reproduction: By seed formed through cross-pollination between different cultivars. Seeds are primarily dispersed by birds that eat the fruits. Can begin flowering and setting fruit after only three years.

Impact: This species invades disturbed areas and can form dense thickets. It outcompetes native species and can disrupt succession in old fields. Wild *P. calleryana* trees produce copious seed from a young age and can spread rapidly.

Control: Do not plant new cultivars of *P. calleryana,* even those marketed as sterile. Shallow-rooted seedlings and saplings can be dug out or pulled when the soil is moist. Larger trees should be cut down and the stumps treated with glyphosate or triclopyr.

Pyrus calleryana leaves. Image by Chuck Bargeron, University of Georgia, Bugwood.org.

Pyrus calleryana fruits. Image by Leslie J. Mehrhoff, University of Connecticut, Bugwood.org.

Pyrus calleryana flowers. Image by Leslie J. Mehrhoff, University of Connecticut, Bugwood.org.

Pyrus calleryana seeds. Image by Steve Hurst, hosted by the USDA-NRCS PLANTS Database.

Pyrus calleryana leaves and thorn. Image by Kathy Smith, Ohio State University Extension, Bugwood. org.

Pyrus calleryana tree in spring. Image by Leslie J. Mehrhoff, University of Connecticut, Bugwood.org.

Pyrus calleryana tree in summer. Image by David Stephens, Bugwood.org.

Pyrus calleryana branch with thorns. Image by Nancy Loewenstein, Auburn University, Bugwood. org.

Rutaceae

Phellodendron amurense

Amur corktree · Chinese corktree

Origin: Native to northeastern China, Japan, and Korea. This species was introduced to the United States in 1856 as a landscape plant.

Description: A medium-sized deciduous tree with distinctive corky bark, a short trunk, and an open, spreading crown. Many different cultivars are available, including males that are sometimes marketed as sterile because they are fruitless. Only female plants produce fruit, which gives them the greatest potential to spread into natural areas and cause problems. However, because male trees can still provide pollen to female trees growing nearby and give rise to fruits on these neighboring plants, it is best to avoid planting *Phellodendron amurense* altogether.

Habitat: Primarily found in disturbed forests in urban and suburban areas. This species prefers full sun and moist, well-drained soils. It is successful in a range of soil types and can tolerate drought and pollution.

Height: Typically 10–14 m (30–46 ft.), occasionally to 18 m (60 ft.).

Foliage: Opposite, pinnately compound with 5–13 leaflets, each 6–12 cm (3–5 in.) long, oval-shaped with a pointed tip. The upper surface of leaves is dark green, pale underneath, leaves turning yellow in the autumn. Crushed leaves smell like citrus or turpentine.

Flowers: Dioecious, with separate male and female flowers on separate plants. Individual flowers are small, 3 mm (⅛ in.) wide, with five to eight yellowish-green or maroon petals, borne in upright clusters. Male flowers have five or six exerted stamens with bright yellow anthers. Each female flower has a large maroon pistil overtopping an ovoid green ovary. May–June.

Fruit: Spherical fruits, 6–13 mm (¼–½ in.) in diameter, are borne on female trees. Fruits are bright green during development, turning black when mature, and have a strong odor when crushed. Each fruit usually contains two or three viable seeds, about 5 mm (¼ in.) long and 2 mm (¹⁄₁₆ in.) wide. Fruits ripen in September–October and persist on trees until winter.

Stems: The outer bark is thick and corky, grayish brown and furrowed, with bright yellow inner bark.

Root system: Shallow root system spreads widely, occasionally forming large surface roots.

Reproduction: Each female tree can produce thousands of seeds, dispersed by birds and mammals. Regeneration occasionally occurs through sprouting from roots, but it is not the primary cause of spread in this species.

Impact: Dense stands of seedlings can establish in the understory in disturbed forests. These suppress the seedlings of native tree species and displace native shrubs and herbaceous plants. There is some evidence that *P. amurense* is allelopathic and can alter the soil microorganisms and surrounding vegetation. Research has shown that it can decrease acorn and nut production in neighboring oak and hickory trees, affecting the wildlife that depends on these food sources.

Native Alternatives

Carya illinoinensis (pecan)
Cladrastis kentukea (American yellowwood)
Diospyros virginiana (common persimmon)
Gymnocladus dioicus (Kentucky coffeetree)
Ptelea trifoliata (hop tree)

305

Control: It is best to avoid planting any new specimens of this tree, including male plants that are marketed as sterile. Seedlings can be pulled by hand. For adult trees, removal of females should be prioritized. Cutting or girdling can be used in combination with spot treatment by glyphosate or triclopyr. Stumps can also be ground. Because this species is capable of resprouting, cutting alone is not recommended.

Phellodendron amurense leaves. Image by Leslie J. Mehrhoff, University of Connecticut, Bugwood.org.

Phellodendron amurense fruits. Image by Patrick Breen, Oregon State University, Bugwood.org.

Male flowers of *Phellodendron amurense.* Image © Polly Ryan.

Phellodendron amurense bark. Image by Troy Kimoto, Canadian Food Inspection Agency, Bugwood.org.

Female flowers of *Phellodendron amurense.* Image © Polly Ryan.

Phellodendron amurense seedlings. Image by Leslie J. Mehrhoff, University of Connecticut, Bugwood.org.

Salicaceae

Populus alba

white poplar · silver poplar · silverleaf poplar · abele

Origin: Native to Europe, western and central Asia, and northern Africa. This species was introduced to North America as an ornamental in 1748. In the past, it was widely planted as a street and shade tree, but it has fallen out of favor because it is susceptible to pests, diseases, and damage by wind. This species is listed as either invasive or as a noxious weed in Pennsylvania, Wisconsin, and Tennessee.

Description: A medium-sized deciduous tree with widely spreading branches that develop into a rounded crown. This species is fast growing, and the wood is weak and brittle. Trees produce root sprouts and can develop into dense colonies that exclude all other vegetation.

Habitat: Found in disturbed forests and forest edges, prairies, agricultural areas, riversides, and roadsides. It prefers full sun and moist soils. It can tolerate air pollution and salt and is adaptable to a wide range of soil pH.

Height: 15–25 m (50–80 ft.).

Foliage: Alternate, simple, with three to seven palmate lobes, 5–13 cm (2–5 in.) long and 4–8 cm (2–3 in.) wide. Young leaves are covered with fine white down, mature leaves are dark green above, bright white and densely hairy on the underside.

Flowers: Dioecious, with separate male and female flowers on different plants. Male flowers are downy gray, borne in 5–8 cm (2–3 in.) long catkins with 10–20 conspicuous dark red stamens. Female flowers are yellow-green, borne in 8–10 cm (3–4 in.) long catkins. Flowers appear before the leaves emerge, March–April.

Fruit: Flowers in female catkins develop into narrow capsules, 4–6 mm (⅛–¼ in.) long, green turning brown at maturity. Each capsule contains two tiny seeds about 2 mm (¹⁄₁₆ in.) long and surrounded by white down. Capsules mature and split open to release the seeds May–June.

Stems: Trunks can grow to a diameter of 60–90 cm (24–35 in.). Young bark is smooth whitish or greenish gray, developing black diamond-shaped marks with age. Older bark becomes ridged and furrowed, especially near the base of the trunk. Young twigs and buds are covered in woolly hairs that can be rubbed off. Twigs have five-angled pith that is star-shaped in cross section.

Root system: The root system is shallow and expansive, sometimes aggressive as the plants invade water and sewer lines. New shoots form on lateral roots.

Reproduction: Abundant seeds are dispersed by wind but have low germination rates. Once established, vegetative reproduction leads to the formation of large colonies.

Impact: Prolific sprouting from roots creates large colonies of these trees that shade out other species.

Control: Removal of seedlings by hand will help prevent the establishment of new clones. The entire root system must be removed to ensure there is no resprouting. For mature trees, any treatment must target all the stems in a clone and follow-up will be necessary to control root sprouts. The main trunks can be cut and glyphosate or triclopyr painted on the cut stumps.

Native Alternatives

Betula lenta (sweet birch)
Betula nigra (river birch)
Carpinus caroliniana (musclewood)
Populus grandidentata (bigtooth aspen)
Populus tremuloides (quaking aspen)

307

Left: *Populus alba* leaves. Image by Robert Vidéki, Doronicum Kft., Bugwood.org. *Right: Populus alba* bark. Image by T. Davis Sydnor, The Ohio State University, Bugwood.org.

Female catkins of *Populus alba*. Image by Leslie J. Mehrhoff, University of Connecticut, Bugwood.org.

Upper and lower surfaces of *Populus alba* leaves. Image by Leslie J. Mehrhoff, University of Connecticut, Bugwood.org.

Developing seed capsules of *Populus alba*. Image by Robert Vidéki, Doronicum Kft., Bugwood.org.

Male catkins of *Populus alba*. Image by Paul Wray, Iowa State University, Bugwood.org.

Stand of *Populus alba* saplings. Image by Robert Vidéki, Doronicum Kft., Bugwood.org.

Salicaceae

Salix alba
white willow · golden willow

Origin: Native to Europe, northern Africa, and western and central Asia. This species was introduced to North America in the 1700s and has since escaped cultivation and become naturalized in many parts of the United States. There are still several popular commercially available cultivars—especially weeping forms.

Description: This medium to large deciduous tree has spreading or pendulous twigs and an irregular crown.

Habitat: This species thrives in habitats with moist soil, including marshes, ditches, river- and streambanks, wetland margins, riverine and riparian forests, roadsides, field edges, and yards. It is not tolerant of heavy shade.

Height: 10–30 m (30–100 ft.).

Foliage: Alternate, simple, narrowly lance-shaped, 4–12 cm (2–5 in.) long and 1–2.5 cm (½–1 in.) wide, with glandular teeth along the margins. Leaves are covered in fine, silky white hairs, and their undersides are a dull, silvery green. There are glands present on the petiole near the base of each leaf.

Flowers: Dioecious, with separate male and female flowers on different plants. Male flowers have two (occasionally three) stamens each with yellow anthers, borne in showy catkins, 3–6 cm (1–3 in.) long. Female flowers are greenish and less conspicuous, in catkins 3–4 cm (1–2 in.) long. Flowering occurs in spring after leaves have emerged, April–May.

Fruit: Female flowers develop into narrow capsules, 3–4.5 mm (⅛ in.) long, green turning brown at maturity. Each capsule contains numerous, minute seeds, 1–1.5 mm ($\frac{1}{32}$ in.) long and 0.4–0.5 mm ($\frac{1}{64}$ in.) wide, surrounded by white down. Fruits mature in midsummer.

Stems: Trunks can grow to a diameter of 1 m (40 in.). Young twigs are bright yellow, turning yellowish brown with age. Mature bark on the trunks is grayish brown and heavily ridged.

Root system: Shallow but widely spreading, can aggressively seek out water in pipes and sewer lines. Lateral roots are capable of developing new stems.

Reproduction: Through seed and root sprouts.

Impact: This species grows rapidly and aggressively in wetland habitats, forming dense stands that limit herbaceous undergrowth and displace native species. It hybridizes freely with both native and nonnative members of the genus, leading to introgression. This has potentially negative consequences for native species.

Control: Seedlings can be pulled or dug out. Because of the species' resprouting capabilities, cutting alone is not likely to kill trees unless it is done repeatedly through the growing season to exhaust root reserves. It is recommended that cutting be combined with treatment by glyphosate, imazapyr, or triclopyr specifically formulated for wet areas, but only a licensed aquatic herbicide applicator should undertake chemical control.

309

Native Alternatives
Salix discolor (pussy willow)
Salix humilis (prairie willow)
Salix nigra (black willow)

Salix alba capsules releasing seeds. Image by Evelyn Simak, used under a CC BY-SA 2.0 license.

Left: Salix alba leaves. *Right:* Developing seed capsules of *Salix alba.* Both images by Robert Vidéki, Doronicum Kft., Bugwood.org.

Salix alba bark. Image by Robert Vidéki, Doronicum Kft., Bugwood.org.

Male catkins of *Salix alba.* Image by Willow, used under a CC BY-SA 2.5 license.

Female catkins of *Salix alba.* Image by Willow, used under a CC BY-SA 2.5 license.

Mature *Salix alba* tree. Image by Robert Vidéki, Doronicum Kft., Bugwood.org.

Salicaceae

Salix fragilis
crack willow · brittle willow

Origin: A native to Europe and western Asia that was introduced to North America as an ornamental during colonial times. This fast-growing species has escaped cultivation and spread across the United States. Although it is not yet listed as invasive in Ohio, it is widespread in the state.

Description: A large deciduous tree with stout twigs and an irregular, often leaning crown. It is often multistemmed and readily sprouts from the roots. The species' common names refer to the fact that the branches break easily, making a cracking sound. The stem fragments can be dispersed downstream, develop roots, and establish new colonies away from the parent plant.

Habitat: Prefers damp habitats such as river- and streambanks, lakesides, marshes, fens, and disturbed areas with moist soils. This species does not tolerate heavy shade.

Height: 10–20 m (30–65 ft.).

Foliage: Alternate, simple, narrowly lance-shaped, 7–15 cm (3–6 in.) long and 1.5–3.5 cm (½–1½ in.) wide. Like *Salix alba* [pp. 309–310], its leaves have glandular teeth along the margins, but the leaves of *Salix fragilis* are more coarsely and sparsely toothed. Leaves become hairless at maturity, dark green above and powdery white underneath. There are glands present on the petiole near the base of each leaf.

Flowers: Dioecious, with separate male and female flowers on different plants. Male flowers have two stamens with yellow anthers, borne in catkins 3–6 cm (1–3 in.) long, at the ends of short, leafy branchlets. Female flowers are green, narrowly conic, borne in catkins 3–6 cm (1–3 in.) long. Flowering occurs when leaves emerge in the spring, April–May.

Fruit: Female flowers develop into narrow capsules, about 4–6 mm (⅛–¼ in.) long and 2.5 mm (¹⁄₁₆ in.) wide, green turning brown at maturity. Each capsule contains numerous, minute seeds, 1–1.5 mm (¹⁄₃₂ in.) long and 0.5 mm (¹⁄₆₄ in.) wide, surrounded by white down. Capsules split open to release the seeds in May–June.

Stems: Trunks can grow to a diameter of 1 m (40 in.), but they are often narrower if multistemmed. Young twigs are greenish to dark red. Mature bark is dark brown to gray and deeply fissured.

Root system: Shallow but widely spreading, often forming dense mats. Roots are aggressive and can grow into waterways, trapping silt and changing the flow of water. Lateral roots are capable of developing new stems.

Reproduction: By seed, root sprouts, and stem fragments.

Impact: Forms pure stands that shade out other plant species. Spreading roots can alter stream hydrology and sedimentation rates. This species hybridizes freely with related willows, including both native and nonnative species.

Control and Native Alternatives: See *Salix alba* [pp. 309–310].

Salix fragilis leaves. Image by Robert Vidéki, Doronicum Kft., Bugwood. org.

Salix fragilis capsules releasing seeds. Image by Robert Vidéki, Doronicum Kft., Bugwood. org.

Male catkins of *Salix fragilis.* Image by Robert Vidéki, Doronicum Kft., Bugwood. org.

Salix fragilis stems and bark. Image by Robert Vidéki, Doronicum Kft., Bugwood. org.

Open *Salix fragilis* capsules. Image by Robert Vidéki, Doronicum Kft., Bugwood.org.

Salix fragilis tree. Image by Robert Vidéki, Doronicum Kft., Bugwood. org.

Female catkins of *Salix fragilis.* Image by Robert Vidéki, Doronicum Kft., Bugwood.org.

Sapindaceae

Acer platanoides
Norway maple

Origin: Native to Europe and western Asia, this species was first introduced to North America in 1756. It is planted widely as an ornamental or specimen tree and has spread into natural areas.

Description: A large, deciduous tree with a widely spreading crown. *Acer platanoides* resembles the native *Acer saccharum* (sugar maple). The two species can be differentiated by the presence of white sap in *A. platanoides*, exuded when the leaf stalks or buds are broken off. In addition, *A. platanoides* has widely spreading wings on its fruits, whereas *A. saccharum*'s wings are smaller and angled or curved downward, giving the paired samaras a horseshoe shape.

Habitat: Yards and gardens, open disturbed areas, roadsides, vacant lots, forested wetlands, early and late successional forest. This species can grow in full sun to deep shade and tolerates a range of temperatures and soil types.

Height: Typically 12–18 m (40–60 ft.), can reach 30 m (100 ft.).

Foliage: Simple, opposite to subopposite, 7–14 cm (3–5 in.) long and 10–18 cm (4–7 in.) wide, with palmate veins and lobes. Leaves have five to seven sharply pointed lobes and a few large, sharp teeth. Leaves are usually dark green, but some cultivars have dark red to purple foliage. The petiole is 8–20 cm (3–8 in.) long and exudes a white milky sap when detached.

Flowers: Small, yellow-green flowers are borne in a flat-topped, branching corymb. Each flower is 5–8 mm (¼ in.) wide, with five petals, five sepals, eight stamens, and a pistil formed from two fused carpels. April–May.

Fruit: Fruit is a double samara with horizontally spreading wings that split down the middle. Each samara develops a single seed and has a flattened wing about 3.5–5 cm (1 ½–2 in.) long, which facilitates dispersal by wind.

Stems: The bark of the tree is grayish brown and shallowly grooved or furrowed. Twigs are stout and smoothly barked, with opposite branches.

Root system: A shallow root system makes this species unstable in high wind. The roots are highly competitive and monopolize soil moisture and nutrients.

Reproduction: By seed.

Impact: This species produces abundant seeds that can germinate rapidly. Seedlings and saplings shade out native understory vegetation such as spring ephemerals. Over time, *A. platanoides* can outcompete and replace native tree species such as *A. saccharum* in the forest canopy, which changes the forest structure. Once established, it creates a canopy of dense shade that prevents regeneration of native seedlings.

Control: Many cultivars are still widely available; it is best to avoid planting any new specimens unless the cultivar has been found to

Native Alternatives
Acer rubrum (red maple)
Acer saccharum (sugar maple)
Liquidambar styraciflua (sweetgum)
Liriodendron tulipifera (tulip poplar)
Nyssa sylvatica (black gum)
Quercus bicolor (swamp white oak)

not be invasive (e.g., 'Crimson King' [p. 360]). Seedlings can be pulled or dug out when the soil is moist. Larger trees should be cut and the stumps painted with glyphosate or triclopyr.

Trees can also be girdled by cutting through the bark and cambium all around the trunk. This is most effective in the spring.

Acer platanoides leaves. Image by Paul Wray, Iowa State University, Bugwood.org.

Mature *Acer platanoides* samaras. Image by Steve Hurst, hosted by the USDA-NRCS PLANTS Database.

Acer platanoides flowers. Image by Jan Samanek, Phytosanitary Administration, Bugwood.org.

Acer platanoides bark. Image by Leslie J. Mehrhoff, University of Connecticut, Bugwood.org.

Acer platanoides samaras developing on tree. Image by Paul Wray, Iowa State University, Bugwood.org.

Planted *Acer platanoides* trees. Image by Leslie J. Mehrhoff, University of Connecticut, Bugwood.org.

Simaroubaceae

Ailanthus altissima

tree-of-heaven · copal tree · Chinese sumac · stinking sumac · stink tree · varnish tree · paradise tree

Origin: A native of northeastern and central China and Taiwan that was first introduced to Philadelphia, Pennsylvania, in 1784 as a horticultural specimen. It was then introduced as an ornamental species in New York in 1820 and as a medicinal plant in California during the Gold Rush of the mid-1800s. The species was widely planted in cities because of its tolerance of pollution and poor soils. From these areas, it has spread in urban, agricultural, and forested areas throughout much of the United States. This well-established invasive is present in every county of Ohio.

Description: A small, rapidly growing, deciduous tree with few branches and coarse twigs. All parts of the tree have a strong, offensive odor, especially the flowers and crushed foliage. It may be confused with native trees and shrubs having compound leaves, such as *Juglans nigra* (black walnut) and *Rhus* spp. (sumacs). *Ailanthus altissima* can be differentiated by its odor, stout twigs, large leaf scars, smooth leaflet margins, and glandular teeth at the base of each leaflet.

Habitat: Disturbed urban areas, roadsides, fields, fencerows, woodland edges, and successional forests. It thrives in poor soils and can tolerate pollution. Does not grow in wetland habitats or sites with dense shade.

Height: Usually 6–10 m (20–30 ft.), but large specimens can grow to 25 m (80 ft.).

Foliage: Alternate, large, pinnately compound leaves that grow to 100 cm (3 ft.) long. Typically 11–41 leaflets per leaf, lance-shaped, 5–18 cm (2–7 in.) long and 2.5–5 cm (1–2 in.) wide, leaflets unlobed except for one or more coarse glandular teeth near the base. Foliage is dark green above and pale green beneath and gives off an unpleasant odor when crushed.

Flowers: Typically, male and female flowers occur on different plants, although there are rare examples of individuals with both male and female flowers. Both flower types are 5 mm (¼ in.) wide, greenish yellow, with five or six petals. Male flowers have 10 stamens, and the pollen has a particularly strong odor. Flowers are borne in many-branched clusters near the ends of branches, 10–30 cm (4–12 in.) long. May–June.

Fruit: A twisted samara with a single seed near the center, 3–5 cm (1–2 in.) long and 7–13 mm (¼–½ in.) wide, yellow-green to orange-red changing to light brown as they mature. Seeds are 5 mm (¼ in.) in diameter. Fruits develop July–August but can persist on the tree over winter.

Stems: Trunks of large trees can grow to 1 m (40 in.) in diameter, although they are typically much smaller. Twigs are stout with slightly hairy reddish-brown bark. Bark on trunks is smooth and light gray, often becoming somewhat rougher with pale stripes as the tree ages.

Root system: Shallow but widely spreading. Root sprouts form readily, and resprouting from the stump and roots occurs after cutting or disturbance.

Native Alternatives

Cladrastis kentukea (American yellowwood)
Carya illinoinensis (pecan)
Carya ovata (shagbark hickory)
Gymnocladus dioicus (Kentucky coffeetree)
Sorbus decora (showy mountain ash)

315

Ailanthus altissima leaves. Image by Robert Vidéki, Doronicum Kft., Bugwood.org.

Female flowers of *Ailanthus altissima.* Image by Doug Goldman, hosted by the USDA-NRCS PLANTS Database.

Glandular teeth at the base of an *Ailanthus altissima* leaflet. Image by James H. Miller, USDA Forest Service, Bugwood.org.

Ailanthus altissima fruits. Image by Chuck Bargeron, University of Georgia, Bugwood.org.

Ailanthus altissima bark. Image by Leslie J. Mehrhoff, University of Connecticut, Bugwood.org.

Male flowers of *Ailanthus altissima.* Image by Jan Samanek, Phytosanitary Administration, Bugwood. org.

Ailanthus altissima seedlings. Image by Leslie J. Mehrhoff, University of Connecticut, Bugwood.org.

Reproduction: Primarily through seed, which is prolific. It is estimated that a single female tree can produce over 300,000 seeds per year. Seeds are spread by wind, water, and other vectors. Root sprouting also contributes to the local spread and persistence of stands.

Impact: Forms dense thickets and displaces native vegetation. It also produces ailanthone, an allelopathic chemical that inhibits the growth of other plants. The aggressive root system can damage pavement, sewers, and building foundations.

Control: Wear protective clothing and gloves, because chemicals in the leaves and flowers sometimes result in contact dermatitis. Pull seedlings by hand before the taproot develops; any root fragments left behind can generate new plants. In mature plants, cutting alone is likely to encourage vigorous resprouting and will need to be repeated for several years to deplete the energy stored in the roots. Foliar treatment using glyphosate, triclopyr, or imazapyr can be done when leaves have emerged in the spring. Cut stump or basal bark treatment with triclopyr is most effective in summer or early autumn. Any control method will require follow-up monitoring and repeated applications.

Ulmaceae

Ulmus pumila

Siberian elm · Asiatic elm · dwarf elm ·
littleleaf elm

318

Origin: Native to Siberia, northern and north-eastern China, and Korea. It was first introduced to North America as a horticultural species in the 1860s. This species, resistant to Dutch elm disease, was planted widely as a replacement for *Ulmus americana* (American elm) following the outbreak of the disease. It has escaped cultivation and spread into natural habitats.

Description: This small- to medium-sized deciduous tree has an open, rounded crown and slender, spreading branches. Occasionally grows in a shrub form. It is extremely fast growing, but its branches are brittle and prone to breakage, which, combined with the lack of ornamental features, makes it an undesirable horticultural specimen. Despite being considered an invasive species in many states, it is still sold commercially as a shelterbelt and windbreak tree. It is sometimes erroneously referred to as Chinese elm, which is the common name of the closely related species *Ulmus parvifolia*.

Habitat: Found in disturbed areas with ample sun, such as pastures, meadows, prairies, roadsides, streambanks, open woodlands, forest margins, vacant lots, and fencerows. This species tolerates a wide range of growing conditions.

Height: Typically 15–20 m (50–65 ft.) at maturity, although sometimes to 25 m (80 ft.) or taller.

Foliage: Simple, alternate, elliptic with a pointed tip, 3–8 cm (1–3 in.) long and 2–3 cm (¾–1 in.) wide, dark green and smooth. The leaves of this species are small compared to others in the genus, and the nearly symmetrical leaf base and single teeth along the leaf margin distinguish it from congeners.

Flowers: Small, drooping clusters of 3–15 flowers develop along the branches. Individual flowers are 3 mm (⅛ in.) across, with a green base, four or five stamens with reddish anthers, a green pistil with a two-lobed style, and no petals. The flowers are cross-pollinated by wind but are also self-compatible. Flowers emerge before the leaves, March–April.

Fruit: A flat, circular samara, 1–2 cm (½–¾ in.) long and 1–1.5 cm (½ in.) wide, deeply notched at the outer tip. Fruit is smooth, light green becoming light tan at maturity. Each fruit contains a single seed in the center, about 4 mm (⅛ in.) long and 2 mm ($\frac{1}{16}$ in.) wide. April–May.

Stems: Trunks can grow to 80 cm (32 in.) in diameter. Twigs are very slender and abundant, light brown to reddish brown and nearly hairless. Bark on mature trunks is gray, rough-textured, and shallowly furrowed.

Root system: Deep, branching, and widely spreading.

Reproduction: Primarily by abundant seed, dispersed by wind. Also capable of sprouting from stumps or roots.

Impact: This species can spread rapidly into disturbed areas with poor soils and low moisture. *Ulmus pumila* often forms dense thickets that displace native vegetation. It can also invade open areas and change the structure to closed woodland.

Native Alternatives

Betula lenta (sweet birch)
Carpinus caroliniana (musclewood)
Celtis occidentalis (hackberry)
Ptelea trifoliata (hop tree)
Ostrya virginiana (hop-hornbeam)
Ulmus rubra (slippery elm)
Ulmus thomasii (rock elm)

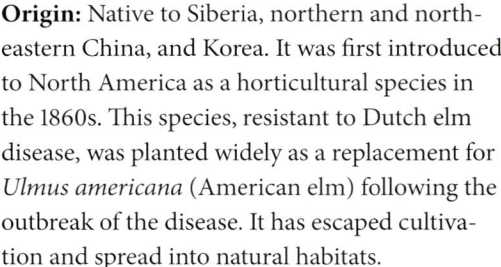

Control: Avoid planting new specimens of *U. pumila*. Seedlings can be pulled by hand. For mature individuals, the recommended management procedure is girdling the trunks from late spring to midsummer. The trees will typically die over a one- to two-year period. To avoid resprouting, cut stumps can be treated with systemic herbicides such as glyphosate and triclopyr. For effective long-term management, it is essential to reduce or eliminate the seed source.

Ulmus pumila leaves. Image by Melburnian, used under a CC BY 2.5 license.

Mature *Ulmus pumila* fruits. Image by Steve Hurst, USDA NRCS PLANTS Database, Bugwood.org.

Left: Ulmus pumila bark. *Right: Ulmus pumila* twigs. Both images by Robert Vidéki, Doronicum Kft., Bugwood.org.

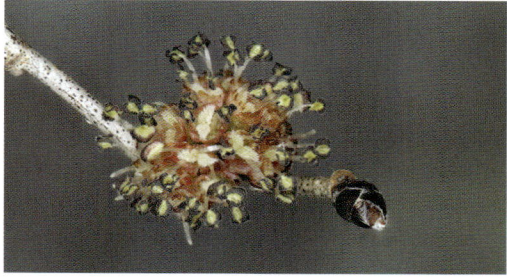

Ulmus pumila flower cluster. Image © Minnesota Wildflowers.

Ulmus pumila fruits developing on tree. Image © Robert L. Carr.

Ulmus pumila tree. Image by Patrick Breen, Oregon State University, Bugwood.org.

Aquatic and Wetland Plants

True aquatic plants are those that grow in the open water of lakes, ponds, rivers, streams, and springs. Many aquatic plants are rooted in the bottoms of water bodies, but some are not anchored to the substrate and are classified as free-floating. Aquatic plants may grow submersed (or submerged), with the entire plant growing below the water level, or they may be emergent, with parts above the water. Different types of aquatic plants grow in distinctly different ways: emergent species have stronger stems and broader leaves, and submersed species typically have weak stems and thin or finely dissected leaves.

Not all wetland plants are true aquatic plants, but they still require habitats inundated by water. Many wetland plants live in saturated soils, bogs, marshes, and wet meadows, while others live at the periphery of open waters. Every species in this section is either a true aquatic species or an obligate wetland species, which means that they almost always occur in wetlands.

Once they enter watercourses, aquatic and wetland plants are often able to invade large sections, particularly areas downstream of the invasion point. A major problem in Ohio's water bodies is that many people use boats to access the water. The movement of boats from one body of water to another can facilitate the dispersal of aquatic and wetland plants. When boats and trailers are removed from the water, great care should be taken to examine them for "hitchhikers" to ensure that aquatic invasive species—plant or animal—are not being moved between areas. An additional source of invasion is the practice of emptying aquariums into water bodies.

Only licensed and certified commercial applicators can apply herbicides to control aquatic and wetland plants. Any herbicide will kill all aquatic plants, including native species, so this is not a safe or practical option in most sites. Furthermore, because of the connectedness of aquatic systems, using chemicals in or adjacent to water bodies has major risks. Misuse of herbicides could contaminate the water table, endangering natural habitats and drinking water. Thus, any chemical applications to aquatic habitats require a permit from the Ohio Environmental Protection Agency (OEPA). OEPA statute 3745–1–01(E)(1) requires that the OEPA be given notice before chemicals are applied to water bodies within one mile of public water supplies, recreational areas, and wetlands. Applications of aquatic herbicides to river or stream banks and township or county drainage ditches also require OEPA notification prior to application. Only chemicals approved for aquatic use by the US Environmental Protection Agency and labeled for aquatic application may be used in and around water bodies. For further details, please see appendix D [pp. 363–365].

Acoraceae

Acorus calamus

sweet flag · sweet cane · calamus · single-veined sweetflag

Origin: Native to central and western Asia and parts of eastern Europe. This species has been used as a medicinal plant for thousands of years and was introduced to North America by early European settlers in the 1600s. *Acorus calamus* has become widespread and abundant in the United States but typically escapes detection because of its extremely close resemblance to the native species *Acorus americanus* (American sweet flag), which it frequently displaces.

Description: An emergent, grasslike, perennial forb that grows along the margins of wetlands. Recent genetic studies have confirmed that *A. calamus* is a sterile triploid (having three sets of chromosomes instead of two). The name *A. calamus* has been incorrectly applied to *A. americanus* (also sometimes referred to as *Acorus calamus* var. *americanus*), further adding to confusion in identification. Leaf venation is the most reliable way to differentiate these species. In *A. calamus,* each leaf has a single distinct primary vein with smaller secondary veins, whereas *A. americanus* has two to six additional large primary veins that run parallel to the midrib. These veins are most visible when the leaf is held up to the light. Leaves of *A. calamus* are typically wider than 1 cm (½ in.), while those of *A. americanus* are narrower. In addition, *A. calamus* is sterile and does not set seed, while *A. americanus* produces a spike of reddish-brown berries that contain two or three seeds each.

Habitat: Found on the margins of standing or slow water, typically along the edges of canals, lakes, ponds, rivers and streams and in marshes or swales. It can grow in wet soil or in water to a depth of 30 cm (1 ft.).

Height: 30–175 cm (1–6 ft.).

Foliage: Tufts of basal leaves emerge directly from the rootstock. They are upright and sword-shaped, 30–175 cm (1–6 ft.) long and 1–2 cm (½–¾ in.) wide, tapering to a sharp point. Leaves are pinkish near the base and bright green above. One or both margins can be wavy or crimped. There is a single distinct midrib along the leaves, often slightly off-center, along with several narrower secondary veins. When crushed, foliage produces a sweet fragrance reminiscent of citrus.

Flowers: Tiny, yellowish-green flowers are borne on a fleshy, cylindrical spadix, 5–10 cm (2–4 in.) long and 5–10 mm (¼–½ in.) wide. A green spathe extends up to 30 cm (12 in.) beyond the spadix. Individual flowers are 3–4 mm (⅛ in.) in diameter, consisting of six tepals (three petals and three modified sepals that look like petals), six stamens, and a green pistil in the center. May–June.

Fruit: This triploid species is sterile and does not produce fruit.

Stems: Flowering stems are smooth, upright, and triangular or rhomboid in cross-section.

Root system: The shallow, branching rhizomes are stout and knobby, with a brown exterior and white interior. Rhizomes give off the same sweet, citrus fragrance as foliage does when crushed. Coarse fibrous roots develop below the rhizomes.

Reproduction: Reproduction is vegetative, through spreading rhizomes and rhizome fragments. Plants are also moved via the aquatic plant trade and escape cultivation.

Impact: This species is very aggressive and can spread into high-quality wetlands. It forms dense colonies that displace native species.

Control: If there is any uncertainty about the proper identification of this species, please consult a botanist before undertaking control efforts. There have been cases of *A. americanus* being removed from wetlands in a misguided attempt to control the nonnative species. For small stands, dig out the plants and rhizomes and remove them from the site. Aquatic formulations of glyphosate and triclopyr amine can control this species. Only a licensed aquatic herbicide applicator should undertake chemical management.

Acorus calamus plants. Image by Mokkie, used under a CC BY-SA 3.0 licence.

Leaf margins and venation on *Acorus calamus*. Image © Arthur Haines, Native Plant Trust.

Developing spadix of *Acorus calamus*. Image by Dũng Nguyễn Việt, used under the Public Domain Mark 1.0.

Rhizomes and roots of *Acorus calamus*. Image by Dũng Nguyễn Việt, used under the Public Domain Mark 1.0.

Acorus calamus flowers. Image by Dũng Nguyễn Việt, used under the Public Domain Mark 1.0.

Acorus calamus stand. Image by Panek, used under a CC BY-SA 4.0 license.

Boraginaceae

Myosotis scorpioides

common forget-me-not · true forget-me-not ·
yelloweye forget-me-not · water forget-me-not ·
water scorpion-grass

Origin: Native to Europe and the western part
of Asia. It was introduced to North America for
ornamental and medicinal purposes. This species had invaded the Great Lakes Basin by 1886.

Description: A perennial forb species. The
generic name *Myosotis* is derived from "mouse
ear" in classical Greek, for its hairy, rounded
leaves. The specific epithet *scorpioides* is a reference to its inflorescence; in bud, the flower
cluster coils around like a scorpion tail. *Myosotis scorpioides* can form dense mats that outcompete most other species.

Habitat: Prefers shallow water and wet soil.
Frequently found along stream banks, springs,
ditches, damp meadows, and rich swamps.

Height: 20–60 cm (8–24 in.).

Foliage: Simple, alternate leaves 2–8 cm (¾–3
in.) long by 7–20 mm (¼–1 in.) wide. Leaves
mostly oblong or lance-shaped, with entire margins and no petiole. Each leaf has a prominent
central vein. Both sides of the leaves have hairs
that are flush with the surface. Plants overwinter
as rosettes of sterile, evergreen leaves.

Flowers: Inflorescence a terminal, curving,
branched cyme with alternate flowers. Flowers
blue (rarely white or pink) with a yellow center, five petals, 6–9 mm (¼–⅜ in.) in diameter.
May–September.

Fruit: The calyx that surrounds the base of
the flower remains after the flower has bloomed.
Within this calyx, four small 1-seeded nutlets
develop, each about 2 mm (¹⁄₁₆ in.) long. The
nutlets are glossy brown and egg-shaped.

Stems: Hairy, circular in cross-section. Stems
often creep along the ground near the base.

Root system: Fibrous rhizomes. Stolons also
develop shallow roots at the nodes where they
touch the ground.

Reproduction: Reproduction is mainly by
seed but also occurs through rhizomes and stolons. Seeds are readily dispersed by water and
moved between sites by animals or humans. It
is not known how long the seeds remain viable
in the soil.

Impact: This species forms dense monocultures, potentially displacing native plant species that specialize in marginal wetland environments. It is widely planted in gardens,
which gives it the opportunity to escape into
natural habitats. Although *M. scorpioides* is not
invasive in Ohio, it is considered a prohibited
noxious weed in Massachusetts and is listed as
invasive in Connecticut and Wisconsin.

Control: Control techniques are not well documented for this species. For small populations, removal by hand is recommended, several times a year.

Native Alternatives

Geranium maculatum (wild geranium)
Lobelia siphilitica (great blue lobelia)
Mertensia virginica (Virginia bluebells)
Myosotis laxa (smaller forget-me-not)
Verbena stricta (hoary vervain)

Myosotis scorpioides plants. Image by Patrick J. Alexander, hosted by the USDA-NRCS PLANTS Database.

Leaf and stem of *Myosotis scorpioides*. Image © Katy Chayka, Minnesota Wildflowers.

Myosotis scorpioides flowers. Image © Robert L. Carr.

Myosotis scorpioides nutlets developing within the persistent calyx. Image by Jouko Lehmuskallio, www.NatureGate.net.

Myosotis scorpioides inflorescence. Image by AnR00002, used under a CC0 1.0 license.

Myosotis scorpioides infestation. Image by Leslie J. Mehrhoff, University of Connecticut, Bugwood.org.

Brassicaceae

Nasturtium officinale

watercress · greencress

Origin: Native distribution is widespread throughout Europe and Asia. This species was deliberately introduced for cultivation as a leafy vegetable. The first record of *Nasturtium officinale* in North America is from 1831 in Connecticut, although it was likely being grown as early as the mid-1700s.

Description: A perennial aquatic or semi-aquatic forb. This plant has been used medicinally in the past and it is now being investigated as an anticarcinogen. The spicy, peppery leaves are commonly used in salads and cooking.

Habitat: Grows in cold, clear, shallow freshwater, particularly along the edges of lakes or springs, in gently flowing rivers and streams, and in wet ditches.

Height: Stems can grow to 100 cm (40 in.).

Foliage: Leaves alternate, pinnately compound, 4–15 cm (2–6 in.) long, with three to nine oval leaflets. The end leaflet is almost round and much larger than the lateral leaflets. Leaves hairless, leaf margins slightly wavy. Leaves can remain green over winter in mild climates or protected sites.

Flowers: Compact terminal clusters. Each flower is 5 mm (¼ in.) wide with four rounded, white petals, six yellow stamens, a short central style, and a purplish-green ovary. Flowering April–October.

Fruit: Slender silique, 1–2 cm (½–¾ in.) by 2 mm ($\frac{1}{16}$ in.) wide, tapering to 1 mm ($\frac{1}{32}$ in.) at the tip. Seeds form in two rows in each chamber; they are reddish brown, ovoid, covered with an intricate netlike pattern, 1 mm ($\frac{1}{32}$ in.) long by 0.7–0.9 mm ($\frac{1}{32}$ in.) wide. Seeds can germinate immediately upon maturation.

Stems: Stems succulent, hairless, many branched, and hollow. Submersed or partly floating, can be erect or creeping. Form tangled mats with other stems.

Root system: Plants have a taproot. Thin, fibrous roots also develop at the nodes along the stem, allowing for vegetative reproduction.

Reproduction: By seed and plant fragments. Seeds are dispersed by gravity, wind, water, animals, and human activity.

Impact: This species grows rapidly and forms impenetrable mats on the surface of waterways. It can displace or outcompete native species, affecting the structure and composition of natural communities. In addition, *N. officinale* can alter stream flow with its dense growth.

Control: There is little information available about the management or control of this species. It is recommended that removal be performed by hand, several times a year.

Nasturtium officinale leaves. Image by Stefan Lefnaer, used under a CC BY-SA 4.0 license.

Nasturtium officinale fruits. Image by Stefan Lefnaer, used under a CC BY-SA 4.0 license.

Nasturtium officinale stem. Image by Stefan Lefnaer, used under a CC BY-SA 4.0 license.

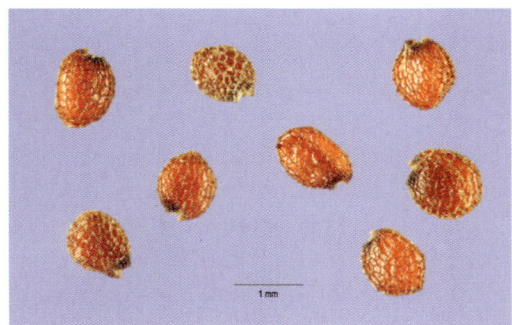

Nasturtium officinale seeds. Image by Steve Hurst, hosted by the USDA-NRCS PLANTS Database.

Nasturtium officinale flowers. Image by Paul Venter, used under a CC BY-SA 4.0 license.

Nasturtium officinale infestation. Image by Leslie J. Mehrhoff, University of Connecticut, Bugwood.org.

Butomaceae

Butomus umbellatus

flowering rush · grass rush · grassy rush · water gladiolus

ODA Invasive
OIPC Invasive

Origin: Native to Europe and Asia. This plant has attractive flowers and was most likely imported to North America as an ornamental. The starchy rhizome is edible, so it may also have been introduced as a food plant. Escaped populations were first detected along the Saint Lawrence River in the late 1800s. The species has since spread across the continent to both the Atlantic and Pacific coasts.

Description: This emergent, perennial forb is not a true rush, but has a rushlike growth form. It will not flower in deep water, often making it difficult to identify.

Habitat: Margins of freshwater wetlands, lakes, and slow-moving rivers or streams. Frequently planted in water features. Tolerant of a wide range of temperatures and soil types. Can grow in water to a depth of 2 m or more.

Height: Flowering stalks 1–1.5 m (3–5 ft.).

Foliage: Leaves long and narrow with a pointed tip, developing from the base of the plant. May be erect or floating on the water surface, up to 1 m (40 in.) long and 5–10 mm (¼–½ in.) wide. The fleshy leaves have parallel veins, a prominent midrib on the underside, and are *V*-shaped in cross-section. Leaves have a purple tinge when new but are green when mature. Emergent leaves sometimes develop a spiral twist.

Flowers: Inflorescence is a many-flowered umbel borne at the top of a single flower stalk. Each flower is borne on its own stalk about 5–10 cm (2–4 in.) long. Flowers are pink to white with red centers, 2–2.5 cm (¾–1 in.) wide. Flowers have three petals, three slightly smaller petal-like sepals, nine stamens, and six pistils that are fused together at the base. June–August.

Fruit: Each pistil develops into a brownish or maroon capsule, about 1 cm (½ in.) long, with six teardrop-shaped sections fused at the base. Capsules contain many small brown seeds, each about 1.5 mm long (⅝ in.) and 0.5 mm (¹⁄₆₄ in.) wide, distinctly ridged on the surface. The capsule splits open to release the seeds, which can float in the water. August–September.

Stems: Smooth and green, triangular in cross section. Can be erect, floating, or submersed in water.

Root system: Short, fleshy rhizomes that can form small secondary bulbs.

Reproduction: Spreads clonally through the branching and fragmentation of rhizomes. Bulblets and other plant parts can be dispersed by water, animals, and boats. Also spreads through seed, although the seeds are not always viable.

Impact: Can form dense stands that displace native riparian vegetation, obstruct boat traffic, and alter water flow.

Control: Cutting below the water surface several times per growing season can decrease abundance. Hand digging or pulling are rec-

Native Alternatives

Chelone glabra (turtlehead)
Hibiscus laevis (smooth rose mallow)
Hibiscus moscheutos (swamp rose mallow)
Lobelia cardinalis (cardinal flower)
Lobelia siphilitica (great blue lobelia)
Physostegia virginiana (obedient plant)
Tradescantia ohiensis (Ohio spiderwort)
Verbena hastata (blue vervain)

ommended for isolated plants. Care must be taken to remove all root material and to ensure that bulblets and root fragments are not spread during removal. No chemical control is recommended, because herbicides wash off the leaves easily and pose risks to nontarget species.

Butomus umbellatus leaves. Image © Peter M. Dziuk, Minnesota Wildflowers.

Flowering *Butomus umbellatus*. Image by Christian Fischer, used under a CC BY-SA 3.0 license.

Butomus umbellatus flower. Image by Leslie J. Mehrhoff, University of Connecticut, Bugwood.org.

Butomus umbellatus fruits. Image by Leslie J. Mehrhoff, University of Connecticut, Bugwood.org.

Butomus umbellatus roots and secondary bulbs. Image by Leslie J. Mehrhoff, University of Connecticut, Bugwood.org.

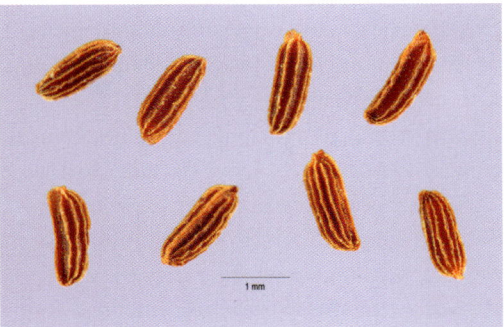

Butomus umbellatus seeds. Image by Steve Hurst, USDA NRCS PLANTS Database.

Haloragaceae

Myriophyllum aquaticum

parrot feather · water feather · Brazilian water-milfoil · diamond milfoil

Origin: Native to the Amazon River basin in South America—including parts of Brazil, Bolivia, Ecuador, Peru, and Colombia—as well as Argentina, Chile, Paraguay, and Uruguay. The earliest record of this species in North America is from 1890. It was introduced for use in aquariums and water gardens and has escaped from cultivation. Populations are now well established in the southern United States, and this species is rapidly expanding its range northward.

Description: A rooted, perennial aquatic plant that can be submersed or emergent. Its submersed leaves are similar to those of *Myriophyllum spicatum* (Eurasian watermilfoil) [pp. 333–334]. The presence of emergent stems—with a growth form somewhat resembling small fir trees—distinguishes *Myriophyllum aquaticum* from *M. spicatum*. In the absence of emergent stems, they can be differentiated by their submersed stems; those of *M. aquaticum* have internodes half the length of the leaves, whereas those of *M. spicatum* have internodes one quarter the length of the leaves. There are few records of *M. aquaticum* in Ohio, but this species is considered invasive in the state.

Habitat: Found in shallow water and margins of ponds, lakes, reservoirs, canals, rivers, streams, and ditches. Most successful in areas with high nutrients.

Height: To 2 m (6 ft.) long; emergent stems can grow up to 30 cm (1 ft.) above the water surface.

Foliage: Leaves are pinnately dissected and borne in whorls of four to six, but their form varies depending on whether they are submersed or emergent. Leaves on emergent stems are featherlike and grayish green, 2.5–3.5 cm (1–1 ½ in.) long and 7–8 mm (¼ in.) wide, with 24–36 uniform linear divisions, each lobe 5 mm (¼ in.) long. The leaves on submersed stems are reddish, more finely textured, 3.5–4.5 cm (1 ½–2 in.) long and 8–15 mm (¼–⅝ in.) wide, with 25–30 uniform linear divisions, each lobe 7 mm (¼ in.) long.

Flowers: The species is dioecious, with separate male and female flowers borne on different plants, but only female plants are found in North America. Female flowers are small and white, 2–3 mm (1/16–⅛ in.) in diameter, borne singly in the axils of emergent leaves. Female flowers lack petals and have four tiny sepals, 0.4–0.5 mm (1/64 in.) long and 0.3 mm (1/64 in.) wide. Each flower has four pistils, and its stigmas, the most visible parts of the flowers, are white and densely fringed. July–August, with occasional flowering into the autumn.

Fruit: No fruit development or seed set has been recorded in the United States. In its native range, the species produces dry capsules that split into four single-seeded segments when ripe, each ovoid and about 1.8 mm (1/16 in.) long by 0.6 mm (1/64 in.) in diameter.

Stems: Stout, 4–5 mm (⅛–¼ in.) in diameter, trailing along the bottom of water bodies or emerging above the water surface. The emergent stems are blue-green and grow upright.

Root system: Plants are rooted in the sediment at the bottom of water bodies. Plants produce rhizomes; most vegetation dies back in autumn, and individuals overwinter as rhizomes or as submersed stems.

Reproduction: In North America, reproduction and dispersal are entirely through vegetative means. Reproduction occurs via fragmentation

Emergent leaves of *Myriophyllum aquaticum*. Image © Gerald D. Carr.

Female flowers of *Myriophyllum aquaticum.* Image by Clinton Morse, University of Connecticut, Ecology and Evolutionary Biology Greenhouses.

Whorl of emergent leaves of *Myriophyllum aquaticum.* Image by André Karwath, used under a CC BY-SA 2.5 license.

Myriophyllum aquaticum stand. Image by Leslie J. Mehrhoff, University of Connecticut, Bugwood.org.

Submersed and emergent *Myriophyllum aquaticum* plant. Image by Leslie J. Mehrhoff, University of Connecticut, Bugwood.org.

Myriophyllum aquaticum infestation. Image by Alison Fox, University of Florida, Bugwood.org.

of emergent and/or submersed shoots, roots, or rhizomes. Plant fragments are often transported inadvertently on boats and boat trailers.

Impact: This species is a particular problem in small water bodies because it grows so densely that it obstructs water flow, impedes recreational activities such as boating and fishing, and causes deoxygenation of water. In addition, dense stands of *M. aquaticum* displace native aquatic plants and provide breeding areas for mosquitoes.

Control: Once established, this species is extremely difficult to eradicate. The first step should be to prevent establishment and spread. It is illegal to buy specimens for aquariums or water gardens, and unwanted plants should not be disposed of near natural water bodies. It is recommended that all boating equipment be cleaned, drained, and dried before moving to new locations. Where populations have established, mechanical harvesters or hand pulling can be used with some success. Aquatic formulations of 2,4-D with imazapyr, diquat with glyphosate, triclopyr, and endothall have been shown to offer partial control with repeated treatments. The thick cuticle on emergent leaves necessitates the use of a surfactant. Herbicide application should be focused in the early part of the growing season. Only licensed aquatic herbicide applicators should undertake chemical management.

Haloragaceae

Myriophyllum spicatum
Eurasian watermilfoil · spiked water-milfoil

Origin: Native to Europe, Asia, and northern Africa. There is not consensus on the date of introduction to North America, but it was likely sometime in the late 1800s or early 1900s. Introduction may have been intentional, through the aquarium trade, or accidental, through ship ballast.

Description: This submersed, perennial aquatic plant is invasive throughout much of the United States. It is extremely successful in highly disturbed lakebeds and water bodies with excess nitrogen and phosphorous.

Habitat: Found in lakes, ponds, reservoirs, and slow-moving rivers and streams. Typically grows at depths of 1–4 m (3–12 ft.) but can be found up to 10 m (30 ft.) deep.

Height: Variable depending on water depth, usually around 3 m (10 ft.), but stems can grow up to 10 m (30 ft.).

Foliage: Finely dissected, pinnately compound leaves with 12–24 leaflets. The feathery leaves are arranged in whorls, usually with four leaves (can be between three and six) per node. Leaves approximately 1–5 cm (½–2 in.) long.

Flowers: Inflorescence is a terminal spike 5–10 cm (2–4 in.) long, borne above the surface of the water. Separate male and female flowers occur on the same inflorescence, with male flowers above and female flowers below. Flowers occur in whorls in the leaf axils or at nodes. Male flowers have small pinkish petals and eight stamens. Female flowers are inconspicuous and yellow-green, each with a four-lobed pistil lacking sepals and petals. Flowering occurs twice per year, June–August.

Fruit: A hard, green-brown capsule with four lobes that split apart at maturity. Each segment contains one seed, about 2 mm (¹⁄₁₆ in.) long. A single plant can produce around 100 seeds per year.

Stems: Slender, typically green to reddish brown. Stems are leafless closer to the base and branch prolifically near the water surface.

Root system: Shallow and fibrous roots. Adventitious roots arise along lower parts of the stem where they come into contact with the substrate. Plant fragments readily develop roots at the nodes, contributing to the species' vegetative spread.

Reproduction: Primarily vegetative, through runners and fragmentation. After flowering, the stem breaks apart, sending fragments into the water, which then root on contact with suitable substrate. Dispersal by seed is considered of minor importance. Unlike many other submersed aquatic plants, this species does not produce turions but dies back to the root crown during winter.

Impact: *Myriophyllum spicatum* displaces other aquatic plants and has less value as a food source for waterfowl than the native species it replaces. This species can also affect recreational activities such as boating, swimming and fishing. Large infestations alter the hydrology and water quality of water bodies.

Control: Measures to minimize the spread of *M. spicatum* between water bodies have been effective, particularly the installation of stations for washing boats and trailers. Removal of invasive milfoils by trained divers has shown much success as a management technique in areas where disturbance must be minimized. Care must be taken to ensure the complete removal of plant material so that plant fragments do not disperse. Aquatic herbicides such as diquat, paraquat, flu-ridone, and the aquatic version of triclopyr have been used to control *M. spicatum*. Only licensed aquatic herbicide applicators should undertake chemical management.

Whorl of leaves in *Myriophyllum spicatum*. Image © Robert L. Carr.

Left: Male flowers of *Myriophyllum spicatum*. *Right:* Female flowers of *Myriophyllum spicatum.* Both images by Stefan Lefnaer, used under a CC BY-SA 4.0 license.

Myriophyllum spicatum leaves and inflorescence. Image by Leslie J. Mehrhoff, University of Connecticut, Bugwood.org.

Submersed *Myriophyllum spicatum* plants. Image by Leslie J. Mehrhoff, University of Connecticut, Bugwood.org.

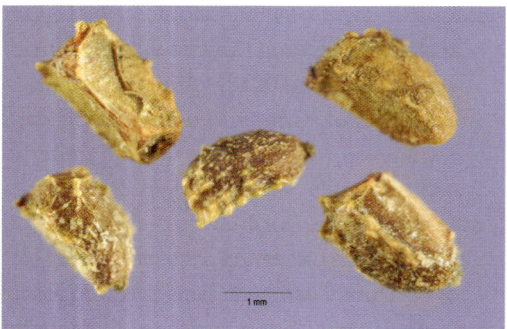

Myriophyllum spicatum seeds. Image by Steve Hurst, USDA-NRCS PLANTS Database.

Myriophyllum spicatum infestation. Image by Chris Evans, University of Illinois, Bugwood.org.

334

Hydrocharitaceae

Egeria densa

Brazilian waterweed · Brazillian elodea · South American waterweed · dense waterweed · common waterweed · leafy elodea · large-flowered waterweed

Synonym: *Elodea densa*

Origin: Native to South America, specifically southeastern Brazil and coastal parts of Argentina and Uruguay. This species was introduced to North America in 1893 for use in aquariums and water gardens and has escaped from cultivation. It has a scattered but widespread distribution throughout North America due to its use as an aquarium plant.

Description: A submersed, perennial aquatic plant that is usually rooted to the substrate but can break off to form free-floating mats. Can be confused with the invasive *Hydrilla verticillata* (hydrilla) [pp. 337–338], but *Egeria densa* has larger leaves and no spines on the midrib on the underside of the leaf. It is also similar to the native *Elodea canadensis,* but *E. densa* typically has four leaves per whorl, whereas *El. canadensis* has three smaller leaves per whorl.

Habitat: Ponds, lakes, streams, and ditches. Found in shallow areas in both still and slow-moving water.

Height: Typically 30–60 cm (1–2 ft.) but can grow to 3 m (10 ft.).

Foliage: Leaves are straplike with pointed tips, 1–4 cm (½–2 in.) long and 2–5 mm (¹⁄₁₆–¼ in.) wide. Arranged in whorls of four to six, most commonly with four leaves per whorl. Leaf margins are finely toothed along the edges. Internodes between leaf whorls are 3–24 mm (⅛–1 in.). Leaves are mostly green, but lower leaves can be yellow or brown.

Flowers: This species is dioecious, with separate male and female flowers borne on different plants, but the introduced populations in North America have only male plants. Flowers arise from reproductive nodes near the top the stalks, each node producing two to four male flowers surrounded by a spathe. Each flower is held on a threadlike hypanthium that elevates it 2–3 cm (¾–1 in.) above the water level. Male flowers are showy, 18–25 mm (¾–1 in.) in diameter, with three white petals and nine yellow stamens surrounding a small central nectary. June–September.

Fruit: No fruits or seeds have been observed in North America.

Stems: Typically bright green, round in cross-section with short internodes, becoming highly branched near the water surface. Lateral branches only grow from specialized reproductive nodes (double nodes), spaced every 6–12 nodes along the stem. Stems are delicate and break easily, giving rise to free-floating mats or stem fragments that disperse and root to form new colonies.

Root system: Plants are rooted in the substrate. Adventitious roots develop on stems or stem fragments where double nodes occur.

Reproduction: By fragmentation of stems and creeping roots. Plant parts are dispersed by water, birds, and boating equipment.

Impact: Dense stands of *E. densa* outcompete native aquatic plants, interfere with recreational activities, reduce populations of fish and waterfowl, and provide breeding ground for mosquitoes.

Control: For small populations, mechanical removal through cutting or hand pulling is feasible, but it is essential that all stem fragments are removed. Copper compounds and the herbicides diquat, acrolein, and fluridone are effective in controlling this species. Only licensed aquatic herbicide applicators should undertake chemical management.

Egeria densa leaves. Image by Robert Vidéki, Doronicum Kft., Bugwood.org.

Egeria densa turion. Image by Robert Vidéki, Doronicum Kft., Bugwood.org.

Double node of *Egeria densa.* Image by U.S. Geological Survey, Bugwood.org.

Egeria densa plants in an infested waterway. Image by Graves Lovell, Alabama Department of Conservation and Natural Resources, Bugwood.org.

Male flower of *Egeria densa.* Image by Leslie J. Mehrhoff, University of Connecticut, Bugwood.org.

Egeria densa infestation. Image by Graves Lovell, Alabama Department of Conservation and Natural Resources, Bugwood.org.

Hydrocharitaceae

Hydrilla verticillata

hydrilla · waterthyme · Esthwaite waterweed ·
Florida elodea

Origin: Native to Europe, Asia, Africa, and Australia. This species was brought to North America as an aquarium plant and first detected in waterways in Florida in the 1950s.

Description: This submersed, perennial aquatic plant is one of the most problematic aquatic invasive plants in North America, although it is not yet widely distributed in Ohio. It grows up to the water surface and forms dense mats that block waterways and crowd out other species. This species has characteristics similar to those of other aquatic plants, including native *Elodea canadensis* and introduced *Egeria densa* (Brazilian waterweed) [pp. 335–336], but both these species lack spines on the midribs that are present on *Hydrilla verticillata*.

Habitat: Rivers, lakes, springs, ponds, marshes, ditches, reservoirs, and low-salinity tidal zones. Tolerant of a wide range of growing conditions.

Height: Stems can grow up to 10 m (30 ft.) long.

Foliage: Leaves are straplike with pointed tips, 6–20 mm (¼–1 in.) long by 2–5 mm (¹⁄₁₆–¼ in.) wide. Arranged in whorls of three to eight, most often with five leaves per whorl. Leaf margins have spines along the edges, giving them a toothed appearance. Leaves have prominent midribs, with one or more spines along the undersides. Color ranges from translucent to green, yellow, or brown.

Flowers: Female and male flowers are separate. Some plants have female flowers only, while others have both female and male flowers on the same plant. Female flowers are tiny, 4–8 mm (⅛–¼ in.) wide, with three white or translucent petals and three green sepals; borne on threadlike stalks that emerge from the leaf axils and extend until the flowers are at or above the water surface. Male flowers are greenish and look like inverted bells; they are closely attached to leaf axils near the stem tips and break loose to float freely on the surface. Fertilization occurs when male flowers bump into the female flowers. June–August.

Fruit: A cylindrical capsule, 7 mm (¼ in.) long by 1.5 mm (⅝ in.) wide, containing two to seven elliptic seeds, each 2–3 mm (¹⁄₁₆–⅛ in.) long.

Stems: Slender, around 1 mm (¹⁄₃₂ in.) wide, with 3–50 mm (⅛–2 in.) between nodes. Stems branch profusely near the water surface.

Root system: Rhizomes are rooted in the substrate. Stolons and stem fragments form adventitious roots where they contact substrate. Small tubers develop at the root tips.

Reproduction: Vegetative reproduction occurs via rhizomes, rooting stems, tubers, and stem fragments. Tubers can remain dormant in the substrate for many years, allowing reestablishment even after plants have been completely eradicated. This species also produces turions in the leaf axils; these break off from the stem and form new plants.

Impact: Displaces native vegetation, clogs irrigation and flood control canals, blocks intakes at water treatment and hydroelectric facilities, and interferes with recreation. Dense infestations can alter water chemistry and decrease dissolved oxygen levels, reducing populations of fish and other aquatic species.

Control: This species is not yet widespread in Ohio, so it is important to prevent its movement and establishment. All boating equipment should be cleaned, drained, and dried before being moved to new locations. In small populations, manual or mechanical harvesting of plants is effective. Because stem and root fragments readily produce new plants, it is critical to ensure that all plant parts are removed. Registered aquatic herbicides including copper compounds, diquat, endothall, and fluridone have been used to control this species. Given the similarity in appearance between *H. verticillata* and native *Elodea*, ensure proper identification before starting any control measures. Only licensed aquatic herbicide applicators should undertake chemical management.

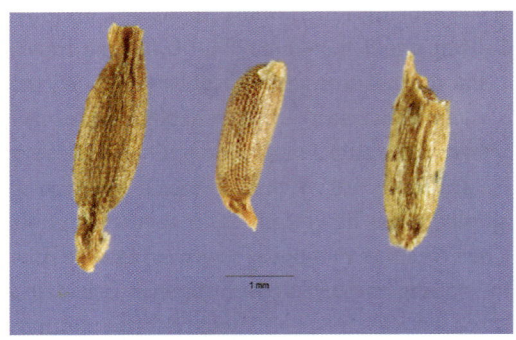

Left: Hydrilla verticillata leaves. *Right: Hydrilla verticillata* tubers. Both images by Robert Vidéki, Doronicum Kft., Bugwood.org.

Hydrilla verticillata seeds. Image by Steve Hurst, USDA NRCS PLANTS Database, Bugwood.org.

Hydrilla verticillata stem. Image by Robert Vidéki, Doronicum Kft., Bugwood.org.

Whorl of *Hydrilla verticillata* leaves. Image by Robert Vidéki, Doronicum Kft., Bugwood.org.

Hydrilla verticillata infestation. Image by Robert Vidéki, Doronicum Kft., Bugwood.org.

Hydrocharitaceae

Hydrocharis morsus-ranae

common frogbit · European frogbit · water frogbit · water poppy

Origin: Native to Europe and temperate parts of western Asia. This species was first introduced to North America in 1932 for use in water gardens. It has escaped cultivation and spread from Canada to the United States through the Saint Lawrence River and the Great Lakes.

Description: A free-floating, perennial aquatic plant that resembles a small water lily. This species is rarely rooted to the substrate and generally floats in mats of tangled stems and roots. It resembles the native American frogbit, *Limnobium spongia.* The two related species can be differentiated, however, by presence of large air pockets across the entire underside of the leaf in *L. spongia,* whereas in *Hydrocharis morsus-ranae* they are confined to the midvein region. There are very few records of this species in Ohio although it is regulated as invasive.

Habitat: Found in shallow, quiet, or slow-moving water, including swamps, marshes, ditches, and along the edges of ponds, lakes, rivers, and streams. It prefers calcium-rich water.

Height: Plants typically grow from the water surface to a depth of 2–10 cm (¾–4 in.).

Foliage: Leaves are round with a heart-shaped base, 2–6 cm (¾–3 in.) long and wide, leathery, held on slender petioles 6–14 cm (3–6 in.) long. Leaf margins are smooth; their undersides are often dark purple and have conspicuous air pockets around the midvein. Primary veins arise from the base of the leaf and broadly curve, running parallel to the leaf margin. Leaves typically float on the water surface, but in very dense stands they can be emergent.

Flowers: This species has separate male and female flowers. In most cases, the male and female flowers are on different plants (dioecious) but some plants have both on a single individual (monoecious). Flowers are white and have three rounded petals, 15 mm (⅝ in.) in diameter, with yellow centers. Male flowers are in groups of one to five, each flower on a stalk up to 4 cm (2 in.) long, with 9–12 stamens per flower. Female flowers are solitary, on stalks up to 9 cm (4 in.) long, with six styles. June–August.

Fruit: A green, spherical or ovoid berry, about 7 mm (¼ in.) long and 5 mm (¼ in.) wide, that splits longitudinally to release numerous tiny, ovoid seeds into the water. There are typically 15–20 seeds per fruit, but up to 74 have been recorded. Seeds are 1 mm ($\frac{1}{32}$ in.) long and 0.5 mm ($\frac{1}{64}$ in.) wide, covered in a viscous material that promotes adhesion to surfaces and potentially aids in dispersal to other locations. July–September.

Stems: Plants develop as rosettes up to 30 cm (12 in.) wide, which spread via stolons to develop new juvenile plants. An individual plant composed of multiple rosettes connected by stolons can be up to 1.5 m (5 ft.) across.

Root system: Typically free-floating but occasionally rooted in the substrate. Roots range from green to white and can grow up to 50 cm (20 in.) long.

Reproduction: Primarily through turions. Turions develop on the stolons in autumn, then detach and sink to the substrate where they remain dormant over winter, and in spring float to the surface and mature. Turions are ovoid, 5–9 mm (¼–⅜ in.) long, and a single plant is capable of producing up to 100 each year. Sexual reproduction is less common but should not be discounted, particularly as seeds are easily dispersed by recreational equipment and wildlife.

Impact: Forms large colonies of dense, floating mats that limit native submersed aquatic plants by reducing available light and nutrients. This species can also decrease water quality and flow, impede boat traffic, and disrupt the movement of wildlife.

Control: This species is not yet widespread in Ohio, so it is important to prevent its movement and establishment. All boating equipment should be cleaned, drained, and dried before being moved to a new location. Most management programs have focused on manual removal by raking plants after overwintering turions have initiated growth on the water surface but before plants have started to spread. Diquat is the only approved herbicide that has been tested and shown effective against this species. Only licensed aquatic herbicide applicators should undertake chemical management.

Hydrocharis morsus-ranae leaves. Image by Christian Fischer, used under a CC BY-SA 3.0 license.

Hydrocharis morsus-ranae flowers. Image by Christian Fischer, used under a CC BY-SA 3.0 license.

Underside of *Hydrocharis morsus-ranae* leaf, which lacks large air pockets. Image by Jouko Lehmuskallio, www.NatureGate.net.

Hydrocharis morsus-ranae turion. Image by Christian Fischer, used under a CC BY-SA 3.0 license.

Hydrocharis morsus-ranae rosette. Image by Leslie J. Mehrhoff, University of Connecticut, Bugwood.org.

Dense cover of *Hydrocharis morsus-ranae*. Image by Christian Fischer, used under a CC BY-SA 3.0 license.

Iridaceae

Iris pseudacorus

yellow iris · pale yellow iris · European yellow iris · yellow flag · water flag

⚠ Do Not Touch

Origin: Native to Europe, western Asia, and northwest Africa. Introduced to North America as an ornamental plant and escaped from cultivation by the early 1900s. This species has also been used for sewage treatment and erosion control.

Description: This emergent, perennial forb has a showy yellow flower and forms dense colonies in wetlands. *Iris pseudacorus* is the only yellow-flowered iris known to occur in natural areas in Ohio, making it easy to identify.

Habitat: Bogs, swamps, marshes, open and forested floodplains, and shallow water along the margins of rivers, streams, lakes, and ponds. Can survive in water to 25 cm (10 in.) deep.

Height: 50–100 cm (20–40 in.).

Foliage: Flat, sword-shaped leaves, growing from the base of the plant in a fanlike arrangement. Leaves are 50–100 cm (20–40 in.) long and 1–3 cm (½–2 in.) wide, stiff and erect, with raised midribs.

Flowers: Bright yellow or cream colored, 7–9 cm (¼–⅜ in.) wide. Each flower has three downward-spreading sepals with brown markings and three smaller erect petals. Typically two or three flowers per stalk. May–July.

Fruit: An ellipsoid capsule, 5–8 cm (2–3 in.) long, green turning to brown. The three segments of the capsule open along the seams and spread widely when mature, revealing flattened brown seeds, 6–7 mm (¼ in.) in diameter, arranged in rows in the chambers of the capsule. Each plant can produce several hundred seeds. August–October.

Stems: Erect flower stalks, shorter than or equaling the length of the leaves.

Root system: This species has fleshy roots 10–30 cm (4–12 in.) long. It also produces pink-fleshed, stout rhizomes, 1–4 cm (½–2 in.) in diameter. The rhizomes branch prolifically and form extensive clumps.

Reproduction: By rhizomes. Seed is viable, although seedlings are not common in the wild. Both rhizome fragments and floating seeds can be dispersed by water. This species is still widely planted in gardens, facilitating its spread.

Impact: Forms dense stands that displace native species and alter wildlife habitat. Stands of *I. pseudacorus* can change the hydrology of a water body by trapping sediment. This species contains glycosides, making it poisonous to animals.

Control: Do not plant new specimens of *I. pseudacorus* in gardens or water features. The glycosides in this plant cause skin irritation, so wear protective clothing and gloves when undertaking control methods. Hand pulling or digging the rhizomes is effective for controlling small populations. Cutting or mowing in the early season can reduce spread. If aboveground parts are cut back repeatedly, the energy reserves in the rhizomes will be exhausted. Leaves of actively growing plants or recently cut leaves can be treated with aquatic glyphosate or imazapyr. Only licensed aquatic herbicide applicators should undertake chemical management.

Native Alternatives

Iris brevicaulis (zigzag iris)
Iris fulva (copper iris)
Iris versicolor (blue flag iris)
Iris virginica (southern blue flag)
Pontederia cordata (pickerelweed)
Sagittaria latifolia (arrowhead)
Saururus cernuus (lizard's tail)

341

Iris pseudacorus leaves. Image by Leslie J. Mehrhoff, University of Connecticut, Bugwood.org.

Iris pseudacorus fruit. Image by Leslie J. Mehrhoff, University of Connecticut, Bugwood.org.

Iris pseudacorus flower. Image by Nancy Loewenstein, Auburn University, Bugwood.org.

Iris pseudacorus seeds. Image by Steve Hurst, USDA-NRCS PLANTS Database.

Iris pseudacorus rhizome and roots. Image by Leslie J. Mehrhoff, University of Connecticut, Bugwood.org.

Iris pseudacorus fruit. Image by Leslie J. Mehrhoff, University of Connecticut, Bugwood.org.

Iris pseudacorus infestation. Image by Leslie J. Mehrhoff, University of Connecticut, Bugwood.org.

Lythraceae

Trapa natans

European water chestnut · horned water chestnut · water nut · bull nut · water caltrop

Origin: This species has a discontinuous native range throughout Europe, Asia, and Africa. Records indicate that it was intentionally planted in the United States sometime before 1879 and escaped cultivation.

Description: A rooted, annual aquatic plant with leaves that float on the water surface. Plants are rooted in the substrate but often break off and become free-floating. This species has a long history of cultivation for its edible nut, although this is not the water chestnut used in Asian cooking (those are the tubers of a sedge species). There are currently no records of this species in Ohio, but it is highly likely that this species occurs here and has simply not yet been reported. It grows in neighboring New York and Pennsylvania, including areas near the eastern end of Lake Erie.

Habitat: Still or slow-moving water of rivers, streams, lakes, ponds, and canals. Prefers high-nutrient systems and full sun, in shallow to deep water.

Height: Typically 2–3 m (6–10 ft.), but can be up to 5 m (15 ft.).

Foliage: This species has two types of leaves. Leaves on the submersed stem are opposite or alternate, linear, falling early to be replaced by whorls of featherlike adventitious roots that grow up to 15 cm (6 in.) long. Floating leaves are alternate and borne in a rosette at the water surface, rhomboid or triangular, 2–5 cm (¾–2 in.) long and wide, sharply toothed along the upper margins. The floating leaves have swollen petioles, 5–9 cm (2–4 in.) long, filled with spongy tissue and air to provide buoyancy for the emergent portion of the plant.

Flowers: Small, solitary flowers develop in the axils of floating leaves. Each flower has four white petals and four green sepals and is about 16 mm (⅝ in.) in diameter. July–September.

Fruit: Fruit is a large, woody nut, 3–4 cm (1–2 in.) long and wide, with four sharp, barbed spines. Each nut contains a single ovoid seed, 15 mm (⅝ in.) long and 10 mm (½ in.) wide. August–October.

Stems: The spongy, cordlike stems can grow to 5 m (15 ft.) in length.

Root system: Plants are typically anchored in the mud by fine roots, although the stems can break off to become free-floating. There are also featherlike adventitious roots that develop along the length of the stem.

Reproduction: Plants die back at the end of the growing season, and the nuts sink to the substrate and overwinter in a dormant state. Seeds can remain viable up to 12 years, although most germinate within the first two years. There is also vegetative reproduction via fragmentation. Dispersal occurs when rosettes detach from stems and float to new areas or when nuts are spread by water or attached to wildlife or equipment.

Impact: Forms dense, floating mats that shade out native aquatic plants, disrupt the movement of wildlife, restrict boating and other recreational activities, and provide habitat for mosquito breeding. The dieback of large stands in the autumn can reduce dissolved oxygen in the water, leading to fish kills. The spiny nuts are also problematic; they can cause injury if stepped on.

Control: Do not purchase or plant in water gardens. To prevent this species from becoming established, early detection is critical. Small populations can be eradicated with manual or mechanical harvesting. The aquatic formulation of 2,4-D can control this species. Only licensed aquatic herbicide applicators should undertake chemical management.

Trapa natans leaves. Image by Agnieszka Kwiecień, used under a CC BY-SA 4.0 license.

Developing fruits on *Trapa natans*. Image by Leslie J. Mehrhoff, University of Connecticut, Bugwood.org.

Trapa natans flower. Image by Georg Schramayr, used under a CC BY-SA 1.0 license.

Inflated petioles on *Trapa natans* leaves. Image by Leslie J. Mehrhoff, University of Connecticut, Bugwood.org.

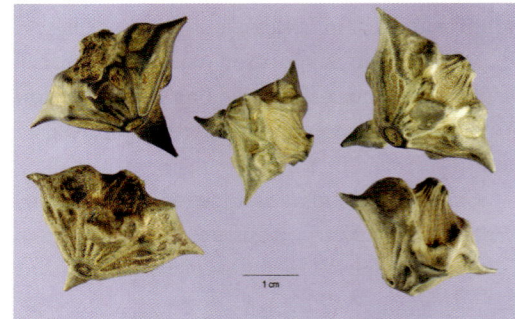

Trapa natans fruits. Image by Steve Hurst, hosted by the USDA-NRCS PLANTS Database.

Trapa natans infestation. Image by Leslie J. Mehrhoff, University of Connecticut, Bugwood.org.

Menyanthaceae

Nymphoides peltata

yellow floating heart · fringed water lily · water fringe · fringed buckbean · marshflower · entire marshwort

Origin: Native to Europe and Asia, this species was introduced to North America in 1882 for use in ornamental water gardens. There are scattered populations established throughout the United States as a result of intentional plantings or escape from cultivation.

Description: A perennial aquatic plant rooted in the substrate, whose leaves and flowers float on the water surface. It superficially resembles a water lily but its fringed yellow flowers differentiate it from native water lily species. *Nymphoides peltata* is still available through the aquatic nursery trade, but the Ohio Department of Agriculture has classified it as invasive, and it should not be sold in the state.

Habitat: Commonly found in still bodies of water such as ponds, reservoirs, and swamps, or in slow-moving areas of rivers, lakes, and canals. Optimal habitats are alkaline and have high nutrients, with a water depth of 1–4 m (3–12 ft.). It cannot grow in shade.

Height: Varies from 1–4 m (3–12 ft.), depending on water depth.

Foliage: Leaves arise singly or—on flowering stems—in an opposite pair of unequal leaves that attach below the flowers. Leaves are ovate or circular with a rounded, heart-shaped base, 3–11 cm (1–5 in.) long and 2–10 cm (¾–4 in.) wide, leathery. The leaves' undersides are frequently purplish or reddish, and the leaves have slightly wavy, shallowly scalloped margins.

Flowers: Flowers are in clusters of two to five in the axil between a pair of leaves. Individual flowers are bright yellow, 3–4 cm (1–2 in.) in diameter when fully open, with five petals that have distinctively fringed margins. There

are five stamens and one pistil. The flowers are held above the water surface on stalks 3–10 cm (1–4 in.) long. June–September.

Fruit: A green capsule, narrowly ovoid with a strong beak, 12–25 mm (½–1 in.) long and 8–11 mm (¼–½ in.) wide, containing numerous seeds. Seeds are yellowish to brown, flat, oval, about 4–5 mm (⅛–¼ in.) long and 3 mm (⅛ in.) wide, with hairs along the margins. August–October.

Stems: Green stolons run horizontal to the substrate up to a length of 3 m (10 ft.). The stolons are about 2–3 mm (1/16–⅛ in.) thick and branching, with nodes that give rise to leaves and flowering stems.

Root system: Plants develop stout white rhizomes.

Reproduction: By seeds, which attach to animals or are moved by water. Vegetative reproduction also occurs via plant fragments and creeping rhizomes.

Impact: Can form dense mats of leaves that can reduce water flow and light penetration, decrease oxygen levels, and alter nutrient cycling. This has negative impacts on native plant and animal communities. In addition, the species can affect fishing, boating, and swimming in recreational areas.

Control: Do not plant specimens of *N. peltata,* and remove any existing populations. Mechanical dredging or raking twice per growing season has been shown to control small populations. It is important to ensure that all plant material is removed from the water, both so decaying plants do not deplete the oxygen in the water and so plant fragments do not reestablish. The leaf surfaces can be treated with aquatic formulations of herbicides, but several applications may be necessary. Only licensed aquatic herbicide applicators should undertake chemical management.

Nymphoides peltata leaves. Image by Leslie J. Mehrhoff, University of Connecticut, Bugwood.org.

Nymphoides peltata fruits. Image by Leslie J. Mehrhoff, University of Connecticut, Bugwood.org.

Nymphoides peltata flowers. Image by Leslie J. Mehrhoff, University of Connecticut, Bugwood.org.

Nymphoides peltata cover. Image by Leslie J. Mehrhoff, University of Connecticut, Bugwood.org.

Nymphoides peltata flower. Image © Gerald D. Carr.

Nymphoides peltata infestation. Image by Rob Andress, Department of Conservation & Natural Resources, Bugwood.org.

Najadaceae

Najas minor

lesser naiad · brittle naiad · slender naiad · spiny-leaf naiad · European naiad · bushy naiad · brittle waternymph

Origin: Native to Europe, Asia, and northern Africa. Genetic studies have determined that this species was introduced to North America at least two separate times. Introduction was likely accidental, through either contamination of cultivated species, the aquarium trade, or shipping. The oldest known record for this species in North America is from 1932 in Lake Cardinal, Ohio (Les et al. 2015).

Description: An annual, submersed aquatic plant. Typically rooted to the substrate but occasionally in free-floating mats.

Habitat: Ponds, lakes, reservoirs, canals, and slow-moving rivers and streams. Typically found in water with depths of up to 4 m (12 ft.). Tolerant of high turbidity and eutrophic conditions.

Height: To 120 cm (4 ft.).

Foliage: Leaves are opposite, 0.5–3.5 cm (¼–1 ½ in.) long and 0.3–0.5 mm (¹⁄₃₂ in.) wide, linear, becoming recurved with age. There are prominent teeth along the leaf margins, with 7–15 multicellular teeth on each side.

Flowers: Plants have separate male and female flowers, greenish, 1–2 mm (¹⁄₃₂–¹⁄₁₆ in.) in size. Flowers occur in the axils of the leaves, with one or two flowers per axil. Male flowers consist of single stamens and are located toward the tips of the stems. Female flowers consist of single pistils and are located below the male flowers. Flowers are water pollinated. July–August.

Fruit: Fruits are ellipsoid, about 1.5–3 mm (¹⁄₁₆–⅛ in.) long, and develop in the leaf axils. Each fruit contains a single, spindle-shaped seed, slightly curved at the tip, with longitudinal ribbing. September–October.

Stems: Slender, branching stems, up to 1 mm (¹⁄₃₂ in.) in diameter. As they age, stems become extremely brittle and fragment easily.

Root system: Shallow and fibrous.

Reproduction: This species produces abundant seeds, dispersed by water and birds. Seed germination occurs in the early spring. Fragmentation also contributes to the species' vegetative spread.

Impact: *Najas minor* grows and reproduces rapidly, forming dense stands that outcompete native plant species and reduce habitat quality for wildlife. In addition, this species can impede recreational activities and block water flow in inlets, streams, and channels.

Control: Mechanical removal or hand pulling can reduce the biomass of infestations. Because fragments can establish new plants, care must be taken to remove all plant parts. Benthic barriers—mats that lie on top of the substrate to keep sunlight from penetrating to the lake bottom—have been used to restrict upward growth, but these can have negative effects on nontarget organisms. Aquatic herbicides containing endothall, diquat, and fluridone have been found to control infestations of *N. minor,* but these chemicals are nonselective and can also affect native plants. Only licensed aquatic herbicide applicators should undertake chemical management.

Najas minor plants. Image by Leslie J. Mehrhoff, University of Connecticut, Bugwood.org.

Najas minor seed. Image by Donald Les, University of Connecticut.

Najas minor stems and leaves. Robert H. Mohlenbrock, hosted by the USDA-NRCS PLANTS Database.

Najas minor cover. Image by Graves Lovell, Alabama Department of Conservation and Natural Resources, Bugwood.org.

Najas minor fragment. Image by Graves Lovell, Alabama Department of Conservation and Natural Resources, Bugwood.org.

Najas minor infestation in Roaming Shores Lake, Ohio. Image by Donald Les, University of Connecticut.

Potamogetonaceae

Potamogeton crispus

curly pondweed · curly-leaf pondweed · crispy-leaved pondweed · curly cabbage · river weed

Origin: Native to Europe, Asia, Africa, and Australia. The first verified record in North America is from the mid-1800s. The source of introduction is unclear, but it is likely that it was introduced multiple times, either intentionally or accidentally.

Description: This submersed, perennial aquatic plant grows rapidly in the early spring and dies in midsummer. It produces turions that remain dormant in the sediment through the summer and germinate in autumn. It overwinters as small plants that grow rapidly in the early spring before other aquatic plant species begin growing.

Habitat: Found in freshwater rivers, streams, ponds, lakes, and in slightly brackish waters. This species can grow in alkaline or high-nutrient conditions. It is tolerant of low light and low water temperatures, allowing it to invade both shallow and deeper water.

Height: Typically 30–80 cm (1–3 ft.); can grow to 4 m (12 ft.) in deep water.

Foliage: Leaves alternate, oblong, 3–8 cm (1–3 in.) by 5–12 mm (¼–½ in.), with a rounded tip. The reddish-green leaves are finely toothed and have distinctive, wavy margins. Leaves attach directly to the stem and have a prominent midrib. Turions develop in the axils of leaves or at the apex of stems, typically 3 cm (1 in.) long by 2 cm (¾ in.) wide.

Flowers: Inflorescence is a short axillary or terminal spike, 2–5 cm (¾–2 in.) long, emerging above the water surface. Flowers are small, 3 mm (⅛ in.) wide, with four greenish-red sepals. Flowering May–June.

Fruit: Fruits are reddish-brown achenes that each contain a single seed. The body of the achene is ovoid, 6 mm (¼ in.) by 2.5 mm (⅛ in.), with a small, recurved beak that measures 2–3 mm ($\frac{1}{16}$–⅛ in.). June–July.

Stems: Flattened and sparingly branched.

Root system: Elongate and slender rhizomes are rooted in the substrate, pale yellow or reddish. Creeping rhizomes and basal stems can root at the nodes.

Reproduction: Both vegetative and through seed. Turions, which are released in midsummer when plants die back, are the main source of vegetative reproduction. Stem fragmentation and creeping rhizomes also contribute to spread.

Impact: This species forms dense mats in the spring and early summer. It is known to displace native species, reduce biodiversity, and impede recreational activities. The dieback in midsummer can alter water quality and lead to anoxic conditions.

Control: Because *Potamogeton crispus* begins growing before other aquatic plant species, it is recommended that control techniques be used in early spring, to reduce impacts on the native plant community. Cutting at the sediment surface has been found to prevent turion production. Herbicides such as copper chelate, endothall, fluridone, and diquat have been used for control. Only licensed aquatic herbicide applicators should undertake chemical management.

349

Potamogeton crispus leaves. Image © Gerald D. Carr.

Potamogeton crispus stem. Image by Chris Evans, University of Illinois, Bugwood.org.

Potamogeton crispus flowers. Image © Robert L. Carr.

Potamogeton crispus plants from an infested waterway. Image by Chris Evans, University of Illinois, Bugwood.org.

Potamogeton crispus turion. Image by Leslie J. Mehrhoff, University of Connecticut, Bugwood.org.

Typhaceae

Typha angustifolia

narrow-leaved cattail · lesser bulrush · lesser
reed-mace · nail-rod · small reed-mace

ODA Invasive
OIPC Invasive

Origin: There is some dispute over the native distribution of this species, but it is often considered an introduced species in North America. The native range is thought to include Europe, temperate Asia, and Northern Africa. Introduction to the United States was likely accidental, perhaps through the dry ballast of ships.

Description: An emergent, grasslike, perennial forb. It is highly competitive and tends to displace the native cattail *Typha latifolia.* As a whole, the genus *Typha* has considerable economic value to humans. The plants are useful in stabilizing shorelines and removing water pollutants, the starchy rhizomes and pollen are edible, and the leaf fibers and seed fluff can be used for many purposes, including insulation and stuffing for pillows, bedding, or life preservers.

Habitat: Found in wet or intermittently wet habitats such as fens, marshes, roadside ditches, and depressions that collect water; also found along the margins of lakes, ponds, streams, reservoirs, and canals. Can tolerate high nutrient levels and moderate salt levels. The species typically grows in water 30–60 cm (1–2 ft.) deep, but it can be in a depth up to about 1 m (3 ft.).

Height: 1–3 m (3–10 ft.).

Foliage: Sword-shaped, 1 m (3 ft.) long, dark green, parallel veined. Leaf width is 3–8 mm (⅛–¼ in.), narrower than both the native *T. latifolia* and invasive *Typha* x *glauca* (hybrid cattail) [p. 353]. About 10 leaves emerge from the base of each stem. The leaf surface is relatively flat on the front and rounded on the back, giving the leaf a distinctive *D*-shape in cross-section. Leaves contain hollow chambers and have a spongy feel.

Flowers: Inflorescence contains tiny, separate male and female flowers on a densely crowded, cylindrical spike. Male flowers are yellowish to brown, in the top part of the spike. Female flowers are reddish to dark brown, in the bottom part of the spike, about 10–15 cm (4–6 in.) long and 1–2 cm (½–¾ in.) thick at maturity. The male and female flowers are separated from each other on the spike by a gap of 1–12 cm (½–5 in.). This feature differentiates *T. angustifolia* from the native *T. latifolia,* because the latter has no gap between the male and female flowers. May–July.

Fruit: Fruit is a tiny achene, 5–8 mm (¼ in.) long, surrounded by hairs. A single plant can produce 20,000–700,000 seeds that are dispersed by wind. Some seeds are released in autumn, while others remain on the dry stalk until the following spring. Seeds are viable for 70–100 years.

Stems: Flowering stems are smooth, 5–12 mm (¼–½ in.) in diameter, narrowing to 2–3 mm ($^1/_{16}$–⅛ in.) near the inflorescence.

Root system: Branching rhizomes, 2–4 cm (¾–2 in.) in diameter. Fibrous root masses develop at the base of stems and at rhizome nodes.

Reproduction: Produces abundant seed and can grow from rhizome fragments, both of which contribute to the establishment of new colonies. Vegetative spread occurs through rhizomes.

Impact: This species displaces less competitive wetland plants. Although *T. angustifolia* provides food and shelter for marsh animals such as muskrats and red-winged blackbirds, areas invaded by dense monospecific stands tend to have reduced biodiversity.

Control: Due to the extensive rhizomes and persistent soil seed bank, once established, this species is difficult to eradicate. Management techniques are primarily aimed at reducing the density of established colonies and controlling the spread into new populations. Mechanical control options include cutting shoots (including any dead leaves) below the waterline several times during the growing season. If there is no tissue above the water surface, then gas diffusion cannot take place and the rhizomes die. Foliar application of dalapon, glyphosate, fluridone, imazapyr, diquat, and glufosinate have all been shown to effectively control *T. angustifolia.* Treatment in late summer or early autumn is recommended. Only licensed aquatic herbicide applicators should undertake chemical management.

Typha angustifolia leaves and flower spikes. Image by Radio Tonreg, used under a CC BY 2.0 license.

Male flowers of *Typha angustifolia.* Image © Gerald D. Carr.

Curved back and spongy interior of *Typha angustifolia* leaf. Image © Robert L. Carr.

Typha angustifolia seed heads. Image by Johann Jaritz, used under a CC BY-SA 3.0 AT license.

Male inflorescence (above) and female inflorescence (below) on *Typha angustifolia.* Image by Rob Routledge, Sault College, Bugwood.org.

Dense stand of *Typha angustifolia* encroaching on a wetland. Image © Robert L. Carr.

Typhaceae

Typha x *glauca*

hybrid cattail · white cattail

Origin: This is a common hybrid of the nonnative species *Typha angustifolia* (narrow-leaved cattail) [pp. 351–352] and the native species *Typha latifolia.*

Description: An emergent, grasslike, perennial forb. As with *T. angustifolia,* the hybrid cattail is highly invasive and can form extensive monocultures. Plants of *Typha* x *glauca* often grow taller than either parent species.

Habitat: Freshwater marshes, fens, roadsides, ditches, shallow ponds, stream margins, and lake shores. This plant thrives in high-nutrient conditions and can withstand long periods of flooding.

Height: 2–3.5 m (6–12 ft.).

Foliage: Sword-shaped, blue-green, parallel veined. The flat leaf blades are 7–15 mm (¼–⅝ in.) wide and generally longer than those of the parent species. The backs of the leaves are convex. About 15 leaves emerge from the base of the stem.

Flowers: Inflorescence composed of tiny, separate male and female flowers densely packed into a cylindrical spike at the end of the stem. Male flowers are yellow, in the top part of the spike. Female flowers are greenish to brown, in the bottom part of the spike, 15 cm (6 in.) long or more and about 14–27 mm (⅝–1 in.) wide at maturity. The gap between the male and female sections is about 1–4 cm (½–2 in.). May–July.

Fruit: Seeds are tiny, 1 mm ($\frac{1}{32}$ in.) in length and width, surrounded by hairs. Unlike those of the parent species, seeds of this hybrid are typically sterile.

Stems: Flowering stems are smooth, 1–2 cm (½–¾ in.) in diameter.

Root system: Thick, starchy rhizomes form large clones.

Reproduction: Seed viability is low; this species spreads primarily by rhizomes.

Impact: The replacement of native wetland plant communities by dense monospecific stands of *T.* x *glauca* reduces biodiversity. The taxonomic confusion with its parent species and common misidentification of this hybrid has likely led to an underestimate of both its distribution and its impact on natural communities.

Control: See *Typha angustifolia* [pp. 351–352].

353

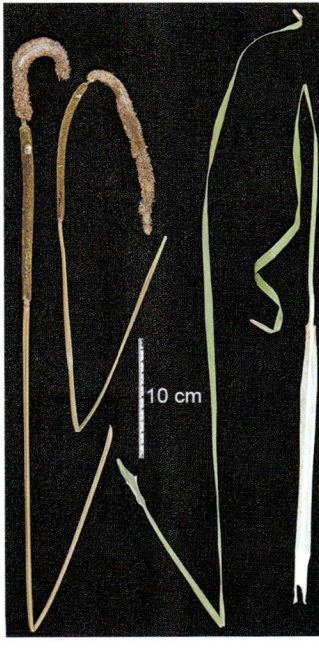

Typha x *glauca* leaves and flowers. Image © Gerald D. Carr.

Acknowledgments

In Ohio, invasive species are designated and controlled under the auspices of the Ohio Department of Agriculture. We acknowledge the important work being done by the ODA, as well as support that the Ohio Department of Natural Resources has provided in educating the public about invasive species through the years. However, much of the work on the frontlines of invasive species research and education in the state is being done by the Ohio Invasive Plants Council. This independent group is working tirelessly to study the impacts of invasive plants, teach people how to control these species, and promote native alternatives that can be used in gardens. The council has taken the critical step of bridging the divide between conservation and horticulture, and we applaud it for helping to make Ohio a better, greener place.

The seed for this book was planted while Megan Griffiths and David Ward still lived in South Africa. Clive Bromilow's *Problem Plants and Alien Weeds of South Africa* (2010) crossed our desks, and we were inspired by its concise descriptions of plants, useful distribution maps, and carefully selected photographs that aided in identification. When David was invited to join the Department of Biological Sciences at Kent State University, we searched for a similar resource to use as he established a new research program on a new continent. Only one book for the region existed, the incredibly important *Invasive Plants of the Upper Midwest* by Elizabeth J. Czarapata (2005). That publication remains the foremost authority on invasive plants for the region, but it is not specific to Ohio and does not contain distribution maps, making it difficult to immediately differentiate the most problematic species in any given location. We believed the state deserved its own treatment of the many invasive plants and weeds found within its borders. Once on the ground in Ohio, we joined forces with Melissa Davis, whose local wisdom, classical botanical training, mapping skills, horticultural knowledge, and herbicide expertise allowed us to produce the complete book you now hold in your hands.

Creating this work has been an extraordinary learning experience for all of us. We have come to this project not as experts but as interested individuals who wanted to know more about Ohio's invasive plant and weed species. We readily acknowledge the many resources in print and online that already cover much of this information, but that information is not all organized in one place. We saw a gap and have tried to compile information into a single resource to make it easier for others to access. This is basically the book we wished we had four years ago. During the writing process, we have become more knowledgeable about the issues and species covered here, but we are all still novices in many ways.

A book like this does not come to be on its own. We would like to express our sincere gratitude to the many individuals, organizations, and agencies that have made this work possible. We specifically thank Dylan Stover at the Kent State University Herbarium for his invaluable help with evaluating species for inclusion, sourcing photographs, and for compiling much of the distribution data used to make the species maps. We also thank Christian Combs—lab manager extraordinaire—for additional technical support whenever and wherever we needed it. We are grateful to Katie Johnson for helping explore funding sources for this project. We thank the herbariums at The Ohio State University and Miami University for sharing species distribution data. Louis Iverson and Matthew Peters at the USDA Forest Service Northern Research Station helped immensely in creating the maps with projected changes in forest type distributions and mean annual temperature included in the introduction.

At the book's inception, we had intended to photograph all the plant species ourselves. However, we quickly realized that amassing a collection of usable photographs would have taken hundreds of hours and multiple growing seasons; we simply did not have the time to both produce the manuscript and take the photographs. In addition, we discovered that the University of Georgia's Bugwood Database already had a huge collection of relevant photographs. It seemed unnecessary to duplicate that effort. Many of the pictures in this volume have come from that database, and we are incredibly grateful to all the photographers who through the Bugwood Database have either made their images available within the public domain or granted us permission to use them for the book. There are more individuals than we can list here, but we have been both humbled and encouraged by the generosity shown by all the photographers and research units whose work is showcased here. We acknowledge specifically the work of the late Leslie Mehrhoff, whose photographs feature heavily in the book. He has left an incredible legacy, and we are beyond grateful that his photographs have been made available for public use. Other photographers whose work we have accessed through Bugwood and on which we have relied heavily include Robert Vidéki, Jan Samanek, Chris Evans, Steve Dewey, Rob Routledge, Forest Starr and Kim Starr, and Bruce Ackley. Terrence Walters and colleagues at the USDA APHIS Plant Protection Quarantine program have allowed use of several of their exquisite seed photos. Catherine Herms, Ken Chamberlain, and colleagues at The Ohio State Weed Lab have generously shared many of their images and continue to provide an invaluable resource to the residents of Ohio.

We are also grateful for the existence of the USDA PLANTS Database, a necessary first stop when determining the distribution of any species and whether it is native or nonnative. The USDA image gallery was critically important for the completion of this book, particularly the nearly exhaustive library of seed images by Steve Hurst and plant photos by Patrick Alexander and Doug Goldman.

Three additional sources proved invaluable in filling in the gaps in our photographic coverage: Gerald Carr and Robert Carr of the Oregon Flora Image Project, Katy Chayka and her collaborators at Minnesota Wildflowers, and the Native Plant Trust (formerly known as the New England Wildflower Society). We are indebted to them (and photographers they have showcased) for generously and freely sharing selected photos, and we encourage our readers to support these organizations if they have the opportunity. In particular, the Native Plant Trust's Go Botany key and the Minnesota Wildflowers plant search function are incredibly useful identification tools; we cannot recommend them highly enough.

While we have made every effort to obtain permission for the use of photographs and images, we may have made mistakes or miscommunications along the way. We apologize for any errors and would appreciate being notified of any corrections to be incorporated in reprints or future editions.

We thank the Native Plant Society of Northeastern Ohio for their help in funding the publication of this book. The manuscript has been greatly improved by several expert reviewers and by our copyeditor, Erin Holman. We are eternally grateful to Susan Wadsworth-Booth, Will Underwood, Mary Young, Christine Brooks, Richard Fugini, and the rest of the team at the Kent State University Press for their patience, expertise, and support of this project at every stage. Thank you for helping us grow this seed of an idea into something bigger and more beautiful than we ever imagined possible.

Appendix A

Plant Species Designated as Invasive in Ohio

List of plant species designated as invasive by the Ohio Department of Agriculture under law ODA 901:5-30-01 (available online at the Ohio Administrative Code website: http://codes.ohio.gov/oac/901:5-30-01). According to the law, no person shall sell, offer for sale, propagate, distribute, import, or intentionally cause the dissemination of any invasive plant as defined in paragraph (A) of this rule in the state of Ohio. This list is effective as of January 7, 2018, and will be subject to a five-year review on January 7, 2023.

Paragraph A	Scientific name	Common name
(1)	*Ailanthus altissima*	tree-of-heaven
(2)	*Alliaria petiolata*	garlic mustard
(3)	*Berberis vulgaris*	common barberry
(4)	*Butomus umbellatus*	flowering rush
(5)	*Celastrus orbiculatus*	oriental bittersweet
(6)	*Centaurea stoebe* ssp. *micranthos*	spotted knapweed
(7)	*Dipsacus fullonum*	common teasel
(8)	*Dipsacus laciniatus*	cutleaf teasel
(9)	*Egeria densa*	Brazilian elodea
(10)	*Elaeagnus angustifolia*	Russian olive
(11)	*Elaeagnus umbellata*	autumn olive
(12)	*Epilobium hirsutum*	hairy willow herb
(13)	*Frangula alnus*	glossy buckthorn
(14)	*Heracleum mantegazzianum*	giant hogweed
(15)	*Hesperis matronalis*	dame's rocket
(16)	*Hydrilla verticillata*	hydrilla
(17)	*Hydrocharis morsus-ranae*	European frog-bit
(18)	*Lonicera japonica*	Japanese honeysuckle
(19)	*Lonicera maackii*	Amur honeysuckle
(20)	*Lonicera morrowii*	Morrow's honeysuckle

Paragraph A	Scientific name	Common name
(21)	*Lonicera tatarica*	Tatarian honeysuckle
(22)	*Lythrum salicaria*	purple loosestrife
(23)	*Lythrum virgatum**	European wand loosestrife
(24)	*Microstegium vimineum*	Japanese stiltgrass
(25)	*Myriophyllum aquaticum*	parrotfeather
(26)	*Myriophyllum spicatum*	Eurasian water-milfoil
(27)	*Nymphoides peltata*	yellow floating heart
(28)	*Phragmites australis*	common reed
(29)	*Potamogeton crispus*	curly-leaved pondweed
(30)	*Pueraria montana* var. *lobata*	kudzu
(31)	*Pyrus calleryana***	callery pear
(32)	*Ranunculus ficaria*	fig buttercup/lesser celandine
(33)	*Rhamnus cathartica*	European buckthorn
(34)	*Rosa multiflora*	multiflora rose
(35)	*Trapa natans*	water chestnut
(36)	*Typha angustifolia*	narrow-leaved cattail
(37)	*Typha* x *glauca*	hybrid cattail
(38)	*Vincetoxicum nigrum*	black dog-strangling vine, black swallowwort

357

*until one year after the effective date of this rule
**until five years after the effective date of this rule

Ohio Invasive Plants Council Plant Assessment Results

The Ohio Invasive Plants Council (OIPC) has developed an objective, science-based process to identify invasive plants that threaten the health and diversity of Ohio's natural ecosystems. Details of this assessment protocol and interpretation of the resulting numerical scores can be found on the organization website (https://www.oipc.info/assessment-protocol.html). Updates are regularly made to the species list, so refer to the assessment results page for the most current information (https://www.oipc.info/assessment-results.html). Species assessed as invasive reproduce and expand into natural or relatively natural areas, negatively affecting existing species. Species assessed as pending further review are those that have not yet been assessed as invasive, largely because additional scientific investigation is still needed. Species not known to be invasive are those not considered invasive at the time the assessment was conducted, based on available information.

Scientific name	Common name	Score	OIPC assessment
Lythrum salicaria	purple loosestrife	77	Invasive
Phalaris arundinacea	reed canary grass	74	Invasive
Typha angustifolia	narrow-leaved cattail	73	Invasive
Typha x *glauca*	hybrid cattail	73	Invasive
Phragmites australis ssp. *australis*	common reed	70	Invasive
Myriophyllum spicatum	Eurasian water-milfoil	69	Invasive
Lonicera maackii	Amur honeysuckle	65	Invasive
Alliaria petiolata	garlic mustard	63	Invasive
Cirsium arvense	Canada thistle	63	Invasive
Elaeagnus umbellata	autumn olive	63	Invasive
Lonicera morrowii	Morrow's honeysuckle	63	Invasive
Frangula alnus	glossy buckthorn	61	Invasive
Fallopia japonica	Japanese knotweed	60	Invasive
Microstegium vimineum	Japanese stiltgrass	60	Invasive
Celastrus orbiculatus	oriental bittersweet	59	Invasive
Centaurea stoebe ssp. *micranthos*	spotted knapweed	59	Invasive

Scientific name	Common name	Score	OIPC assessment
Bromus inermis	smooth bromegrass	58	Invasive
Rhamnus cathartica	common buckthorn	58	Invasive
Rosa multiflora	multiflora rose	58	Invasive
Ailanthus altissima	tree-of-heaven	56	Invasive
Butomus umbellatus	flowering rush	56	Invasive
Pueraria lobata	kudzu	56	Invasive
Berberis thunbergii	Japanese barberry	54	Invasive
Pyrus calleryana	callery pear	54	Invasive
Conium maculatum	poison hemlock	53	Invasive
Fallopia sachalinense	giant knotweed	53	Invasive
Fallopia x *bohemica*	Bohemian knotweed	53	Invasive
Lonicera tatarica	Tatarian honeysuckle	51	Invasive
Melilotus officinalis	yellow sweet clover	51	Invasive
Melilotus alba	white sweet clover	50	Invasive
Dipsacus laciniatus	cutleaf teasel	49	Invasive
Lonicera japonica	Japanese honeysuckle	49	Invasive
Miscanthus sinensis	Chinese silvergrass	49	Invasive
Pastinaca sativa	wild parsnip	49	Invasive
Potamogeton crispus	curly-leaf pondweed	49	Invasive
Elaeagnus angustifolia	Russian olive	48	Invasive
Dipsacus fullonum	common teasel	47	Invasive
Ficaria verna	lesser celandine	47	Invasive
Saponaria officinalis	bouncing bet	47	Invasive
Vincetoxicum nigrum	black swallowwort	47	Invasive
Dioscorea polystachya	air potato	46	Invasive
Epilobium hirsutum	hairy willow-herb	45	Invasive
Euonymus fortunei	wintercreeper	45	Invasive
Hesperis matronalis	dame's rocket	45	Invasive
Ipomoea purpurea	morning glory	45	Invasive
Phellodendron amurense	Amur corktree	45	Invasive
Ulmus pumila	Siberian elm	45	Invasive
Ligustrum vulgare	common privet	44	Pending
Persicaria perfoliata	mile-a-minute	43	Pending
Achyranthes japonica	Japanese chaff flower	42	Pending
Arthraxon hispidus	small carpgrass	42	Pending
Convolvulus arvensis	field bindweed	42	Pending
Najas minor	lesser naiad	42	Pending
Arctium minus	common burdock	41	Pending
Ampelopsis brevipedunculata	porcelainberry	40	Pending
Morus alba	white mulberry	40	Pending
Barbarea vulgaris	garden yellowrocket	37	Pending

Scientific name	Common name	Score	OIPC assessment
Hemerocallis fulva	daylily	37	Pending
Acer platanoides	Norway maple	36	Pending
Paulownia tomentosa	princess tree	36	Pending
Rubus phoenicolasius	wineberry	36	Pending
Tripidium ravennae	Ravenna grass	36	Pending
Rosa canina	dog rose	35	Pending
Ligustrum obtusifolium	border privet	30	Not invasive
Taraxacum officinale	dandelion	29	Not invasive
Ornithogalum umbellatum	star-of-Bethlehem	28	Not invasive
Plantago major	broad leaf plantain	23	Not invasive
Aegopodium podagraria	goutweed	20	Not invasive
Koelreuteria paniculata	golden raintree	18	Not invasive
Acer ginnala	Amur maple	15	Not invasive
Eleutherococcus sieboldianus	aralia	15	Not invasive
Acer platanoides 'Crimson King'	Norway maple 'Crimson King'	11	Not invasive
Acer campestre	hedge maple	11	Not invasive
Liriope muscari	big blue lilyturf	7	Not invasive

Appendix C

Plant Species Designated as Prohibited Noxious Weeds in Ohio

The Ohio Department of Agriculture, under law ODA 901:5-37-01 (available online at the Ohio Administrative Codes website: http://codes.ohio.gov/oac/901:5-37-01), has designated the following plants as noxious weeds. This list is effective as of September 14, 2018, and will be subject to a five-year review on September 14, 2023.

	Scientific name	Common name
(A)	*Sorghum bicolor*	shatter cane
(B)	*Salsola kali* var. *tenuifolia*	Russian thistle
(C)	*Sorghum halepense*	Johnsongrass
(D)	*Pastinaca sativa*	wild parsnip
(E)	*Vitis* spp.*	grapevines
(F)	*Cirsium arvense*	Canada thistle
(G)	*Conium maculatum*	poison hemlock
(H)	*Senecio glabellus*	cressleaf groundsel
(I)	*Carduus nutans*	musk thistle
(J)	*Lythrum salicaria*	purple loosestrife
(K)	*Polygonum perfoliatum*	mile-a-minute weed
(L)	*Heracleum mantegazzianum*	giant hogweed
(M)	*Nicandra physalodes*	apple of Peru
(N)	*Conyza canadensis*	marestail
(O)	*Bassia scoparia*	kochia
(P)	*Amaranthus palmeri*	Palmer amaranth
(Q)	*Pueraria montana* var. *lobata*	kudzu
(R)	*Polygonum cuspidatum*	Japanese knotweed
(S)	*Phyllostachys aureasculata***	yellow groove bamboo
(T)	*Convolvulus arvensis*	field bindweed
(U)	*Lepidium draba* ssp. *draba*	heart-podded hoary cress
(V)	*Lepidium appelianum*	hairy whitetop or ballcress

	Scientific name	Common name
(W)	*Sonchus arvensis*	perennial sowthistle
(X)	*Acroptilon repens*	Russian knapweed
(Y)	*Euphorbia esula*	leafy spurge
(Z)	*Calystegia sepium*	hedge bindweed
(AA)	*Nassella trichotoma*	serrated tussock
(BB)	*Sorghum x almum*	Columbus grass
(CC)	*Carduus nutans*	musk thistle
(DD)	*Bassia prostrata*	forage kochia
(EE)	*Amaranthus tuberculatus*	water hemp

*when growing in groups of one hundred or more and not pruned, sprayed, cultivated, or otherwise maintained for two consecutive years

**when the plant has spread from its original premise of planting and is not being maintained

Herbicides

These are the most common herbicides available commercially. Before undertaking a chemical control program, always consult your local extension service and refer to product label and Material Safety Data Sheets (MSDS). Premix products are available whenever combinations of two or more herbicides are recommended. **Always consult chemical manufacturer and follow label instructions for use guidelines.**

Active ingredient	Trade name	Family name	MOA	SOA	Usage notes
2,4-D (2,4-dichloro-phenoxyacetic acid)	various	phenoxy	1	A	BL, WO, PO, S
2,4-D (granular)	Aquaside, Aqua-Kleen,	phenoxy	1	A	AQ, BL, PO, S
acrolein	Aqualin	aldehyde	6	M	AQ, BL, PO, S, CH, R
chlorsulfuron	Glean, Telar	sulfonylurea	2	B	BL, GR, PR, PO, S, CH
clopyralid	Lontrel, Transline, Stinger	carboxylic acid (pyridine)	1	A	BL, WO, S, PO, †
copper chelate	Cutrine	copper compound	8	M	AQ, PO, S
copper sulfate	Blue Vitriol, Bluestone	copper compound	8	M	AQ, PO, S
dalapon	Revenge, Diamond	aliphatic acid	6	I	GR, PR, PO, S
DCPA (dimethyl tetra-chloroterephthalate)	Dacthal	benzoic acid	4	E	GR, BL, PR, S
dicamba[a]	Vanquish, Clarity, Distinct, Banvel	benzoic acid	1	A	BL, PR, PO, S, †
diquat	Reward	bipyridylium	6	I	AQ, BL, PO, N, CH
diuron	Karmex	ureas	5	H	BL, WO, GR, PE, PO, N
endothall	Aquathol	unknown	8	M	AQ, BL, S, CH

Active ingredient	Trade name	Family name	MOA	SOA	Usage notes
fluazifop	Fusilade	aryloxyphenoxy propionate	3	D	GR, PO, S
fluridone	Sonar, Whitecap, Avast!	organic compound	5	L	AQ, GR, BL, PR, S
glufosinate	Finale, Rely, Ignite, Liberty	phosphinic acid	7	K	BL, GR, PO, N
glyphosate (dry sites)	Roundup, Honcho, various	glycine	2	C	BL, WO, GR, PO, N
glyphosate (wet sites)	Rodeo, Aqua Neat, Accord	glycine	2	C	AQ, BL, GR, PO, N
haloxyfop	Verdict	pyridine	4	E	GR, PR, PO, S
hexazinone	Velpar, Pronone	triazinone	5	G	BL, WO, GR, PO, N, †
imazapic	Plateau	imidazolinone	2	B	GR, BL, PR, PO, S
imazapyr	Arsenal	imidazolinone	2	B	BL, WO, GR, PR, PO, S
isoxaben	Gallery	benzamide	4	F	BL, GR, PR, S
MCPA (2-methyl-4-chlorophenoxyacetic acid)	various	phenoxy	1	A	BL, PO, S, R
MCPP (Mecoprop)	various	dimethylamine salt	1	A	BL, PO, S
metsulfuron	Cimarron, Escort	sulfonylurea	2	B	BL, WO, GR, PR, PO, S
paraquat	Gramoxone, Boa	bipyridylium	6	I	AQ, BL, PO, S, R
picloram	Tordon	caroxylic acid (pyridines)	1	A	BL, WO, PO, S, R, †
sethoxydim	Poast, Vantage	cyclohexanedione	3	D	GR, PO, S, CH
triclopyr	Garlon	carboxylic acid (pyridine)	1	A	BL, WO, PO, S
trifluralin	Treflan, Tri-4	dinitroaniline	4	E	GR, PR, S

Notes:

[a]Only certified applicators are permitted to use this herbicide in Ohio.

Mode of Action (MOA):
1. plant growth regulators
2. amino acid biosynthesis inhibitors
3. fatty acid biosynthesis inhibitors
4. seedling growth inhibitors (root and shoot)
5. photosynthesis inhibitors
6. cell membrane disruptors
7. phosphorylated amino acid
8. not classified / unknown

Site of Action (SOA): repeated use of herbicides with the same site of action can cause herbicide-resistant weeds.
A. auxin receptors (indole-3-acetic acid-like synthetic auxins)
B. ALS (acetolactate synthesis) enzyme inhibitor
C. EPSP (5-enolpyruvyl-shikimate-3-phosphate) enzyme inhibitor
D. ACCase (acetyl CoA carboxylase) inhibitors
E. microtubule inhibitor
F. cell wall biosynthesis inhibitor
G. photosystem II inhibitor (binds to the D-1 quinone protein)

H. photosystem II inhibitor (blocks the plastoquinone binding site)
I. PPG (propargylglycine) enzyme inhibitor
J. photosystem I electron diverter
K. glutamine synthesis inhibitor
L. phytoene desaturase inhibitor
M. not classified / unknown

Usage Notes: PO = postemergence; PR = preemergence; BL = broadleaf weeds; GR = grasses; WO = woody species; AQ = aquatic use; R = restricted use; S = selective; N = nonselective; CH = contact herbicide; † = groundwater or surface water advisory

For further information, we recommend the following resources:

Pesticide Safety Education Program, Ohio State University Extension, https://pested.osu.edu/.

2019 Ohio, Indiana and Illinois Weed Control Guide, The Ohio State University, https://extensionpubs.osu.edu/2019-weed-control-guide-for-ohio-indiana-and-illinois-pdf/.

Midwest Invasive Plant Network (MIPN), https://mipncontroldatabase.wisc.edu/.

Introduction to Weeds and Herbicides, Penn State Extension, https://extension.psu.edu/introduction-to-weeds-and-herbicides

Herbicide Mode of Action Chart - Purdue Agriculture, https://ag.purdue.edu/btny/.../Herbicide_MOA_CornSoy_12_2012%5B1%5D.pdf.

WSSA: Herbicide Site of Action (SOA) Classification List, wssa.net/wssa/weed/herbicides.

Additional Synonyms of Scientific Names For Problem Plant Species

Scientific name	Synonym(s)[a]
Amaranthaceae	
Bassia scoparia	*Bassia sieversiana, Kochia alata, Kochia sieversiana, Kochia trichophila*
Salsola tragus	*Salsola iberica, Salsola kali, Salsola pestifer, Salsola ruthenica*
Apocynaceae	
Vincetoxicum nigrum	*Asclepias nigra, Cynanchum nigrum*
Asteraceae	
Arctium minus	*Lappa minor*
Centaurea stoebe ssp. *micranthos*	*Centaurea biebersteinii*
Cirsium arvense	*Carduus arvensis, Cirsium incanum, Cirsium setosum*
Cirsium vulgare	*Carduus lanceolatus, Carduus vulgaris, Cirsium lanceolatum, Cirsium lanceolatus*
Conyza canadensis	*Conyza parva, Erigeron pusillus, Leptilon canadense, Leptilon pusillum*
Hieracium caespitosum	*Hieracium pratense, Pilosella caespitosa*
Lactuca serriola	*Lactuca scariola*
Leucanthemum vulgare	*Chrysanthemum leucanthemum, Leucanthemum leucanthemum*
Packera glabella	*Senecio lobatus*
Sonchus arvensis	*Sonchus uliginosus*
Taraxacum officinale	*Leontodon taraxacum, Taraxacum dens-leonis, Taraxacum palustre, Taraxacum vulgare*
Betulaceae	
Alnus glutinosa	*Alnus alnus, Betula alnus* var. *glutinosa, Betula glutinosa*
Betula pendula	*Betula verrucosa*
Boraginaceae	
Myosotis scorpioides	*Myosotis palustris*
Brassicaceae	
Alliaria petiolata	*Alliaria alliaria, Alliaria officinalis, Erysimum alliaria, Sisymbrium alliaria*

Barbarea vulgaris	*Barbarea arcuata, Barbarea barbarea, Barbarea stricta, Campe barbarea, Campe stricta, Crucifera barbarea*
Nasturtium officinale	*Nasturtium nasturtium-aquaticum, Rorippa nasturtium-aquaticum, Sisymbrium nasturtium-aquaticum*
Thlaspi arvense	*Crucifera thlaspi, Lepidium thlaspi, Teruncius arvensis, Thlaspi strictum, Thlaspidea arvensis, Thlaspidium arvense*

Butomaceae

Butomus umbellatus	*Butomus junceus*

Cannabaceae

Humulus japonicus	*Humulus scandens*

Caprifoliaceae

Lonicera japonica	*Caprifolium japonicum, Nintooa japonica*
Lonicera morrowii	*Lonicera insularis*
Lonicera tatarica	*Lonicera sibirica, Lonicera micrantha*

Caryophyllaceae

Cerastium fontanum	*Cerastium vulgatum*
Saponaria officinalis	*Bootia saponaria, Bootia vulgaris, Lychnis officinalis, Lychnis saponaria, Silene saponaria*

Celastraceae

Celastrus orbiculatus	*Celastrus articulatus*
Euonymus alatus	*Celastrus alatus, Celastrus striata, Euonymus alata, Euonymus striata*
Euonymus europaeus	*Euonymus vulgaris*
Euonymus fortunei	*Elaeodendron fortunei*

Convolvulaceae

Calystegia sepium	*Convolvulus sepium*
Convolvulus arvensis	*Convolvulus ambigens, Convolvulus incanus, Strophocaulos arvensis*
Ipomoea purpurea	*Convolvulus purpureus, Ipomoea hirsutula, Pharbitis purpurea*

Dipsacaceae

Dipsacus fullonum	*Dipsacus sylvestris*

Elaeagnaceae

Elaeagnus angustifolia	*Elaeagnus argentea, Elaeagnus orientalis*

Euphorbiaceae

Euphorbia esula	*Euphorbia intercedens, Euphorbia podperae, Euphorbion esulum, Galarhoeus esula, Tithymalus esula*

Fabaceae

Lotus corniculatus	*Lotus caucasicus, Lotus filicaulis*
Melilotus albus	*Melilotus alba, Melilotus argutus, Melilotus leucanthus, Melilotus melanospermus, Melilotus vulgaris*
Melilotus officinalis	*Melilotus arvensis, Melilotus luteus*
Pueraria montana var. *lobata*	*Dolichos lobatus, Pueraria hirsuta, Pueraria lobata, Pueraria thunbergiana*

Haloragaceae

Myriophyllum aquaticum	*Enydria aquatica, Myriophyllum brasiliense, Myriophyllum proserpina-coides*

Hydrocharitaceae

Egeria densa   *Anacharis densa, Philotria densa*

Hydrilla verticillata   *Serpicula verticillata*

Hypericaceae

Hypericum perforatum   *Hypericum assurgens, Hypericum deidesheimense, Hypericum lineolatum, Hypericum marylandicum*

Lamiaceae

Glechoma hederacea   *Nepeta glechoma, Nepeta hederacea*

Lythraceae

Trapa natans   *Trapa bispinosa, Trapa quadrispinosa*

Malvaceae

Hibiscus syriacus   *Althaea frutex, Hibiscus rhombifolius, Ketmia syriaca*

Menyanthaceae

Nymphoides peltata   *Limnanthemum peltatum, Menyanthes nymphoides, Nymphoides flava, Nymphoides nymphaeoides, Nymphoides orbiculata, Villarsia nymphoides*

Moraceae

Morus alba   *Morus tatarica*

Najadaceae

Najas minor   *Caulinia minor*

Onagraceae

Epilobium hirsutum   *Chamaenerion hirsutum, Epilobium tomentosum, Epilobium velutinum, Epilobium villosum*

Papaveraceae

Chelidonium majus   *Chelidonium grandiflorum*

Paulowniaceae

Paulownia tomentosa   *Paulownia imperialis*

Plantaginaceae

Linaria vulgaris   *Antirrhinum linaria, Linaria linaria*

Plantago major   *Plantago asiatica* (misapplied), *Plantago halophila*

Poaceae

Arthraxon hispidus   *Arthraxon ciliaris, Arthraxon hispidus, Arthraxon micans, Arthraxon quartinianus, Digitaria hispida, Lasiolytrum hispidum, Phalaris hispida*

Bromus inermis   *Festuca inermis, Poa bromoides, Schedonorus inermis*

Cynodon dactylon   *Agrostis bermudiana, Capriola dactylon, Chloris cynodon, Digitaria dactylon, Panicum dactylon, Paspalum umbellatum, Phleum dactylon, Vilfa stellata*

Dactylis glomerata   *Bromus glomeratus, Festuca glomerata, Limnetis glomerata, Phalaris glomerata, Trachypoa vulgaris*

Echinochloa crus-galli   *Milium crus-galli, Oplismenus crus-galli, Panicum crus-galli, Panicum hispidulum, Pennisetum crus-galli, Setaria muricata*

Elymus repens   *Agropyron repens, Elytrigia repens, Elytrigia vaillantiana, Triticum repens, Triticum vaillantianum*

Festuca arundinacea   *Festuca elatior, Lolium arundinaceum, Schedonorus arundinaceus, Schedonorus phoenix*

Festuca pratensis	*Bromus pratensis, Lolium pratense, Schedonorus pratensis, Tragus pratensis*
Holcus lanatus	*Aira holcus-lanatus, Avena lanata, Ginannia lanata, Holcus argenteus, Notholcus lanatus*
Microstegium vimineum	*Andropogon vimineus, Eulalia viminea, Pollinia imberbis, Pollinia viminea*
Miscanthus sinensis	*Erianthus japonicus, Eulalia japonica, Saccharum japopnicum*
Phalaris arundinacea	*Arundo colorata, Baldingera arundinacea, Calamagrostis colorata, Digraphis arundinacea, Phalaroides arundinacea, Typhoides arundinacea*
Phleum pratense	*Phleum nodosum, Plantinia pratensis, Stelephuros pratensis*
Phragmites australis ssp. *australis*	*Arundo australis, Arundo phragmites, Phragmites communis, Phragmites phragmites*
Poa annua	*Aira pumila, Catabrosa pumila, Ochlopoa annua, Poa aestivalis, Poa algida*
Setaria pumila	*Setaria glauca, Setaria lutescens, Setaria pallide-fusca*
Setaria viridis	*Chaetochloa viridis, Chamaeraphis viridis, Ixophorus viridis, Panicum viride, Pennisetum viride*
Sorghum bicolor	*Andropogon sorghum, Holcus bicolor, Holcus sorghum, Milium bicolor, Setaria vulgare, Sorghum vulgare*
Sorghum halepense	*Andropogon halepensis, Holcus halepensis, Milium halepense, Sorghum miliaceum*
Polygonaceae	
Fallopia convolvulus	*Bilderdykia convolvulus, Fagopyrum convolvulus, Tiniaria convolvulus*
Persicaria perfoliata	*Ampelygonum perfoliatum*
Potamogetonaceae	
Potamogeton crispus	*Potamogeton crenulatus, Potamogeton crispum, Potamogeton tuberosus*
Primulaceae	
Lysimachia nummularia	*Ephemerum nummularia, Lysimachusa nummularia*
Ranunculaceae	
Clematis terniflora	*Clematis dioscoreifolia, Clematis maximowicziana, Clematis paniculata*
Rhamnaceae	
Rhamnus cathartica	*Cervispina cathartica, Rhamnus cathaticus*
Rosaceae	
Potentilla recta	*Potentilla sulphurea*
Rhodotypos scandens	*Corchorus scandens, Kerria tetrapetala, Rhodotypos tetrapetalus, Rhodotypos kerrioides*
Rosa canina	*Rosa corymbifera, Rosa dumetorum*
Rosa multiflora	*Rosa cathayensis, Rosa japonica, Rosa polyanthus, Rosa watsoniana*
Rubiaceae	
Galium mollugo	*Galium erectum*
Salicaceae	
Salix alba	*Salix vitellina*
Simaroubaceae	
Ailanthus altissima	*Ailanthus glandulosa, Toxicodendron altissimum*

Solanaceae

Nicandra physalodes   *Atropa physalodes, Physalis daturaefolia, Physalodes physalodes*

Typhaceae

Typha x glauca   *Typha angustifolia x latifolia, Typha latifolia var. elongata*

Vitaceae

Ampelopsis brevipedunculata   *Ampelopsis brevipedunculata var. maximowiczii, Ampelopsis glandulosa var. brevipedunculata, Ampelopsis heterophylla*

Xanthorrhoeaceae

Hemerocallis fulva   *Gloriosa luxurians, Hemerocallis crocea, Hemerocallis kwanso, Hemerocallis maculata*

Note:

[a]Where synonyms are commonly used or species have been newly renamed, those synonyms are included in the species descriptions in the text.

Glossary

achene: A small, dry fruit that encloses a single seed and does not open at maturity

adventitious: Describes structures that arise in areas other than where they normally occur

aerial roots: Roots produced above the ground

allelopathy: Release of chemicals by plants that inhibit growth of other plants

anther: Part of a stamen that contains pollen

auricle: Small appendage at the base of the leaf blade

awn: Slender bristles, particularly at the tips of glumes and lemmas (Poaceae)

biennial: Plant that lives for two years, vegetative in the first and reproductive in the second

biternate: Compound leaf with three groups of three leaflets

bract: Modified leaf or leaflike structure at the base of a flower or inflorescence

bulblet: Small bulb either produced on a larger bulb or on an aerial portion of a plant

bulbil: Vegetative bulb that typically develops in the leaf axils and can give rise to a new plant

bunchgrass: A grass that grows in clumps

calyx: Collective term for all the sepals of a flower

cambium: Living tissue on a woody plant that produces xylem and phloem

caryopsis: Dry fruit in which a single seed is fused to the embryo wall (Poaceae)

catkin: Spikelike inflorescence of unisexual flowers, in which petals are reduced or absent

composite flower: Flower head that includes numerous ray and disc florets (Asteraceae)

cool-season grasses: Grasses with optimum growth between 60°F and 75°F (15°C and 24°C)

corolla: Collective term for all the petals of a flower

corymb: Branching inflorescence with a flat top

culm: Main stalk of a grass, jointed and hollow except at the nodes

cultivar: Named and registered plant variety that has been cultivated through selective breeding

cyathium: Small inflorescence that resembles a single flower (Euphorbiaceae)

cyme: Inflorescence with a flat or rounded top, in which the central flower blooms first

dextrorse: Turned or spirally arranged to the right, twining upward from left to right

dioecious: Species with male and female flowers on separate individuals

disc floret: Small tubular flower typically in the central portion of a composite flower head

elaiosome: Fleshy structure attached to a seed, containing lipids and proteins to attract ants

emergent: Aquatic plant with most vegetative parts growing above the water surface

filament: Stalk portion of a stamen which bears the anther

follicle: Dry dehiscent fruit, splitting down one side to expose seeds at maturity

glume: Lowermost bracts at the base of a spikelet in the inflorescence of a grass

haustorium/haustoria: In parasitic plants, a rootlike structure that draws water and nutrients from host

hypanthium: Structure where the base of the sepals and petals fuse into cup-shaped tube

inflorescence: Group or cluster of many flowers, with varying arrangement of branches

infructescence: Group or cluster of many fruits

introgression: Movement of genes from one species into the gene pool of another

lenticels: Linear or circular corky elevations on bark that allow for oxygen exchange

ligule: Appendage in the inner base of the leaf and sheath in grasses and some sedges

lobe: Divided or projected segment along the edge of a leaf

monocarpic perennial: Plant that lives for more than two years, flowers once, then dies

monoecious: Species with separate male and female flowers borne on the same plant

node: Area of a stem where leaves or branches originate

ocrea/ocreae: Sheath(es) around stem arising from fused stipules (Polygonaceae)

ovate: Egg-shaped, broader at the base and narrower at the tip

ovoid: Egg-shaped in three dimensions

palate: The projecting part on the lower lip of a two-lobed floral tube that closes the throat

palmate: Veins or leaflets radiating from a single point, usually at the top of a petiole

panicle: Spreading flower cluster that branches from a central axis

petiole: Stalk of a leaf that connects the blade to the stem

petiolule: Stalk of a leaflet in a compound leaf

phyllary: One of several bracts surrounding the base of the flower head (Asteraceae)

phytophotodermatitis: Skin hypersensitivity to ultraviolet light caused by chemicals in a plant

pinnate: Veins or leaflets arranged along a central elongate axis

pistil: Female reproductive structure of a flower, consisting of an ovary, style, and stigma

pith: Tissue in the center of a stem, particularly in the twigs of woody plants

polygamo-dioecious: Species with hermaphroditic or unisexual flowers on separate plants

prickle: Sharp projection of the epidermis with no vascular tissue

propagule: Any plant material that is capable of growing a new plant (e.g., seed, spore, bud, cutting)

propagule pressure: A composite measure of the number of individuals released into a region to which they are not native

ray floret: Small flower with a strap-shaped petal in a composite flower head (Asteraceae)

rhizome: Underground horizontal stem that sends out roots and shoots

rootlets: Small roots branching off a main root

rosette: Cluster of leaves radiating in a circle from the center, usually close to the ground

samara: Dry, winged fruit that does not split open

sepal: Leaflike or petallike component that surrounds the outermost part of a flower

silicle: Dry, flattened fruit with two valves, less than twice as long as wide (Brassicaceae)

silique: Dry, long, narrow fruit with two valves, more than twice as long as wide (Brassicaceae)

sinistrorse: Turned or spirally arranged to the left, twining upward from right to left

solarization: Process of killing off vegetation by covering the area with black plastic

spadix: Fleshy spike bearing tiny flowers (Araceae)

spathe: Leaflike or petallike bract partly surrounding a spadix (Araceae)

spine: Stiff, sharply pointed modified leaf or stipule originating from below epidermis

stamen: Male reproductive structure of a flower, consisting of a filament and an anther

sterile: Not capable of sexual reproduction

stigma: The top section of the pistil that is receptive to pollen

stipule: A leafy or spiny appendage at the base of a petiole or node

stolon: Horizontal, aboveground creeping stem, forming roots and shoots at nodes and tip

strobile: Conelike reproductive structure with overlapping scales arranged along a central stem

stump sprouting: Regrowth of stems from a cut stump

style: Elongated portion of the pistil between the ovary and stigma

syncarp: Aggregate fruit derived from an entire inflorescence

tendril: Twining modified stem used for support and climbing

tepal: Collective term for sepals and petals

thorn: Sharply pointed modified stem with vascular tissue

triploid: Having three sets of chromosomes

turion: In aquatic plants, a vegetative bud that is dormant in unfavorable conditions

tussock: Grass that grows in a bunch rather than forming a continuous sod

umbel: Inflorescence in which flower stalks or clusters arise from the same point

umbellet: One cluster of a compound umbel inflorescence

warm-season grasses: Grasses with optimum growth between 75°F and 90°F (24°C and 32°C)

Online Resources

California Invasive Plant Council, https://www.cal-ipc.org/.

Center for Aquatic and Invasive Plants, University of Florida, Institute of Food and Agricultural Sciences, https://plants.ifas.ufl.edu/.

Centre for Agriculture and Biosciences International, Invasive Species Compendium, http://www.cabi.org/isc/.

Cincinnati Wildflower Preservation Society, www.cincywildflower.org/.

EDDMapS—Early Detection and Distribution Mapping System, https://www.eddmaps.org/.

Encyclopedia of Life, http://eol.org/.

Fire Effects Information System, United States Department of Agriculture, Forest Service, https://www.feis-crs.org/feis/.

Flora of China, http://www.efloras.org.

Flora of North America, http://www.efloras.org.

Global Invasive Species Database, 100 of the World's Worst Invasive Alien Species. http://www.iucngisd.org/gisd/100_worst.php.

Global Invasive Species Database, Invasive Species Specialist Group, http://www.iucngisd.org/gisd/.

Go Botany, Native Plant Trust, https://gobotany.newenglandwild.org.

Great Lakes Aquatic Nonindigenous Species Information System, https://www.glerl.noaa.gov/glansis/index.html.

Illinois Wildflowers, http://www.illinoiswildflowers.info/.

Indiana Department of Natural Resources, https://www.in.gov/dnr/3123.htm.

Integrated Taxonomic Information System, https://www.itis.gov/.

Invasipedia, BugwoodWiki, https://wiki.bugwood.org/Invasipedia.

Midwest Invasive Plant Network, https://www.mipn.org/plantlist/.

Midwest Native Plant Society, http://midwestnativeplants.org/.

Minnesota Department of Agriculture, http://www.mda.state.mn.us/plants/pestmanagement/weedcontrol/noxiouslist/.

Minnesota Wildflowers, https://www.minnesotawildflowers.info.

Missouri Botanical Garden Plant Finder, http://www.missouribotanicalgarden.org/plantfinder/plantfindersearch.aspx.

National Invasive Species Information Center, United States Department of Agriculture, https://www.invasivespeciesinfo.gov/.

Native Plant Society of Northeastern Ohio, www.nativeplantsocietyneo.org/.

New York Invasive Species, http://nyis.info/non-native-plant-assessments/.

Nonindigenous Aquatic Species, United States Geological Survey, https://nas.er.usgs.gov/taxgroup/Plants/default.aspx.

Ohio Department of Agriculture, ODA 901:5–30–01, Invasive Plant Species, http://codes.ohio.gov/oac/901:5–30–01.

Ohio Department of Natural Resources, Invasive Plants, http://ohiodnr.gov/invasiveplants.

Ohio Department of Natural Resources, Native Plant List, http://ohiodnr.gov/gonative.

Ohio Department of Natural Resources, Prescribed Fire, http://forestry.ohiodnr.gov/prescribedfire.

Ohio Environmental Protection Agency, Open Burning, https://www.epa.ohio.gov/dapc/general/open burning.

Ohio Invasive Plants Council, https://www.oipc.info/.

Ohio Perennial and Biennial Weed Guide, Ohio Agricultural Research and Development Center, The Ohio State University, http://www.oardc.ohio-state.edu/weedguide/.

Ohio Prescribed Fire Council, https://www.ohiopre scribedfire.org/.

Pennsylvania Department of Conservation and Natural Resources, http://www.dcnr.pa.gov/Conser vation/WildPlants/InvasivePlants/.

Plant Conservation Alliance, Alien Plant Working Group, https://www.nps.gov/plants/alien/fact.htm.

Southeast Exotic Pest Plant Council, https://www.se-eppc.org/.

Southwest Desert Flora, southwestdesertflora.com.

Texas Invasive Species Institute, http://www.tsusinva sives.org/home/database/plants.html.

The Invasive Plant Atlas of New England, http://www.eddmaps.org/ipane/.

The Ohio State University Extension, Locate an Office Page, https://extension.osu.edu/lao.

The Plant List, http://www.theplantlist.org/.

The PLANTS Database, National Plant Data Center, Natural Resources Conservation Service, United States Department of Agriculture, http://plants.usda.gov/.

The University of Georgia, Center for Invasive Species and Ecosystem Health, https://www.invasive.org/species/weeds.cfm.

Tropicos, Missouri Botanical Garden, http://www.tropicos.org/.

University of California Agriculture and Natural Resources, Statewide Integrated Pest Management Program, http://ipm.ucanr.edu/.

Virginia Tech Dendrology, http://dendro.cnre.vt.edu/dendrology/.

Wisconsin Department of Natural Resources, https://dnr.wi.gov/topic/Invasives/.

Amann, Joyce E. 1961. "A Survey of the Vascular Plants of Stark County, Ohio." MA thesis, Kent State University.

Anderson, Sandra M. 1969. "The Vascular Plant Flora of Medina County, Ohio." MA thesis, Kent State University.

Andreas, Barbara K. 1980. "The Flora of Portage, Stark, Summit, and Wayne Counties, Ohio." PhD diss., Kent State University.

———. 1989. *The Vascular Flora of the Glaciated Allegheny Plateau Region of Ohio.* Columbus: Ohio Biological Survey.

Beardslee, Henry C. 1874. *Catalogue of the Plants of Ohio: Including the Flowering Plants, Ferns, Mosses, and Liverworts.* Painesville: Journal Office.

Bojnanský, Vít, and Agáta Fargašová. 2007. *Atlas of Seeds and Fruits of Central and East-European Flora: The Carpathian Mountains Region.* Dordrecht: Springer Science.

Bower, Frederick O. 1919. *Botany of the Living Plant.* London: Macmillan.

Bradley, Bethany A., David S. Wilcove, and Michael Oppenheimer. 2010. "Climate Change Increases Risk of Plant Invasion in the Eastern United States." *Biological Invasions* 12 (6):1855–72.

Bromilow, Clive. 2010. *Problem Plants and Alien Weeds of South Africa.* Pretoria: Briza.

Burns, James F. 1980. "The Flora of Trumbull County, Ohio." MS thesis, Kent State University.

Burrell, C. Colston. 2006. *Native Alternatives to Invasive Plants.* Brooklyn: Brooklyn Botanic Garden.

Callaway, Ragan M., and Erik T. Aschehoug. 2000. "Invasive Plants Versus Their New and Old Neighbors: A Mechanism For Exotic Invasion." *Science* 290 (5491):521–23.

Chace, Teri Dunn. 2013. *How to Eradicate Invasive Plants.* Portland, OR: Timber Press.

Cline, Robert J. 1977. "The Vascular Plants of Perry County, Ohio." MS thesis, Kent State University.

Cooperrider, Tom S., Allison W. Cusick, and John T. Kartesz. 2001. *Seventh Catalog of the Vascular Plants of Ohio.* Columbus: Ohio State Univ. Press.

Crow, Garrett E., and C. Barre Hellquist. 1999a. *Pteridophytes, Gymnosperms, and Angiosperms: Dicotyledons.* Vol. 1 of *Aquatic and Wetland Plants of Northeastern North America.* Madison: Univ. of Wisconsin Press.

———. 1999b. *Angiosperms: Monocotyledons.* Vol. 2 of *Aquatic and Wetland Plants of Northeastern North America.* Madison: Univ. of Wisconsin Press.

Curtis, Virginia L. 1996. "The Vascular Plant Flora of Columbiana County, Ohio." MS thesis, Kent State University.

Cusick, Allison W. 1967. "The Vascular Plants of Jefferson County, Ohio." MA thesis, Kent State University.

Czarapata, Elizabeth J. 2005. *Invasive Plants of the Upper Midwest: An Illustrated Guide to Their Identification and Control.* Madison: Univ. of Wisconsin Press.

Daehler, Curtis C. 2001. "Darwin's Naturalization Hypothesis Revisited." *American Naturalist* 158 (3):324–30.

Darwin, Charles. 1859. *On the Origin of Species by Means of Natural Selection, or Preservation of Favoured Races in the Struggle for Life.* London: John Murray.

Davis, Mark A. 2009. *Invasion Biology.* Oxford: Oxford Univ. Press.

Davis, Melissa A. 2012. "A Floristic Survey of Geauga County, Ohio: 50 Years of Change." MS thesis, Kent State University.

Daws, Matthew I., J. Hall, Sarah Flynn, and Hugh W. Pritchard. 2007. "Do Invasive Species Have Bigger Seeds? Evidence From Intra- and Inter-specific Comparisons." *South African Journal of Botany* 73:138–43.

Dietz, Hansjörg, and Peter J. Edwards. 2006. "Recognition That Causal Processes Change during Plant Invasion Helps Explain Conflicts in Evidence." *Ecology* 87 (6):359–67.

Diez, Jeff M., Ian Dickie, Grant Edwards, Philip E. Hulme, Jon J. Sullivan, and Richard P. Duncan. 2010. "Negative Soil Feedbacks Accumulate Over Time for Non-native Plant Species." *Ecology Letters* 13 (7):803–9.

Dirr, Michael A. 2009. *Manual of Woody Landscape Plants,* 6th ed. Champaign, IL: Stipes.

Duncan, Richard P., and Peter A. Williams. 2002. "Darwin's Naturalization Hypothesis Challenged." *Nature* 417 (6889):608–9.

Edwards, Erika J., and Christopher J. Still. 2008. "Climate, Phylogeny, and the Ecological Distribution of C4 Grasses." *Ecology Letters* 11 (3):266–76.

Emmitt, David P. 1981. "The Vascular Plants of Ashland County." MS thesis, Kent State University.

Gibson, David J. 2009. *Grasses and Grassland Ecology.* Oxford: Oxford Univ. Press.

Gleason, Harry A., and Arthur Cronquist. 1991. *Manual of Vascular Plants of the Northeastern United States and Adjacent Canada,* 2nd ed. Bronx: New York Botanical Garden.

Hatch, Stephan L., Joseph L. Schuster, and D. Lynn Drawe. 1999. *Grasses of the Texas Gulf Prairies and Marshes.* College Station: Texas A&M Univ. Press.

Hawver, William D. 1961. "The Vascular Flora of Geauga County, Ohio." MA thesis, Kent State University.

Intergovernmental Panel on Climate Change. 2013. "Climate Change 2013: The Physical Science Basis." Working Group I Contribution to the Fifth Assessment Report of the Intergovernmental Panel on Climate Change, edited by Thomas F. Stocker, Dahe Qin, Gian-Kasper Plattner, Melinda M. B. Tignor, Simon K. Allen, Judith Boschung, Alexander Nauels, et al. Cambridge: Cambridge Univ. Press.

Kashanski, Susan. 2002. Illustrations for course CSP3130 *Advanced Plant Identification: Grasses, Sedges, Rushes and Composites.* Shepherdstown, WV: National Conservation Training Center, US Fish and Wildlife Service.

Kaufman, Sylvan Ramsey, and Wallace Kaufman. 2007. *Invasive Plants: A Guide to Identification, Impacts, and Control of Common North American Species.* Mechanicsburg, PA: Stackpole.

Keane, Ryan M., and Michael J. Crawley. 2002. "Exotic Plant Invasions and the Enemy Release Hypothesis." *Trends in Ecology and Evolution* 17 (4):165–70.

Kellerman, William Ashbrook. 1899. *The Fourth State Catalogue of Ohio Plants: Consisting of a Serially Numbered Systematic Check-List of the Pteridophytes and Spermatophytes.* Columbus: Ohio State Univ. .

Kellerman, William Ashbrook, and William C. Werner. 1894. "Catalogue of Ohio Plants." *Geological Survey of Ohio Report* 7:56–406.

Kourtev, Peter S., Joan G. Ehrenfeld, and Max Häggblom. 2003. "Experimental Analysis of the Effect of Exotic and Native Plant Species on the Structure and Function of Soil Microbial Communities." *Soil Biology and Biochemistry* 35 (7):895–905.

Les, Donald H., Elena L. Peredo, Nicholas P. Tippery, Lori K. Benoit, Hamid Razifard, Ursula M. King, Hye Ryun Na, et al. 2015. "*Najas minor* (Hydrocharitaceae) in North America: A Reappraisal." *Aquatic Botany* 126: 60–72.

Levine, Jonathan M., Peter B. Adler, and Stephanie G. Yelenik. 2004. "A Meta-Analysis of Biotic Resistance to Exotic Plant Invasions." *Ecology Letters* 7 (10):975–89.

Lockwood, Julie L., Phillip Cassey, and Tim Blackburn. 2005. "The Role of Propagule Pressure in Explaining Species Invasions." *Trends in Ecology and Evolution* 20 (5):223–28.

Lorenzi, Harri J., and Larry S. Jeffery. 1987. *Weeds of the United States and Their Control.* New York: Van Nostrand Reinhold.

Lovett Doust, Lesley, and Jon Lovett Doust. 1982. "The Battle Strategies of Plants." *New Scientist* 95 (1313):81–84.

Mitchell, Charles E., Anurag A. Agrawal, James D. Bever, Gregory S. Gilbert, Ruth A. Hufbauer, John N. Klironomos, John L. Maron, et al. 2006. "Biotic Interactions and Plant Invasions." *Ecology Letters* 9 (6):726–40.

Moles, Angela T., Habacuc Flores-Moreno, Stephen P. Bonser, David I. Warton, Aveliina Helm, Laura Warman, David J. Eldridge, Enrique Jurado, et al. 2012. "Invasions: The Trail Behind, the Path Ahead, and a Test of a Disturbing Idea." *Journal of Ecology* 100 (1):116–27.

Myers, Judith H., and Dawn R. Bazely. 2003. *Ecology and Control of Introduced Plants.* Cambridge: Cambridge Univ. Press.

Narango, Desirée L., Douglas W. Tallamy, and Peter P. Marra. 2018. "Nonnative Plants Reduce Population Growth of an Insectivorous Bird." *Proceedings of the National Academy of Sciences* 115 (45):11549–54.

Newberry, John Strong. 1860. *Catalogue of the Flowering Plants and Ferns of Ohio.* Columbus: Richard Nevins, Printer.

Orians, Colin M., and David Ward. 2010. "Evolution of Plant Defenses in Nonindigenous Environments." *Annual Review of Entomology* 55 (1):439–59.

Pigliucci, Massimo. 2005. "Evolution of Phenotypic Plasticity: Where Are We Going Now?" *Trends in Ecology and Evolution* 20 (9):481–86.

Pimentel, David. 2011. "Environmental and Economic Costs Associated with Alien Invasive Species in the United States." In *Biological Invasions: Economic and Environmental Costs of Alien Plant, Animal, and Microbe Species,* edited by David Pimentel, 411–30. Boca Raton, FL: CRC Press.

Prasad, Anantha M., Louis R. Iverson, Stephen N. Matthews, and Matthew Peters. 2007. *A Climate Change Atlas for 134 Forest Tree Species of the Eastern United States.* Delaware, OH: Northern Research Station, USDA Forest Service. https://www.nrs.fs.fed.us/atlas/tree.

———. 2014. *Climate Change Tree Atlas.* Delaware, OH: Northern Research Station, USDA Forest Service. https://www.nrs.fs.fed.us/atlas/tree/fut_fortypes.html

Pusey, Paul L. 1976. "The Vascular Plants of Knox County, Ohio." MA thesis, Kent State University.

Pyšek, Petr, and David M. Richardson. 2007. Traits Associated with Invasiveness in Alien Plants: Where Do We Stand? In *Biological Invasions,* edited by Wolfgang Nentwig, 97–125. Berlin: Springer-Verlag.

Rejmánek, Marcel. 1998. "Invasive Plant Species and Invasible Ecosystems." In *Invasive Species and Biodiversity Management,* edited by Odd Terje Sandlund, Peter Johan Schei, and Åslaug Viken, 79–101. Dordrecht: Kluwer.

Rejmánek, Marcel, David M. Richardson, and Petr Pyšek. 2005. "Plant Invasions and Invasibility of Plant Communities." In *Vegetation Ecology,* edited by Eddy Van Der Maarel, 332–55. Oxford: Blackwell.

Schaffner, John H. 1914. *Catalog of Ohio Vascular Plants.* Columbus: Ohio Biological Survey.

———. 1932. *Revised Catalogue of Ohio Vascular Plants.* Columbus: Ohio Biological Survey.

Seabloom, Eric W., W. Stanley Harpole, O. J. Reichman, and David Tilman. 2003. "Invasion, Competitive Dominance, and Resource Use by Exotic and Native California Grassland Species." *Proceedings of the National Academy of Sciences* 100 (23):13384–89.

Selby, Augustine D. 1899. "The Flora of Franklin County, Ohio." *Proceedings of the American Association for the Advancement of Science* 48:300–303.

Shea, Katriona, and Peter Chesson. 2002. "Community Ecology Theory as a Framework for Biological Invasions." *Trends in Ecology and Evolution* 17 (4):170–76.

Shen, Xiaoli, Norman A. Bourg, William J. McShea, and Benjamin L. Turner. 2016. "Long-Term Effects of White-Tailed Deer Exclusion on the Invasion of Exotic Plants: A Case Study in a Mid-Atlantic Temperate Forest." *PLoS One* 11 (3):e0151825.

Silberhorn, Gene M. 1970. "The Flora of the Unglaciated Allegheny Plateau of Southeastern Ohio." PhD diss., Kent State University.

Strauss, Sharon Y., Campbell O. Webb, and Nicolas Salamin. 2006. "Exotic Taxa Less Related to Native Species Are More Invasive." *Proceedings of the National Academy of Sciences* 103 (15):5841–45.

Sullivant, William Starling. 1840. *Catalogue of Plants, Native or Naturalized, in the Vicinity of Columbus, Ohio.* Columbus: Charles Scott.

Swearingen, Jil, and Kristen Saltonstall. 2010. Phragmites *Field Guide: Distinguishing Native and Exotic Forms of Common Reed* (Phragmites australis) *in the United States.* Athens: Univ. of Georgia, Plant Conservation Alliance, Weeds Gone Wild. http://www.nps.gov/plants/alien/pubs/index.htm.

Tallamy, Douglas W., and Kimberly J. Shropshire. 2009. "Ranking Lepidopteran Use of Native versus Introduced Plants." *Conservation Biology* 23 (4):941–47.

Teeri, James A., and Lawrence G. Stowe. 1976. "Climatic Patterns and the Distribution of C4 Grasses in North America." *Oecologia* 23 (1):1–12.

Van Wilgen Brian W., Belinda Reyers, David C. Le Maitre, David M. Richardson, and Lucille Schonegevel. 2009. "A Biome-Scale Assessment of the Impact of Invasive Alien Plants on Ecosystem Services in South Africa." *Journal of Environmental Management* 89:336–49.

Vellend, Mark. 2002. "A Pest and an Invader: White-Tailed deer (*Odocoileus virginianus* Zimm.) as a Seed Dispersal Agent for Honeysuckle Shrubs (*Lonicera* L.)." *Natural Areas Journal* 22 (3):230–34.

Vilà, Montserrat, José L Espinar, Martin Hejda, Philip E Hulme, Vojtěch Jarošík, John L Maron, Jan Pergl et al. 2011. "Ecological Impacts of Invasive Alien Plants: A Meta-Analysis of Their Effects on Species, Communities and Ecosystems." *Ecology Letters* 14 (7):702–8.

Weber, Ewald. 2003. *Invasive Plant Species of the World: A Reference Guide to Environmental Weeds.* Wallingford, CT: Centre for Agriculture and Biosciences International Publishing.

Weishaupt, Clara G. 1971. Vascular Plants of Ohio: A Manual for Use in Field and Laboratory, 3rd ed. Dubuque, IA: Kendall Hunt.

Wilson, Hugh D. 1974. "Vascular Plants of Holmes County, Ohio." *Ohio Journal of Science* 74 (5):277–81.

Index

Acer: platanoides, 313–14; *rubrum,* 313; *saccharum,* 313
Acorus: americanus, 322; *calamus,* 322–23
Actinidia arguta, 12
Aegopodium podagraria, 70–72
Aesculus pavia, 255
Agastache scrophulariifolia, 154, 156
Ageratina altissima, 13
Agrostis hyemalis, 29
Ailanthus altissima, 315–17
air potato, 231
alder: black or European, 288–89; smooth, 288
alexanders: golden, 80, 180; heartleaf, 80, 180
allelopathy, 3, 6, 8, 9; in creepers and climbers, 208, 241;
    in forbs, 67, 90, 94–96, 104, 107, 141; in grasses, 28,
    32, 40, 42, 62; in shrubs, 257; in trees, 305, 317
Alliaria petiolata, 120–21
allspice, Carolina, 248
Alnus: glutinosa, 288–89; *serrulata,* 288
alumroot, 70
amaranth, Palmer's, 64–65
Amaranthus: hybridus, 64; *palmeri,* 64–65; *retroflexus,*
    64; *tuberculatus,* 64
Amelanchier: arborea, 302; *laevis,* 302
Ammophila arenaria, 9
Amorpha fruticosa, 272
Ampelopsis: arborea, 204, 245; *brevipedunculata,* 200,
    245–46; *cordata,* 245
Amphicarpaea bracteata, 235
Andropogon gerardii, 26, 45
Anemone: acutiloba, 86; *canadensis,* 70, 86
anemone: Canada, 70, 86; rue, 86
Angelica: atropurpurea, 80; *venenosa,* 70
angelica: hairy, 70; purplestem, 80
Apocynum androsaemifolium, 236

apple-of-Peru, 193–94
Aralia nudicaulis, 70, 169
Arctium: lappa, 88; *minus,* 88–89; *tomentosum,* 88
Arctostaphylos uva-ursi, 202, 220
Aristolochia tomentosa, 217
Aronia melanocarpa, 251, 255, 264, 274, 298
arrowhead, 341
Artemisia vulgaris, 90–91
Arthraxon hispidus, 24–25
Aruncus dioicus var. *dioicus,* 169
Asarum canadense, 70, 178, 210
Asclepias, 9, 10; *incarnata,* 156, 207; *syriaca,* 208; *verti-*
    cillata, 142, 207, 236
aspen: bigtooth, 307; quaking, 307
aster: common bluewood, 97; New England, 97; smooth,
    97
Astragalus canadensis, 142
Atropa belladonna, 196

bachelor's buttons, 94
Baptisia tinctoria, 142, 144
Barbarea: orthoceras, 122; *verna,* 122; *vulgaris,* 122–23
barberry: common, 251, 253–54; Japanese, 5, 9, 247,
    251–52
Bassia scoparia, 66–67
bayberry, northern, 251, 268, 274
bearberry, 202, 220
beard-tongue: foxglove, 164; hairy, 124
bedstraw: fragrant, 187; northern, 187; shining, 187;
    smooth, 187–88
beebalm, spotted, 154
bellwort, large-flowered, 161
Berberis: thunbergii, 5, 9, 247, 251–52, 252; *vulgaris,* 251,
    253–54

bergamot, wild, 154
Betula: lenta, 288, 307, 318; *nigra,* 288, 307; *pendula,* 286, 290–91; *populifolia,* 290
Bignonia capreolata, 204, 214, 229
bindweed: black, 14, 15, 239–40; false, 222; field, 201, 222, 225–26; fringed black, 239; hedge, 222–24, 225
biological control, 14, 141, 158
biotic resistance hypothesis, 7
birch: European, 286, 290–91; gray, 290; river, 288, 307; sweet, 288, 307, 318
bittersweet: American, 217; oriental or Asiatic, 217–19
bladdernut, American, 298
bladderpod, Short's, 38
blazing star: dense, 156; prairie, 156
Blephilia ciliata, 154
bloodroot, 86
bluebells, Virginia, 324
blueberry, highbush, 264, 248
blue-eyed Mary, 164
bluegrass: annual, 53–54; Kentucky, 53; rough, 53
bluejoint, Canada, 26, 36, 46, 50
bluestem: big, 26, 45; little, 26, 36
bouncing bet, 133–34
bower, virgin's, 243
bowman's root, 187
Brassica: arvensis (see *Sinapis: arvensis); juncea,* 127; *kaber* (see *Sinapis: arvensis); napus,* 10, 127; *nigra,* 127
brome, smooth, 26–27
Bromus inermis, 26–27
buckeye, red, 255
buckthorn: European or common, 16, 286, 300–301; glossy or alder, 6, 286, 298–99, 300
buckwheat: climbing false, 239; wild, 239
buffaloberry, russet, 268
burdock: common, 88–89; great, 88; lesser, 88; woolly, 88
burning bush, 247, 264–65
bushclover, round-headed, 144
Butomus umbellatus, 328–29
butterbur, common, 118
butterfly, monarch, 9, 10, 208
buttonbush, 255, 298

C₃ photosynthetic pathway, 23
C₄ photosynthetic pathway, 23
Calamagrostis canadensis, 26, 36, 46, 50
Calamovilfa longifolia, 26, 36
Caltha palustris, 180, 182
Calycanthus floridus, 248
Calystegia: sepium, 222–24, 225; *spithamaea,* 222
Campanula rotundifolia, 97
campion: bladder, 133; white, 133
Cannabis sativa, 185
cardinal flower, 328
Carduus: acanthoides, 102; *nutans,* 92–93
Carpinus caroliniana, 268, 288, 307, 318

Carya: illinoinensis, 305, 315; *ovata,* 292, 315
Castilleja coccinea, 164
Catalpa: bignonioides, 296; *speciosa,* 296
catalpa, northern, 296
cattail: hybrid, 351, 353; narrow-leaved, 351–52
Ceanothus americanus, 251, 274
cedar, eastern red, 12
celandine: greater, 19, 161–63; lesser, 19, 161, 180–82
Celastrus: orbiculatus, 217–19; *scandens,* 217
Celtis occidentalis, 318
Centaurea: cyanus, 94; *stoebe* ssp. *micranthos,* 6, 8, 94–96
Cephalanthus occidentalis, 255, 298
Cerastium fontanum, 131–32
Cercis canadensis, 302
Chaiturus marrubiastrum, 154
Chamerion angustifolium, 124, 156
Chasmanthium latifolium, 46
Chelidonium majus, 19, 161–63
Chelone glabra, 164, 328
cherry, pin, 302
chickweed: common, 131; mouse-ear, 131–32
chicory, 97–98
Chionanthus virginicus, 274, 302
chokeberry, black, 251, 255, 264, 274, 298
chokecherry, 274, 302
Chrysogonum virginianum, 161, 180, 204
cicely, sweet, 70
Cichorium intybus, 97–98
Cicuta maculata, 73, 75
cinquefoil, sulfur, 185–86
Cirsium: arvense, 99–101; *vulgare,* 102–3
Cladrastis kentukea, 305, 315
Clarkia pulchella, 124
Clematis: terniflora, 200, 243–44; *virginiana,* 229, 243
clematis, sweet autumn, 200, 243–44
cliff green, 202, 220
climate change, 11–12; projected changes in dominant forest types, *12*
clover, purple prairie, 236
coffeetree, Kentucky, 305, 315
Collinsia verna, 164
coltsfoot, 118–19
coneflower, purple, 110
Conium maculatum, 3, 73–74, 75
control recommendations for problem plants, 13–17
Convolvulus arvensis, 222, 225–26, 201
Conyza canadensis, 2, 12, 104–5
cool-season grass: control methods, 26; defined, 23; species, 26, 30, 34, 36, 38, 40, 46, 48, 53
coral bells, 70
coralberry, 251, 255
cordgrass, prairie, 46, 50
corktree, Amur, 305–6
Cornus: alternifolia, 296; *amomum,* 255, 264, 278, 298; *racemosa,* 255, 264, 278, 298; *sericea,* 248

379

Coronilla varia. See *Securigera varia*
Corylus americana, 278
cranberry, American highbush, 248
creeping Charlie. *See* ivy: ground
creeping Jenny. *See* moneywort
cress, early winter, 122
crops: transgenic, 10; yield reduction as a result of prob-
    lem plants, 10, 32, 55, 57, 59, 62, 66, 101, 104, 129,
    224, 228, 240
crossvine, 204, 214, 229
crowpoison, 86
Culver's root, 144
Cuscuta epithymum, 200, 227–28
Cynanchum: leave, 207; *louiseae* (see *Vincetoxicum
    nigrum*)
Cynodon dactylon, 28–29

Dactylis glomerata, 30–31
daisy: Michaelmas, 110; oxeye, 110–11; Shasta, 110
Dalea purpurea, 236
dame's rocket, 124–26
dandelion: common, 116–17; red-seeded, 116
Dasistoma macrophylla, 164
Daucus carota, 70, 75–76, 77, 82
daylily: tawny, 197–99; yellow, 197
deer, white-tailed, 5
Delphinium exaltatum, 97
Desmodium, 142, 144; *rotundifolium,* 235
dew flower, March, 12
dextrorse, 200, *201*
Dicanthelium clandestinum, 24
Diervilla lonicera, 251, 255, 298
Dioscorea: batatas (see *Dioscorea: polystachya*); *bulbi-
    fera,* 231; *oppositifolia* (see *Dioscorea: polystachya*);
    polystachya, 200, 231–32
Diospyros virginiana, 305
Dipsacus: fullonum, 135–36, 137; *laciniatus,* 135, 137–38, 138
dodder, clover or common, 200, 227–28
dogbane, spreading, 236
dogwood: gray, 255, 264, 278, 298; pagoda, 296; red
    osier, 248; silky, 255, 264, 278, 298
dropseed, prairie, 26
Duchesnea indica, 183–84
Dutchman's pipe, woolly, 217

Echinacea purpurea, 110
Echinochloa crus-galli, 32–33
ecosystem: negative effects of problem plants, 3; services, 9
Egeria densa, 5, 335–36, 337
Eichhornia crassipes, 5
Elaeagnus: angustifolia, 268–69, 270; *umbellata,* 268,
    270–71
elderberry: common, 298; red, 169, 248, 298
elm: American, 318; Chinese (misapplied), 318; rock,
    318; Siberian, 318–19; slippery, 318
Elodea: canadensis, 335, 337; *densa* (see *Egeria densa*)

Elymus repens, 34–35
empty niche hypothesis, 7–8
encroachment, shrub or tree, 12
enemy release hypothesis, 7
Epilobium hirsutum, 159–60
Eragrostis spectabilis, 26
Erigeron: canadensis (see *Conyza canadensis*); *pulchel-
    lus,* 110
Euonymus: alatus, 247, 264–65; *americanus,* 264; *atro-
    purpureus,* 264; *europaeus,* 266–67; *fortunei,* 220–21
Euphorbia: corollata, 139, 187; *cyparissias,* 139; *esula,* 3,
    139–41; *virgata,* 139
Eutrochium purpureum, 156

Falcaria vulgaris, 12
Fallopia: cilinodis, 239; *convolvulus,* 14, 15, 239–40;
    japonica, 169–71, 172, 174; *sachalinensis,* 171, 172–73,
    174; *scandens,* 239; x *bohemica,* 171, 174–75
fescue: meadow, 38–39; red, 29; tall, 36–37
Festuca: arundinacea, 36–37, 38; *pratensis,* 38–39; *rubra,* 29
feverfew, American, 110
Ficaria verna, 19, 161, 180–82
Filipendula rubra, 156
fire: as control method, 15; as disturbance, 8; hazards,
    68; suppression, 9
fireweed, 124, 156
Firmiana simplex, 12
flag, southern blue, 341
floating heart, yellow, 345
foamflower, heartleaf, 70
forget-me-not: common, 324–25; smaller, 324
foxglove, mullein, 164
foxtail: green, 57–58; yellow, 55–56
Fragaria indica. See *Duchesnea indica*
Frangula alnus, 6, 286, 298–99, 300
fringetree, 274, 302
frogbit: American, 339; common or European, 339–40

Galium: boreale, 187; *concinnum,* 187; *mollugo,* 187–88;
    triflorum, 187
garlic, false, 86
garlic mustard, 120–21
Gaultheria procumbens, 202, 220
geranium, wild, 124, 324
Geranium maculatum, 124, 324
germander, American, 154
Gillenia trifoliata, 187
ginger: Canadian wild, 210; wild, 70, 178
Glechoma hederacea, 152–53, 178
Glyceria grandis, 46
goatsbeard, 169
goat's rue, 142
goldenrod, 104
goutweed, 70–72
grass: barnyard, 32–33; beach or marram, 9; Bermuda,
    28–29; carpet (small), 24–25; cogon, 12; deertongue,

24; gama, 61; Indian, 26, 45, 46, 50; itch, 12; Japanese stilt, 5, 24, 42–43; Johnson, 61–62; manna, 46; orchard, 30–31; porcupine, 36; purple love, 26; quack, 34–35; reed canary, 3, 46–47; silver plume, 45; tickle, 29; Timothy, 48–49; velvet, 40–41; wavy basket, 12

grazing: as control for problem plants, 14; disturbance caused by, 8

green-and-gold, 161, 180, 204

groundsel: cressleaf, 112–13; golden, 180

Guelder rose, 248

gum: black, 313; sweet, 313

Gymnocladus dioicus, 305, 315

gypsyweed, common, 178

hackberry, 318

Hamamelis virginiana, 268, 298

harebell, 97

hawkweed, yellow or meadow, 106–7

hazelnut, American, 278

Hedera helix, 200, 204–6

hedgeparsley, 82–83, 84; Japanese, 82, 84–85

Helianthus: giganteus, 110; *occidentalis,* 110; *tuberosus,* 110

Hemerocallis: fulva, 197–99; *lilioasphodelus,* 197

hemlock: poison, 3, 73–74, 75; water, 73, 75

henbit, 152

hepatica, sharp-lobed, 86

Heracleum mantegazzianum, 63, 77–79, 80

herbicide, 16–17, 363–65

Hesperis matronalis, 124–26

Hesperostipa spartea, 36

Heuchera americana, 70

Hibiscus: laevis, 272, 328; *moscheutos,* 272, 328; *syriacus,* 247, 272–73

hickory, shagbark, 292, 315

Hieracium caespitosum, 106–7

hogweed, giant, 63, 77–79

Holcus lanatus, 40–41

holly, winterberry, 248, 251, 274

honeysuckle: Amur, 255–57; Bell's, 262–63; Japanese, 201, 214–16; Morrow's, 258–59; northern bush, 251, 255, 298; shrub, 5, 247, 255–63; Tatarian, 260–61; trumpet, 214, 217, 229; wild, 214, 229;

honeyvine, 207

hop-hornbeam, 288, 318

hops: common, 212; Japanese, 212–13

hop tree, 305, 318

horsefly weed, 142, 144

Humulus: japonicus, 212–13, 213; *lupulus,* 212

hyacinth, water, 5

hydrangea, wild, 248, 272, 278

Hydrangea arborescens, 248, 272, 278

hydrilla, 337–38

Hydrilla verticillata, 335, 337–38

Hydrocharis morsus-ranae, 339–40

hydrology, effects of problem plants on, 9, 52, 269, 311, 333, 341

Hypericum: hypericoides, 149; *perforatum,* 149–51; *prolificum,* 149; *pyramidatum,* 149

hyssop, purple giant, 154, 156

Ilex verticillata, 248, 251, 255, 264, 274

Imperata cylindrica, 12

Indian paintbrush, 164

indigo bush, false, 272

invasion: general hypotheses, 7–8; stages, 3–5

invasiveness, 2, 6

invasive species: characteristics of successful invaders, 6–7; economic costs, 10

Ipomoea purpurea, 229–30

iris: blue flag, 341; copper, 341; yellow, 341–42; zigzag, 341

Iris: brevicaulis, 341; *fulva,* 341; *pseudacorus,* 341–42; *versicolor,* 341; *virginica,* 341

ironweed, prairie, 156

ivy: English, 200, 204–6; ground, 152–53, 178; poison, 13, 235

Jerusalem artichoke, 110

jetbead, black, 278–79

Joe-Pye weed, sweet, 156

Juglans nigra, 315

Juniperus virginiana, 12

Kali tragus. See *Salsola tragus*

kiwi vine, hardy, 12

knapweed, spotted, 6, 8

knotweed: Bohemian, 171, 174–75; giant, 171, 172–73, 174; Japanese, 169–71, 172, 174

kochia, 66–67

Kochia scoparia. See *Bassia scoparia*

kudzu, 14, 15, 233–35

Lactuca serriola, 108–9, 114

lady's thumb: Oriental, 176–77; spotted, 176

Lamium amplexicaule, 152

Laportea canadensis, 13

larkspur, tall, 97

Lathyrus venosus, 236

Leonurus: cardiaca, 154–55; *sibiricus,* 154

Lespedeza capitata, 144

lettuce: prickly, 108–9, 114; wild, 97

Leucanthemum: vulgare, 110–11; x *superbum,* 110

Liatris: pycnostachya, 156; *spicata,* 156

Ligustrum: lucidum, 12; *obtusifolium,* 274–75; *vulgare,* 274, 276–77

Lilium: canadense, 197; *michiganense,* 197; *philadelphicum,* 197; *superbum,* 197

lily: Canada, 197; Michigan, 197; turk's cap, 197; wood, 197

Limnobium spongia, 339

Linaria: dalmatica, 164; *vulgaris,* 164–66

Lindera benzoin, 255, 274

Liquidambar styraciflua, 313

Liriodendron tulipifera, 296, 313

lizard's tail, 341
Lobelia: cardinalis, 328; *siphilitica,* 124, 324, 328
lobelia, great blue, 124, 324, 328
Lonicera: dioica, 214, 229; *japonica,* 201, 214–16; *maackii,* 255–57; *morrowii,* 258–59, 262; *sempervirens,* 214, 217, 229; *tatarica,* 260–61, 262; x *bella,* 262–63
loosestrife: European wand, 156; purple, 4, 6, 10, 156–58; winged, 156
Lotus corniculatus, 142–43, 236
Lunaria annua, 124
lupine, sundial, 142, 144
Lupinus perennis, 142, 144
Lysimachia nummularia, 178–79
Lythrum: alatum, 156; *salicaria,* 4, 6, 10, 156–58; *virgatum,* 156

magnolia, cucumber tree, 296
Magnolia acuminata, 296
mahogany, Chinese, 12
Maianthemum canadense, 70
mallow: smooth rose, 272, 328; swamp rose, 272, 328
maple: Norway, 313–14; red, 313; sugar, 313
marestail, 2, 12, 104–5
marigold, marsh, 180
mayflower, Canada, 70
meadowsweet, 280
Medicago, 142, 227
Meehania cordata, 154, 178
Melilotus: albus, 144–46, 147; *altissimus,* 147; *officinalis,* 144, 147–48
Mertensia virginica, 324
Microstegium vimineum, 5, 24, 42–43
Microthlaspi perfoliatum, 129
mile-a-minute weed, 200, 241–42
milkweed: common, 208; swamp, 156, 207; whorled, 142, 207, 236
millet, foxtail, 57
mint: creeping, 154, 178; hairy wood, 154; mountain, 154
Miscanthus: giganteus, 10; *sacchariflorus,* 44; *sinensis,* 44–45; x *giganteus,* 44
Mitchella repens, 178, 202, 220
Monarda: fistulosa, 154; *punctata,* 154
moneywort, 178–79
Morella pensylvanica, 251, 268, 274
morning glory, tall or common, 229
Morus: alba, 294–95; *rubra,* 294
motherwort, 154–55; false, 154; Siberian, 154
mountain ash, showy, 315
mugwort, 90–91
mulberry: red, 294; white or common, 294–95
Mulgedium genus, 97
mullein: common or great, 191–92; moth, 189–90
Murdannia keisak, 12
musclewood, 268, 288, 307, 318
mustard: black, 127; brown, 127; white, 127; wild, 127–28

mycorrhizal fungi, 8, 120
Myosotis: laxa, 324; *scorpioides,* 324–25
Myriophyllum: aquaticum, 330–32; *spicatum,* 330, 333–34

naiad, lesser or brittle, 347–48
Najas minor, 347–48
Nasturtium officinale, 326–27
native plants for gardens, 13; aquatic plants, 324, 328; creepers and climbers, 204, 207, 210, 214, 217, 220, 229, 233, 236, 243, 245; forbs, 70, 80, 86, 97, 110, 124, 142, 144, 154, 156, 161, 164, 169, 178, 180, 187, 197, 202; grasses, 26, 29, 36, 45, 46, 50; shrubs, 248, 251, 255, 264, 268, 272, 274, 278, 280; trees, 288, 292, 296, 298, 302, 305, 307, 309, 313, 315, 318
naturalization hypothesis, Darwin, 8
nettle: stinging, 13; wood, 13
New Jersey tea, 251, 274
Nicandra physalodes, 193–94
nightshade: bittersweet, 195–96; deadly, 196
ninebark, 278
Nothoscordum bivalve, 86
novel weapons hypothesis, 8
Nuttallanthus canadensis, 164
Nymphoides peltata, 345–46
Nyssa sylvatica, 302, 313

oak: chinquapin, 292; sawtooth, 292–93; shingle, 292; swamp white, 292, 313
obedient plant, 124, 156, 328
Ohio Department of Agriculture (ODA), 1; invasive species regulation, 2, 3, 356–57; prohibited noxious weed regulation, 3, 361–62
Ohio Department of Natural Resources (ODNR), 13, 15
Ohio Environmental Protection Agency (OEPA), 15, 320, 321
Ohio Invasive Plants Council (OIPC), 2, 3, 13, 14, 358–60
olive: autumn, 268, 270–71; Russian, 268–69
Oplismenus hirtellus ssp. *undulatifolius,* 12
Ornithogalum: nutans, 86; *umbellatum,* 86–87
Osmorhiza longistylis, 70
Ostrya virginiana, 288, 318

Pachysandra: procumbens, 202, 210; *terminalis,* 210–11
Packera: aurea, 180; *glabella,* 112–13
Panicum virgatum, 36, 45, 46
parasol tree, Chinese, 12
parrot feather, 330–32
parsnip, wild, 6, 80–81
Parthenium integrifolium, 110
partridgeberry, 178, 178, 202, 220, 220
Pastinaca sativa, 6, 80–81
pasture degradation, 3; by forbs, 68, 76, 82, 84, 101, 107, 111, 112, 130, 131, 141, 151, 186; by shrubs, 238, 280, 282
pathogens, problem plants as hosts/reservoirs, 3; in creepers and climbers, 206; in forbs, 67, 89, 101, 104, 111, 123, 126, 128, 130, 146, 166, 168,

193, 196; in grasses, 32, 59, 62; in shrubs, 253; in trees, 298, 300

Paulownia tomentosa, 296–97

Paxistima canbyi, 202, 220

pea, veiny, 236

peanut, hog, 235

pear: callery or 'Bradford,' 286, 302–4; common, 302

pecan, 305, 315

pennycress: field, 129–30; perfoliate, 129; roadside, 129

Penstemon: digitalis, 124, 164; *hirsutus,* 124

penstemon, smooth, 124

pepper tree, Peruvian, 5

peppervine, 204, 245; heartleaf, 245

periwinkle: common or lesser, 202–3, 220; greater, 202

Persicaria: arifolia, 241; *longiseta,* 176–77; *maculosa,* 176; *perfoliata,* 200, 241–42, 242; *sagittata,* 241

persimmon, common, 305

pests, problem plants as hosts, 9, 59, 62, 104, 111, 123, 128, 130, 193, 196, 290

Petasites hybridus, 118

Phalaris arundinacea, 3, 46–47

Phellodendron amurense, 305–6

Phleum pratense, 48–49

phlox: creeping, 178; garden, 124; prairie, 124; smooth, 124; wild blue, 124

Phlox: divaricata, 124; *glaberrima,* 124; *paniculata,* 124; *pilosa,* 124; *stolonifera,* 178

Phragmites: australis ssp. *americanus,* 50; *australis* ssp. *australis,* 50–52

Physaria globosa, 38

Physocarpus opulifolius, 278

Physostegia virginiana, 124, 156, 328

pickerelweed, 341

pigweed: redroot, 64; smooth, 64

pine, Virginia, 12

pinkfairies, 124

Pinus virginiana, 12

Plantago: lanceolata, 167; *major,* 167–68

plantain: broadleaf, 167–68; buckhorn or narrowleaf, 167; robin's, 110

plum, American, 302

Poa: annua, 53–54; *pratensis,* 53; *trivialis,* 53

pollination, 8, 9, 13, 247

pollution, 9

Polygonum: cespitosum (see *Persicaria: longiseta*); *convolvulus* (see *Fallopia: convolvulus*); *cuspidatum* (see *Fallopia: japonica*); *longisetum* (see *Persicaria: longiseta*); *perfoliatum* (see *Persicaria: perfoliata*); *schalinense* (see *Fallopia: sachalinensis*); *x bohemicum* (see *Fallopia: x bohemica*)

pondweed, curly, 349–50

Pontederia cordata, 341

poplar, white, 307–8

poppy: celandine, 161; wood, 161

Populus: alba, 307–8; *grandidentata,* 307; *tremuloides,* 307

porcelain berry, 200, 245–46

Potamogeton crispus, 349–50

Potentilla: indica (see *Duchesnea indica*); *recta,* 185–86

princess tree, 296–97

privet: border, 274–75; European or common, 276; glossy, 12

prohibited noxious weed, defined, 2, 3

propagule pressure, 5, 14

Prunus: americana, 302; *pensylvanica,* 302; *virginiana,* 274, 302

Ptelea trifoliata, 305, 318

Puccinia graminis, 253

Pueraria montana var. *lobata,* 14, 15, 233–35

Pycnanthemum virginianum, 154

Pyrus: calleryana, 286, 302–4; *communis,* 302

Queen Anne's lace, 70, 75–76, 77, 82

queen-of-the-prairie, 156

Quercus: acutissima, 292–93; *bicolor,* 292, 313; *imbricaria,* 292; *muehlenbergii,* 292

radish, wild, 127

rangeland, degradation by problem plants, 3, 68, 94, 96, 101, 107, 141, 164, 186

Ranunculus ficaria. See *Ficaria verna*

rapeseed, cultivated, 127

Raphanus raphanistrum, 127

raspberry, purple-flowering, 280

redbud, eastern, 302

reed, common, 50–52

Reynoutria: convolvulus (see *Fallopia: convolvulus*); *japonica* (see *Fallopia: japonica*); *sachalinensis* (see *Fallopia: sachalinensis*); *x bohemica* (see *Fallopia: x bohemica*)

Rhamnus: cathartica, 16, 286, 300–301; *frangula* (see *Frangula alnus*)

Rhodotypos scandens, 278–79

Rhus aromatica, 210, 264

rocket, yellow, 122–23

Rosa: canina, 280–81; *carolina,* 280; *multiflora,* 282–83; *palustris,* 280; *setigera,* 280

rose: climbing prairie, 280; dog, 280–81; multiflora, 282–83; pasture, 280; swamp, 280

rose of Sharon, 247, 272–73

Rottboellia cochinchininsus, 12

Rubus: odoratus, 280; *phoenicolasius,* 5, 284–85

rue: goat's, 144; meadow, 144

rush, flowering, 328–29

Saccharum alopecuroideum, 45

Sagittaria latifolia, 341

Salix: alba, 309–10; *discolor,* 309; *fragilis,* 311–12; *humilis,* 309; *nigra,* 309

Salsola tragus, 68–69

Sambucus: canadensis, 298; *pubens,* 298; *racemosa,* 169, 248

sandreed, prairie, 26, 36

Sanguinaria canadensis, 86
Saponaria officinalis, 133–34
sarsaparilla, wild, 70
Saururus cernuus, 341
Schinus molle, 5
Schizachyrium scoparium, 26, 36
Scutellaria incana, 154
Securigera varia, 236–38, *237*
Senecio glabellus. See *Packera: glabella*
serviceberry: Allegheny, 302; downy, 302
Setaria: italica, 57; *pumila,* 55–56; *viridis,* 57–58
shattercane, 59–60
Shepherdia canadensis, 268
sickle weed, 12
Silene: latifolia, 133; *vulgaris,* 133
silver dollar plant, 124
silvergrass, 10; Chinese, 44–45
Sinapis: alba, 127; *arvensis,* 127–28
sinistrorse, 201, *201*
skullcap, downy, 154
smartweed, 176
snakeroot, white, 13
snowbell, American, 268, 296, 298, 302
snowberry, 251, 255
Solanum dulcamara, 195–96
Sonchus arvensis, 114–15
Sorbus decora, 315
Sorghastrum nutans, 26, 45, 46, 50
Sorghum: bicolor, 59–60, 61; *halepense,* 61–62
sowthistle, perennial, 114–15
Spartina pectinata, 46, 50
spicebush, 255, 274
spiderwort, Ohio, 328
spikenard, American, 169
spindletree, European, 266–67
Spiraea: alba, 280; *tomentosa,* 169, 274
Sporobolus heterolepis, 26
spotted knapweed, 94–96
spurge: Allegheny, 202, 210; cypress, 139; flowering, 139, 187; Japanese, 210–11; leafy, 3, 139–41
St. Andrew's cross, 149
Staphylea trifolia, 298
star-of-Bethehem, 86–87
steeplebush, 169, 274
Stellaria media, 131
St. Johnswort: common, 149–51; great, 149; shrubby, 149; spotted, 149
strawberry: barren, 202; Indian or mock, 183–84; wild, 183
strawberry bush, 264
Stylophorum diphyllum, 161
Styrax americanus, 268, 296, 298, 302
sumac, 315; fragrant, 210, 264
sunflower: giant, 110; oxeye, 110
swallowwort: black or Louis', 207–9; smooth, 207
sweet clover: tall, 147; white, 144–46, 147; yellow, 144, 147–48

sweet flag, 322–23; American, 322
switchgrass, 36, 45, 46
Symphoricarpos: albus, 251, 255; *orbiculatus,* 251, 255
Symphyotrichum: cordifolium, 97; *laeve,* 97; *novae-angliae,* 97; *novi-belgii,* 110

Taraxacum: erythrospermum, 116; *officinale,* 116–17
tearthumb: arrow-leaved, 241; devil's, 241; halberd-leaved, 241
teasel: common, 135–36, 137; cutleaf, 135, 137–38
Tephrosia virginiana, 142, 144
Teucrium canadense, 154
Thalictrum: dasycarpum, 144; *thalictroides,* 86
thistle: bull, 102–3; Canada, 99–101; musk, 92–93; nodding, 92–93; plumeless, 102; Russian, 68–69
Thlaspi: alliaceum, 129; *arvense,* 129–30
Thymus genus, 227
Tiarella cordifolia, 70
tick-trefoil, 142, 144; prostrate, 235
toadflax: blue, 164; Dalmatian, 164; yellow, 164–66
Toona sinensis, 12
Torilis: arvensis, 82–83, 84; *japonica,* 82, 84–85
Toxicodendron radicans, 13, 235
toxic plants: to horses, 152, 238; to humans, 13, 73, 77, 87, 155, 163, 196, 230, 238, 243, 275, 278, 300; to livestock, 3, 36, 38, 59, 62, 64, 67, 80, 87, 146, 151, 182, 226; to mammals, 112, 133, 143, 275, 278, 341; to sheep, 68
Tradescantia ohiensis, 328
trait diversity hypothesis, 7
Trapa natans, 343–44
tree-of-heaven, 315–17
trefoil, birdsfoot, 142–43, 236
tridens, purpletop, 26, 29, 36
Tridens flavus, 26, 29, 36
Trifolium genus, 142, 227
Trillium: cernuum, 161; *grandiflorum,* 161
trillium: great white, 161; nodding, 161
Tripsacum dactyloides, 61
tulip poplar, 296, 313
tupelo, black, 302
turtlehead, 164, 328
Tussilago farfara, 118–19
Typha: angustifolia, 351–52, 353; *latifolia,* 351, 353; x *glauca,* 351, 353

Ulmus: americana, 318; *parvifolia,* 318; *pumila,* 318–19; *rubra,* 318; *thomasii,* 318
United States Department of Agriculture (USDA), 2; PLANTS database, 13, 18
unpalatable plants, to livestock, 92, 96, 101, 166, 185–86, 188, 191
Urticaria dioica, 13
Uvularia grandiflora, 161

Vaccinium corymbosum, 248, 264
Verbascum: blattaria, 189–90; *thapsus,* 191–92

Verbena: hastata, 156, 328; *stricta,* 324

Vernonia: fasciculata, 156; *officinalis,* 178

Veronicastrum virginicum, 144

vervain: blue, 156, 328; hoary, 324

vetch: American, 236; Canada milk, 142; crown, 236–38

Viburnum: acerifolium, 248, 278; *dentatum,* 255, 278; *nudum,* 248, 264; *opulus* var. *americanum,* 248; *opulus* var. *opulus,* 248–49; *prunifolium,* 274

viburnum: arrowwood, 255, 278; blackhaw, 274; cranberry, 248–49; mapleleaf, 248, 278; witherod, 248, 264

Vicia americana, 236

Vinca: major, 202–3, 220; *minor,* 202

Vincetoxicum nigrum, 207–9

virgin's bower, 229

wahoo, eastern, 264

Waldsteinia fragarioides, 202

walnut, black, 315

warm-season grass: defined, 23; species, 24, 28, 32, 42, 50, 55, 57, 59, 61

water chestnut, European, 343–44

watercress, 326–27

waterhemp, tall, 64

watermilfoil, Eurasian, 330, 333–34

waterweed, Brazilian, 5, 335–36, 337

willow: black, 309; crack or brittle, 311–12; prairie, 309; pussy, 309; white, 309–10

willowherb, hairy, 159–60

wineberry, 5, 284–85

winterberry, 255, 264

wintercreeper, 220–21

wintergreen, 202, 220

wisteria, Atlantic, 233

Wisteria frutescens, 233

witch hazel, 268, 298

woodbine, 243

woodoats, Indian, 46

wormwood, common, 90

yam, Chinese, 200, 231–32

yellowrocket, American, 122

yellowwood, American, 305, 315

Zizia: aptera, 80, 180; *aurea,* 80, 180

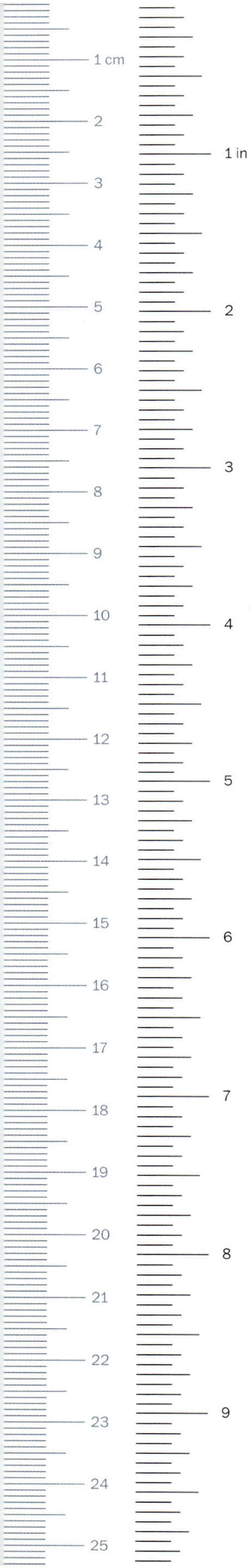

Metric-Imperial Conversions

Millimeters to Inches
1 mm = $^3/_{64}$ in.
2 mm = $^5/_{64}$ in.
3 mm = ⅛ in.
4 mm = $^5/_{32}$ in.
5 mm = $^{13}/_{64}$ in.
6 mm = ¼ in.
7 mm = $^9/_{32}$ in.
8 mm = ⅓ in.
9 mm = $^{23}/_{64}$ in.
10 mm = ⅜ in.

Centimeters to Inches
1 cm = ⅜ in.
2 cm = ¾ in.
3 cm = 1¼ in.
4 cm = 1½ in.
5 cm = 2 in.
6 cm = 2⅜ in.
7 cm = 2¾ in.
8 cm = 3 in.
9 cm = 3½ in.
10 cm = 4 in.
20 cm = 8 in.
30 cm = 12 in.
40 cm = 15¾ in.
50 cm = 19$^{11}/_{16}$ in.
60 cm = 23⅝ in.
70 cm = 27$^9/_{16}$ in.
80 cm = 31½ in.
90 cm = 35$^7/_{16}$ in.
100 cm = 39⅜ in.

Meters to Feet
1 m = 3 ft., 3⅜ in.
2 m = 6 ft., 7 in.
3 m = 9 ft., 10 in.
4 m = 13 ft., 1½ in.
5 m = 16 ft., 5 in.
6 m = 19 ft., 8 in.
7 m = 22 ft., 11 in.
8 m = 26 ft., 3 in.
9 m = 29 ft., 6⅓ in.
10 m = 32 ft., 9¾ in.
15 m = 49 ft., 2½ in.
20 m = 65 ft., 7⅜ in.
25 m = 82 ft.
30 m = 98 ft., 5 in.